Community Ecology

Community Ecology

Gary G. Mittelbach
Michigan State University

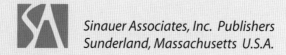

Sinauer Associates, Inc. Publishers
Sunderland, Massachusetts U.S.A.

Cover art by Joanne Delphia

COMMUNITY ECOLOGY
Copyright © 2012 by Sinauer Associates, Inc. All Rights Reserved.
For information address Sinauer Associates, 23 Plumtree Road,
Sunderland, MA 01375 U.S.A.

FAX: 413-549-1118
orders@sinauer.com
publish@sinauer.com
www.sinauer.com

Library of Congress Cataloging-in-Publication Data

Mittelbach, Gary George.
 Community ecology / Gary G. Mittelbach.
    p. cm.
 Includes bibliographical references and index.
 ISBN 978-0-87893-509-3 (paperbound)
 1. Biotic communities. 2. Ecology. I. Title.
 QH541.M526 2012
 577.8'2--dc23
                                        2012005549

Printed in China

5 4 3 2

To my parents, George and Dorothy Mittelbach, and to my mentor,
Earl Werner, and to my loving sons, John and Mark,
and most especially to my wife, Katherine Gross,
who for more than 35 years has been my companion,
scientific collaborator, and great friend.

Brief Table of Contents

Table of Contents

Preface

The community is … the hierarchical level where the basic characteristics of life— its diversity, complexity and historical nature—are perhaps most daunting and challenging.

Michel Loreau, 2010: 50

For communities … the task of choosing which topics to emphasize and which to elide [omit] … is necessarily quirky.

Robert M. May et al., 2007: 111

Well those drifter's days are past me now
I've got so much more to think about
Deadlines and commitments
What to leave in, what to leave out

Bob Seger, *Against the Wind*

"What to leave in, what to leave out"? Much of writing well about any subject comes down to this simple question. (I sometimes think much of life comes down to this simple question.) But, as Robert May notes, the task of choosing which topics to emphasize in community ecology is "necessarily quirky." Our discipline is broad and there is no clear roadmap. The organizational path I have chosen for this book is the one that works best for me when teaching this subject to graduate students. I begin with an exploration of patterns of biodiversity—that is, how does the diversity of life vary across space and time? Documenting and understanding spatial and temporal patterns of biodiversity are key components of community ecology, and recent advances in remote sensing, GIS mapping, and spatial analysis allow ecologists to examine these patterns as never before. Next, with species disappearing from the Earth today at a rate unprecedented since the extinction of the dinosaurs, what are the consequences of species loss to the functioning of communities and ecosystems? This question drives the very active research area of biodiversity and ecosystem functioning discussed in Chapter 2.

With the patterns of biodiversity at local and regional scales firmly in mind and with an appreciation for the potential consequences of species loss, I next shift gears to focus on what I call the "nitty-gritty" of community ecology: population regulation and species interactions, including predation, competition, mutualism, and facilitation. The goal here is to understand in some detail the mechanics of species interactions by focusing on consumers and resources in modules of a few interacting species. From these simple building blocks we can assemble more complex ecological networks, such as food webs and mutualistic networks, which involves exploring the importance of indirect effects, trophic cascades, top-down and bottom-up regulation, alternative stable states, diversity–stability relationships, and much more. In the section on spatial ecology, I focus on the processes that link populations and communities across space (metapopulations and metacommunities) and on the consequences of these local and regional links for species diversity.

The interplay between local and regional processes is a prominent theme throughout the book. Likewise, the interplay between ecology and evolution—what is termed "eco-evolutionary dynamics" or, more broadly, evolutionary community ecology—is an important new area of research. I explore evolutionary community ecology, along with the impacts of variable environments on species interactions, in the book's final section on changing environments and changing species. Applied aspects of community ecology (e.g., resource management and harvesting, invasive species, diseases and parasites, and community restoration) are treated throughout the book as natural extensions of basic theoretical and empirical work. The emphasis, however, is on the basic science. Theoretical concepts are developed using simple equations, with an emphasis on the graphical presentation of ideas.

This is a book for graduate students, advanced undergraduates, and researchers seeking a broad coverage of ecological concepts at the community level. As a textbook for advanced courses in ecology, it is not meant to replace reading and discussing the primary literature. Rather, this book is designed to give students a common background in the principles of community ecology at a conceptually advanced level. At Michigan State University, our graduate community ecology course draws students from many departments (zoology, plant biology, fisheries and wildlife, microbiology, computer science, entomology, and more) and students come into the course with vastly different exposures to ecology. I hope this book helps students from varied academic backgrounds fill in the gaps in their ecological understanding, approach a new topic more easily, and find an entry point into the primary literature. I'd be doubly pleased if it can do the same for practicing ecologists.

When teaching community ecology, I try to show students how seemingly differently ideas in ecology have developed over time and are linked together. This is important and hopefully useful to students who are just beginning to sink their teeth deeply into the study of ecology. At least, I believe it is useful. An early reviewer of this book wrote, "It is obvious that Mittelbach has a deep understanding and respect for the literature." I take this as a great compliment. We do, after all, stand on the shoulders of giants, and it's important to acknowledge where ideas come from. Moreover, an appreciation for the historical development of ideas and for how concepts are linked together helps deter us from recycling old ideas under new guises. However, another early reviewer suggested that students today aren't all that interested in the history of ideas and that a modern textbook on community ecology should focus on what's new, particularly on how community ecology can inform and guide conservation biology and the preservation of biodiversity. I appreciate this advice as well. I have worked to include cutting edge ideas and to provide case studies from the most recent literature, along with some of the classics. Hopefully, the balance between old and new contained herein is one that works. I recognize, however, that more could be done to illuminate the links between community ecology and conservation biology. Perhaps someone else will take up this call.

Finally, I want to say a few words about the use of mathematical models and theory in this book. Robert May (2010: viii) wrote that "mathematics is ultimately no more, although no less, than a way of thinking clearly." May also pointed out that one of the most celebrated theories in all of biology, Darwin's theory of evolution by natural selection, is a verbal theory. In most cases, however, the ability to express an idea mathematically makes crystal clear the assumptions and processes that underlie an explanation. I have used mathematical models here as a way to "think more clearly" about ecological processes and the theories put forth to explain them. The mathematical models in this book are simple, heuristic tools that, combined with graphical analyses, can help guide our thinking. Readers with limited mathematical skills should not be anxious when they see equations. I am a mathematical lightweight myself, and if I can follow the models presented here, so can you. On the other hand, readers with a strong background in mathematics and modeling will quickly recognize that I have stuck to the very basics and that much more sophisticated mathematical treatments of these topics abound. I have tried to point the way to these treatments in the references cited.

This book is a labor of love that has stretched out for over five years. I always knew that the ecological literature was vast, but I never truly appreciated its scale until I started this project. OMG! It's humbling to spend weeks reviewing the literature on a topic, only to stumble across a key paper later (and purely by accident). I know that I have missed much. I apologize in advance to those scientists whose excellent work I passed by (or simply missed in my ignorance) in favor of studies that were better known to me. Please don't be shy in telling me what I missed.

The first ecology textbook I purchased as an undergraduate in the early 1970s (at the hardcover price of $7.00!) was Larry Slobodkin's *Growth and Regulation of Animal Populations*. In his preface to this marvelous little book, Slobodkin wrote, "Every reader will find some material in this book that appears trivially obvious to him. I doubt, however, that all of it will appear obvious to any one person or that any two readers will be in agreement as to which parts are obvious. Bear with me when I repeat, in a naïve-sounding way, things you already know" (Slobodkin 1961: page v). Ditto.

Gary G. Mittelbach
March, 2012

Acknowledgments

I have had the privilege of teaching a graduate course in population and community ecology at Michigan State University for 25 years. At some point it occurred to me that I should take what I have learned from teaching this course and put it in a book. No doubt the thought of leaving something behind drives many an author to write a book, for as Peter Atkins notes, textbooks capture a mode of thinking. I have focused on that part of ecology that I find most exciting—community ecology. It was only after I was well into the writing that I realized how woefully ignorant I was about my chosen field. I still am ignorant, but less so now. Writing this book has made me a better student of ecology and a better teacher. I will count myself lucky if it helps others in the same way.

I am fortunate to have had five excellent co-instructors in our "Pop and Com" course at MSU over the years: Don Hall, Katherine Gross, Doug Schemske, Elena Litchman, and Kyle Edwards. I thank you all for helping make teaching a fun and rewarding experience. Thanks also to our students (500+ and counting). You listened and challenged, and I hope you will recognize your many contributions in these pages.

Special thanks to the many people who read and commented on early drafts of chapters: Peter Abrams, Andrea Bowling, Stephen Burton, Peter Chesson, Ryan Chisholm, Kristy Deiner, Jim Estes, Emily Grman, Jim Grover, Sally Hacker, Patrick Hanly, Allen Hurlbert, Sonia Kéfi, Jen Lau, Mathew Leibold, Jonathan Levine, Nancy McIntyre, Brian McGill, Mark McPeek, Carlos Melián, Sabrina Russo, Dov Sax, Oz Schmitz, Jon Shurin, Chris Steiner, Steve Stephenson, Katie Suding, Casey terHorst, Mark Vellend, Tim Wootton, and Justin Wright. I owe a particularly large debt to Peter Abrams, who piloted an early draft of this book in his graduate course at the University of Toronto and who provided many insightful comments in his usual, no-holds-barred style.

Interactions with Doug Schemske, Kaustuv Roy, Howard Cornell, Jay Sobel, David Currie, Brad Hawkins, and Mark McPeek have been instrumental in helping me think about broad-scale patterns of biodiversity. Likewise, conversions with Kevin Gross, Armand Kirus, Chris Klausmeier, Kevin Lafferty, Jonathan Levine, Ed McCauley, Craig Osenberg, Josh Tewksbury,

Earl Werner, and the "2010–2011 cohort" of postdocs at the National Center for Ecological Analysis and Synthesis (NCEAS) had a significant impact on the book. Thanks to my former graduate students for so many things and to my current graduate students for putting up with an advisor far too preoccupied with writing a book. Thanks also to Colin Kremer and Mark Mittelbach for their mathematical help. I gratefully acknowledge colleagues and staff at the Kellogg Biological Station for many years of support and friendship. I don't dare start naming names now, because there are too many people to thank. You know who you are and you know why you make KBS such a special place and that's enough. How was I ever so lucky to land here and somehow make it stick for a career?

This book has had a long gestation. When I first approached Andy Sinauer with a book proposal, my one request was that he not put time constraints on me, because I knew this would take awhile (and, secretly, I questioned whether I could pull it off at all). Andy graciously agreed, and he and the staff at Sinauer have been extraordinarily encouraging and helpful in every step of the processes. In particular, I thank my terrific editors, Carol Wigg and Norma Sims Roche, as well as art and production director Chris Small. I appreciate that Michigan State University granted me sabbatical leaves in 2001–2002 and 2010–2011, the first of which helped inspire this book; the second allowed me to finish it (almost).

Large parts of both sabbaticals were spent at the National Center for Ecological Analysis and Synthesis (NCEAS), a center funded by the National Science Foundation, the University of California at Santa Barbara, and the State of California. NCEAS provided the ideal environment for thinking and writing. This book would never have happened without its support. I am particularly grateful to Jim Reichman, Ed McCauley, Stephanie Hampton, and the wonderful NCEAS staff for their friendship and support. In spring 2011, Kay and I spent a short but magical time at EAWAG research institute on the shores of Lake Lucerne, Switzerland, where I worked on the final chapters of this book. I thank Ole Seehausen, Carlos Melián, and the scientists and staff at EAWAG Kastenienbaum for their hospitality and for making our brief stay productive and memorable.

I have enjoyed writing this book. My fond hope is that you enjoy reading it and will find it useful.

1 Community Ecology's Roots

Every genuine worker in science is an explorer, who is continually meeting fresh things and fresh situations, to which he has to adapt his material and mental equipment. This is conspicuously true of our subject, and is one of the greatest attractions of ecology to the student who is at once eager, imaginative, and determined. To the lover of prescribed routine methods with the certainty of "safe" results, the study of ecology is not to be recommended.

Arthur Tansley, 1923: 97

If we knew what we were doing, it wouldn't be called research.

Albert Einstein

The diversity of life on our planet is remarkable; indeed, among the biggest questions in all of biology are: How did such a variety of life arise? How is it maintained? What would happen if it were lost? Community ecology is that branch of science focused squarely on understanding Earth's biodiversity, including the generation, maintenance, and distribution of the diversity of life in space and time. It is a fascinating subject, but not an easy one. Species interact with their environment and with one another. As we will see in the pages that follow, these interactions underlie the processes that determine biodiversity. Yet, unlike the interacting particles studied by physicists, species also change through time—they evolve. This continual change makes the study of interacting species perhaps even more challenging than the study of interacting particles.

In his 1959 address to the American Society of Naturalists, G. Evelyn Hutchinson posed a simple question: "Why are there so many kinds [species] of animals?" Hutchinson's question remains as fresh and relevant today as it was half a century ago. In this book we will explore what ecologists understand about the processes that drive the distribution of animal and plant diversity across different spatial and temporal scales. In order to appreciate the current state of community ecology, however, it is important to know something about its history, particularly the development of ideas. This first chapter provides a brief summary of that history. Those of you familiar with the field may skip ahead, while those of you interested in learning more should consult the books and papers by Hutchinson (1978), Colwell (1985), Kingsland (1985), May and Seger (1986), McIntosh (1980, 1985, 1987), and Ricklefs (1987, 2004). Many of these "histories" were written by ecologists actively involved in the field's development, for community ecology is a relatively young science.

What Is a Community?

A **community** is a group of species that occur together in space and time (Begon et al. 2006). This definition is an operational one. Any limits on space and time are arbitrary, as are any limits on the number of species in a community. For example, we might refer to the study of "bird communities" or "fish communities" to delimit the assemblage of interest, recognizing that it is impossible to study all the species that occur together in the same place at the same time. Although most ecologists would be happy with this definition, the concept of what a community is and how is it organized has changed widely through time (Elton 1927; Fauth et al. 1996).

The first community ecologists were botanists who noted what appeared to be repeated associations between plant species in response to spatial and temporal variation in the environment. Frederic Clements, the pioneer of North American plant ecology, viewed these plant associations as a coherent unit—a kind of superorganism—and he presumed that plant communities followed a pattern of succession to some stable climax community (Clements 1916). Limnology, the study of lakes, also adopted a superorganism view of communities. In 1887, the limnologist Stephen Forbes published a famous paper entitled "The lake as a microcosm," in which he stated that all organisms in a lake tend to function in harmony to create a system in balance. Thus, these early ecologists tended to view communities as unique entities, and they became preoccupied with classifying plant and lake communities into specific "types."

This superorganism concept of communities did not sit well with everyone. It was soon questioned by a number of plant ecologists, most notably Henry Gleason (1926) and Arthur Tansley (1939). Gleason asserted that species have distinct ecological characteristics, and that what appear to be tightly knit associations of species on a local scale are in fact the responses of individual species to environmental gradients. Gleason's individualistic concept of communities was ignored at first, but later asserted itself and led to a rejection of Clements's hypotheses in favor of a continuum or gradient theory of plant distributions (Whittaker 1956).

The debate between Clements and Gleason over the nature of communities may seem like a historical footnote today, but at its core is a question that is very much alive: To what extent are local communities—the collections of species occurring together at a site—real entities? Ricklefs (2004) suggested that "ecologists should abandon circumscribed concepts of local communities. Except in cases of highly discrete resources or environments with sharp ecological boundaries, local communities do not exist. What ecologists have called communities in the past should be thought of as point estimates of overlapping regional species distributions." This focus on the interplay between local and regional processes in determining species associations is a theme that we will return to often in this book.

In contrast to plant ecology, the study of animal communities grew out of laboratory and field studies of populations. Animal population biologists, resource managers, and human demographers were concerned with the factors that regulate the abundance of individuals over time (birth, death, migration). Charles Elton, one of the pioneers of animal community ecology, worked for a time as a consultant for the Hudson's Bay Company, and his thinking was strongly influenced by the fluctuations he observed in the abundance of Arctic animals. Elton was opposed to the "balance of nature" concept espoused by Forbes and others, and in a book entitled *Animal Ecology* (1927), he discussed such important ideas as food webs, community diversity and community invasibility, and the niche. In another book, *The Ecology of Animals*, Elton (1950: 22) proposed that communities have limited membership, stating that in any prescribed area, "only a fraction of the forms that could theoretically do so actually form a community at any one time." Elton went on to note that for animals as well as humans, it appears that "many are called, but few are chosen."

Elton's idea of limited membership was a significant insight, and it meshed well with the concurrent development of mathematical theories of population growth and species interactions. In the 1920s, mathematical ecologists Alfred Lotka (1925) and Vito Volterra (1926) independently developed the now famous equations that bear their names, which describe competition and predation between two or more species. These mathematical models showed that two species competing for a single resource cannot coexist. Gause (1934) experimentally tested this theory with protozoan populations growing in small bottles on a single resource. He found that species grown separately achieved stable densities, but that when pairs of species were grown together, one species always won out and the other species went extinct (**Figure 1.1**). Other "bottle experiments" with fruit flies, flour beetles, and annual plants produced similar results. The apparent generality of these results led to the formulation of what became known as Gause's **competitive exclusion principle**, which can be stated as "two species cannot coexist on one limiting resource." The competitive exclusion principle had a profound effect on animal ecology at the time and, in a modified form, became a cornerstone of the developing field of community ecology.

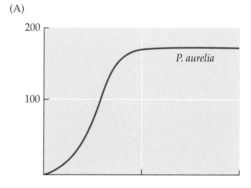

(A)

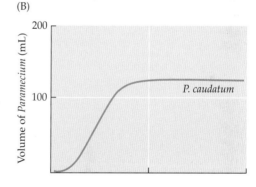

(B)

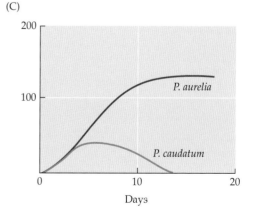

(C)

Figure 1.1   Results of Gause's competition experiments with two *Paramecium* species (*P. aurelia* and *P. caudatum*) grown separately and together in small containers in the laboratory. (A, B) Each species reached a stable population size (carrying capacity) when grown in isolation. (C) When grown together, however, one species always outcompeted and eliminated the other. (After Gause 1934.)

In the 1940s and 1950s, there was vigorous debate over the competitive exclusion principle and whether populations were regulated by density-dependent or density-independent factors. Important figures in this debate were Elton, Lack, and Nicholson on the side of competition and density-dependent regulation and Andrewartha and Birch on the side of density independence. In 1957, a number of ecologists and human demographers met at the Cold Spring Harbor Institute in Long Island, New York, to debate the issues of population regulation, with little consensus. However, this symposium did lead to one remarkable result. At the end of the published conference proceedings is a paper by G. E. Hutchinson (1957), modestly entitled "Concluding Remarks." In this paper, Hutchinson formalized the concept of the niche and ushered in what I consider the modern age of community ecology.

The Ecological Niche

The concept of the **niche** has a long history in ecology (see Chase and Leibold 2003a for an excellent summary). Grinnell (1917) defined the niche of an organism as the habitat or environment it is capable of occupying. Elton (1927) independently defined the niche as the role a species plays in the community. Gause (1934) made the connection between the degree to which the niches of two species overlap and the intensity of competition between them. Each of these concepts of the niche was incorporated into Hutchinson's thinking when he formalized the niche concept and connected it to the problem of species diversity and coexistence (Hutchinson 1957, 1959). In his "Concluding Remarks," Hutchinson showed how we might quantify an organism's niche, including both biotic and abiotic dimensions of the environment, as axes of an *n*-**dimensional hypervolume** (**Figure 1.2**).

Hutchinson (1957) went on to distinguish between an organism's **fundamental** (or *preinteractive*) **niche** and its **realized** (or *postinteractive*) **niche**. The fundamental niche encompasses those parts of the environment a species could occupy in the absence of interactions with other species, whereas the realized niche encompasses those parts of the environment that a species actually occupies in the presence of interacting species (e.g., competitors and predators). In Hutchinson's view, a species' realized niche was smaller than its fundamental niche due to negative interspecific interactions. However, positive interactions between species (mutualisms, commensalisms) can result in a species occupying portions of the environment that were previously unsuitable; in other words, it is possible for the realized niche to be *larger* than the fundamental niche (Bruno et al. 2003). The fact that positive interactions were not explicitly considered in Hutchinson's niche concept shows how completely the ideas of competition and predation permeated ecological thinking at the time.

Hutchinson's definition of the niche provided the framework on which ecologists would build a theory of community organization based on interspecific competition. First, however, they needed to make Hutchinson's concept more workable. An "*n*-dimensional niche" is fine in the abstract, but empirically, it is impossible to measure an undefined number of niche dimensions. It took one of Hutchinson's students, Robert MacArthur, to

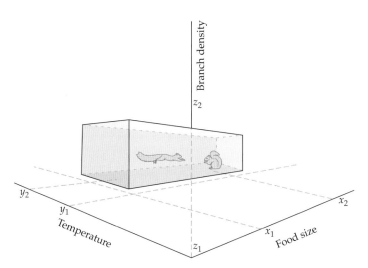

Figure 1.2 Hutchinson's visualization of the niche as an n-dimensional hyper-volume. In this hypothetical example, the fundamental niche of a squirrel species is shown along three environmental dimensions. One axis (here labeled y) might define the range of temperatures tolerated by the species, another dimension (x) might describe the range of seed sizes (e.g., acorns) eaten, and a third axis (z) the range of tree branch densities (diameter, volume) that support this squirrel species. Subscripts 1 and 2 represent the upper and lower limits for each niche dimension. (After Hutchinson 1978.)

make the concept operational. MacArthur's approach (1969) was to focus on only a few critical niche axes—those for which competition occurs. If, for example, interspecific competition for seeds limits the number of seed-eating birds in a community, then we should focus our study on some measure of seed availability to define a species' niche (e.g., seed size). Thus, it became possible to examine the distribution of species in "niche space" within a community (see Figure 8.5). More importantly, MacArthur's approach showed how we might quantitatively link overlap in niche space to the process of competitive exclusion (MacArthur 1972).

Shortly after the publication of his "Concluding Remarks" in 1957, Hutchinson (1959) provided another key insight in his published presidential address to the American Society of Naturalists, entitled "Homage to Santa Rosalia, or why are there so many kinds of animals?" Here, Hutchinson pondered a question that goes a step beyond Gause's competitive exclusion principle: If the competitive exclusion principle is true and interspecific competition limits the coexistence of species within the same niche, then how dissimilar must species be in their niches in order to coexist? Hutchinson suggested that the answer might be found in the seemingly regular patterns of difference in body size among members of an **ecological guild**—co-occurring species that use the same types of resources. He noted that species that were similar in most ways except the sizes of prey eaten tended to differ by a constant size ratio: a factor of about 1.3 in length and 2.0 in body mass. Such regular differences in size among coexisting species

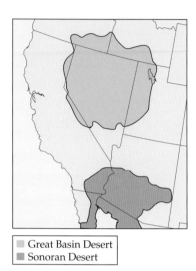

Great Basin Desert
Sonoran Desert

Figure 1.3 Hutchinsonian ratios among desert rodents found in the Great Basin and Sonoran Deserts of the western United States. Differences in body size reflect differences in diet and habitat use (niche differences) between these species. The pattern of body size spacing observed in these two desert rodent ecological guilds is more regular than would be expected by chance. (From Brown 1975.)

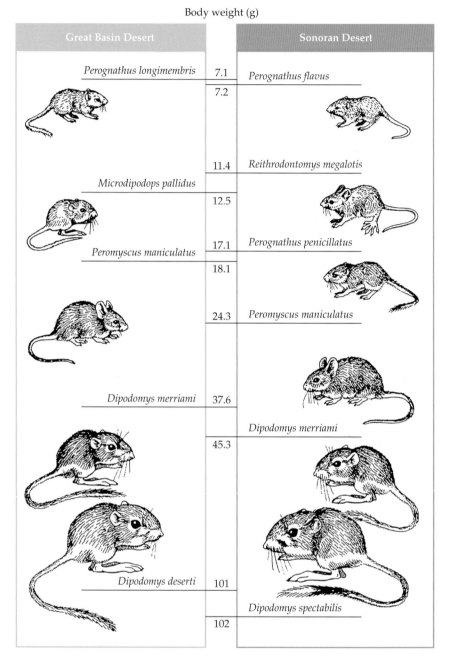

Body weight (g)

Great Basin Desert		Sonoran Desert
Perognathus longimembris	7.1	Perognathus flavus
	7.2	
	11.4	Reithrodontomys megalotis
Microdipodops pallidus	12.5	
Peromyscus maniculatus	17.1	Perognathus penicillatus
	18.1	
	24.3	Peromyscus maniculatus
Dipodomys merriami	37.6	
	45.3	Dipodomys merriami
Dipodomys deserti	101	
	102	Dipodomys spectabilis

were found in many ecological guilds (one example is shown in **Figure 1.3**) and became known as **Hutchinsonian ratios**.

MacArthur and Levins (1967) built on these ideas and introduced the concept of **limiting similarity**, which specified the minimal niche difference between two competing species that would allow them to coexist. In MacArthur and Levins's theory, species are arrayed linearly along a

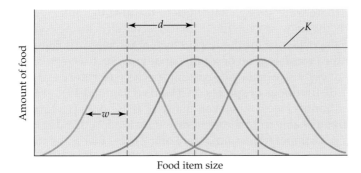

Figure 1.4 The concept of limiting similarity illustrated for three species utilizing a continuum of food resources. *K* represents a resource continuum (for example, the amount of food as a function of food size). Each species' niche is represented by the mean and the standard deviation (*w*) of its resource utilization curve, and *d* is the distance between the mean resource uses of the closest pair of species. MacArthur and Levins (1967) and May and MacArthur (1972) showed that the minimal niche separation required for the coexistence of competing species (under very specific conditions) is *d/w* ≈ 1. (After May and MacArthur 1972.)

resource (niche) axis, and each species' resource use is represented by a normal (bell-shaped) utilization curve (**Figure 1.4**). The overlap between adjacent utilization curves can be used (under specific assumptions) as a measure of the competition coefficients (α's) in the Lotka–Volterra model of interspecific competition (described in Chapter 7). Using this model of competition, MacArthur and Levins (1967) were able to specify the minimum niche difference required for two species to coexist. Later, May and MacArthur (1972) and May (1973b) used a different approach, based on species in fluctuating environments, to arrive at a very similar outcome: the limiting similarity between two competing species is reached when d/w ≈ 1, where d is the separation in mean resource use between species and w is the standard deviation in resource use (see Figure 1.4).

Whither Competition Theory?

In less than 50 years, animal community ecology progressed from the simple recognition that species too similar in their niches cannot coexist to the development of a theoretical framework poised to predict the number and types of species found in natural communities based on a functional limit to the similarity of competing species. This was an enormous leap forward, and community ecology seemed well on its way to becoming a more quantitative and predictive science. The heady optimism of the times is reflected in Robert May's (1977a: 195) comment that "the question of the limits to similarity among coexisting competitors is ultimately as deep as the origin of species itself: although undoubtedly modified by prey–predator and mutualistic relations, such limits to similarity are probably the major factor determining how many species there are." In the end, however, the theory of limiting similarity failed to achieve its promise. What happened?

First, there were strong challenges to the idea that interspecific competition is the only, or even the primary, factor structuring communities. Much of the evidence for the importance of interspecific competition in communities was based on descriptive patterns, such as regularly spaced patterns of body size among coexisting species (see Figure 1.3) or "checkerboard" distributions of species on islands (Diamond 1973, 1975). When examined more closely, however, many of these patterns turned out to be indistinguishable from those predicted by models that did *not* include interspecific competition as an organizing force—that is, by null models (Strong et al. 1979; Simberloff and Conner 1981; Gotelli and Graves 1996). Second, predictions of limiting similarity between species turned out to be model-dependent. That is, even though most mathematical models of interspecific competition predict some limit to how similar species may be in their resource use and still coexist, that limit varies widely depending on the assumptions and structure of the model. Thus, as noted by Abrams (1975, 1983b), there is no hard or fixed limit to similarity between species. These challenges caused ecologists to look beyond interspecific competition and to consider the plurality of factors that might determine species diversity. In contrast to the unbridled optimism that characterized community ecology in the 1960s and early 1970s, the next decade was a period of soul searching as ecologists struggled to find a conceptual framework to replace what had seemingly been lost (McIntosh 1987). In the end, however, the idea that communities are organized around strong interspecific interactions was not so much wrong as it was overly simplistic.

New Directions

The "failure" of simple competition-based models to explain community diversity led to important new directions in community ecology, and many of these directions continue to influence how we study ecology. For example, the "null model debate" of the 1970s led directly to an increased emphasis on using field experiments to test ecological hypotheses. Many of these field experiments focused on studying interspecific competition. However, pioneering experiments by Paine (1966), Dayton (1971), and Lubchenco (1978), all working in the marine intertidal zone, also showed that the presence or absence of predators could have dramatic effects on species diversity. These experiments set the stage for a wealth of future work on food webs, trophic cascades, and top-down effects. We will consider these topics in detail in subsequent chapters, as well as more recent approaches to characterizing food webs and other types of ecological networks (see Chapters 10 and 11).

The experimental studies cited above demonstrated how competition and predation may interact to affect species diversity and composition (for example, diversity is increased when predators feed preferentially on a competitive dominant), again setting the stage for subsequent empirical and theoretical work on keystone predation and competition–predation trade-offs. Over time, the accumulation of results from multiple field experiments fostered the application of **meta-analysis** in ecology, in which the outcomes of many experiments are combined and synthesized to arrive at general conclusions (Gurevitch et al. 1992; Osenberg et al. 1999). Today, ecologists rely heavy on meta-analysis, and there will be many times in this book when we will look to the results of meta-analyses to evaluate the importance of a process in ecology.

The "failure" of a single process (interspecific competition) to account for many of the patterns in species diversity observed in the 1960s and 1970s led ecologists to take a more pluralistic approach to their science (Schoener 1986; McIntosh 1987). A pluralistic ecology recognizes that multiple factors may interact to determine the distribution and abundance of species. The difficulty with pluralism, however, is that it can quickly lead us into a morass. In a recent, thought-provoking article, Vellend (2010: 183) suggested that "despite the overwhelmingly large number of mechanisms thought to underpin patterns in ecological communities, all such mechanisms involve only four distinct kinds of processes: selection, drift, speciation, and dispersal." In Vellend's framework, "selection" encompasses the processes that determine the relative success of species within a local community (e.g., competition, predation, disease), whereas "drift" refers to changes in species' relative abundances due to chance or other random effects, and "dispersal" is the movement of individuals and species into and out of local communities. "Speciation" operates over spatial scales larger than the local community, and it is the process that ultimately generates diversity in regional species pools. Vellend (2010) suggests that conceptual synthesis in community ecology can be achieved by focusing on these four major drivers of species diversity patterns at different spatial and temporal scales (**Figure 1.5**). I agree with Vellend's suggestion. Moreover, I think we are further challenged as community ecologists to illuminate the interior of the "black box" in Figure 1.5, and to better understand how the four basic processes of community ecology interact to determine patterns of biodiversity.

The recognition that local communities bear the footprint of historical and regional processes (Ricklefs and Schluter 1993) is an important insight that grew out of the narrow, local community focus of the 1960s and 1970s. Interestingly, MacArthur (1972) anticipated this paradigm shift, but he

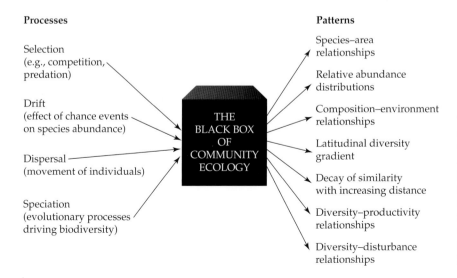

Figure 1.5 A conceptual view of the functioning of community ecology, in which four basic processes (selection, drift, speciation, and dispersal) combine to determine the biodiversity patterns listed on the right. The "black box of community ecology" refers to the fact that there are many ways in which the four processes listed at the left may combine to produce the patterns listed at the right. (After Vellend 2010.)

died too young to be a part of it (see discussion in McIntosh 1987). Simply put, few communities exist in isolation. Instead, the diversity of species within a community is a product of their biotic and abiotic interactions (i.e., species sorting or "selection," along with drift), the dispersal of species between communities, and the composition of the regional species pool (a function of biogeography and evolutionary history). Therefore, we need to consider the processes that regulate diversity at a local scale as well as the processes that link populations and communities into metapopulations and metacommunities and the processes that ultimately generate diversity at regional scales. This is a tall order. In the following chapter, I will use broad-scale diversity gradients, particularly the latitudinal diversity gradient, as a vehicle to begin to think about the processes that generate regional diversity. At geographic scales of regions or continents, biodiversity is a function of evolutionary processes that may play out over millions of years. In addition, chance events in Earth's history can influence a region's size, geomorphology, climate, and the amount of time available for speciation. These historical factors conspire to make our study of the processes that determine regional biodiversity challenging. Of course, there is also little opportunity to do experiments at such vast scales of time and space. But, as we will see, recent advances in molecular biology, phylogenetics, paleontology, and biogeography have greatly facilitated the study of broad-scale diversity patterns, and these new tools are providing the keys to understanding the factors that generate biodiversity at regional scales.

The Big Picture

Patterns, Causes, and Consequences of Biodiversity

2 Patterns of Biological Diversity

Biodiversity, the variety of life, is distributed heterogeneously across the Earth.
Kevin Gaston, 2000: 1

To do science is to search for repeated patterns, not simply to accumulate facts.
Robert MacArthur, 1972: 1

Why do different regions of Earth vary so dramatically in the number and types of species they contain? This question has challenged ecologists and evolutionary biologists for a very long time. The early naturalists were struck by the remarkable diversity of life they found in the humid tropics (e.g., von Humboldt 1808; Darwin 1859; Wallace 1878). As an example, there are approximately 600 tree species in all of North America (Currie and Paquin 1987), whereas a tropical rainforest may contain 600 tree species in just a few hectares (Pitman et al. 2002), and the New World tropics as a whole support an estimated 22,500 tree species (Fine and Ree 2006). For the past 200 years ecologists have sought to explain why the species richness of most taxa increases dramatically from the poles to the equator (**Figure 2.1**). In his review of almost 600 published studies, Hillebrand (2004) documents that this **latitudinal diversity gradient** exists for nearly all taxonomic groups. The strength of the latitudinal diversity gradient does not differ between plants and animals, nor between marine and terrestrial environments, nor between warm- and cold-blooded organisms (**Figure 2.2**).

The striking, repeatable increase in species richness from the poles to the tropics is without a doubt the prime example of what Robert MacArthur (1972) referred to when he said that ecologists should "look for repeated patterns" in nature. Robert Ricklefs called it "the major, unexplained pattern in natural history … one that mocks our ignorance" (quoted in Lewin 1989: 527). In this chapter, we will first examine the factors thought to drive variation in species diversity across broad spatial scales, using the latitudinal diversity gradient as a focus. We will then shift our attention to patterns of biodiversity at smaller spatial scales, exploring the relationships between productivity and species richness and between area and species richness; the distribution of species abundances; and finally the relationship between

(A) Bird species

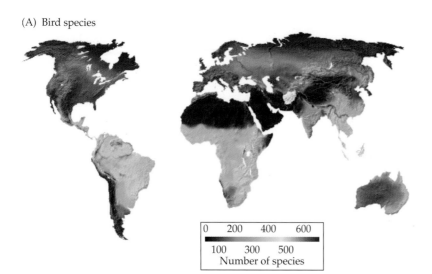

(B) Angiosperm families

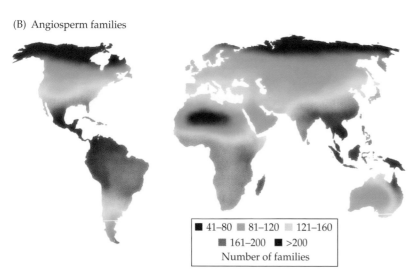

(C) Shallow marine gastropods

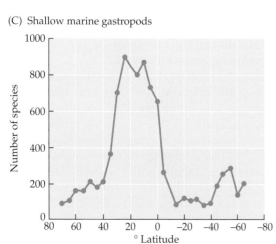

(D) Freshwater phytoplankton

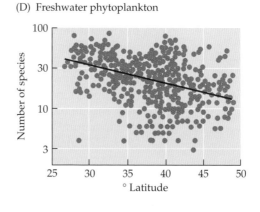

Figure 2.1 Examples of latitudinal gradients in biodiversity. Most biologists are familiar with polar–tropical gradients in large, conspicuous organism such as birds, mammals, and trees. However, recent work documents that latitudinal diversity gradients exist for a wide variety of taxa, including microorganisms, and that these gradients occur in terrestrial, marine, and freshwater habitats. (A) Bird species richness. (B) Angiosperm family richness. (C) Marine gastropod species richness. (D) Freshwater phytoplankton species richness. (A from Hawkins et al. 2008; B from Francis and Currie 2003; C after Rex et al. 2005; D after Stomp et al. 2011.)

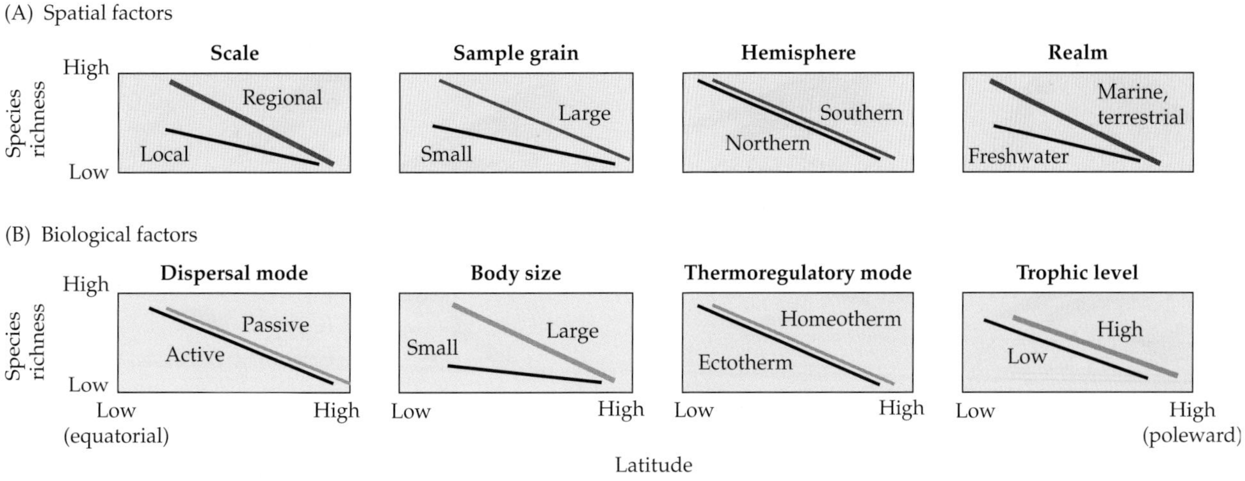

(A) Spatial factors

(B) Biological factors

Figure 2.2   Data compiled from nearly 600 studies of species richness show the trend of a latitudinal gradient—i.e., increasing species richness closer to the equator—for spatial (A) and organismal (B) factors. The graphs illustrate in stylized fashion the observed change in species richness from low to high latitude for eight variables, as well as the relative impact (shown by the difference in the two slopes) and correlational strength (shown by the relative line thickness) of each variable with the latitudinal gradient. (After Hillebrand 2004.)

local and regional species richness. These and other empirical patterns of biodiversity constitute much of what community ecology seeks to explain about nature (see Figure 1.5). An awareness of how species diversity varies across space and time also provides a foundation from which to explore the mechanisms of species interactions and to understand the processes that drive variation in species numbers and their distribution.

Before we delve into patterns of species diversity, however, we will take a moment to consider some important aspects of assessing species diversity at different spatial scales.

Assessing Species Diversity in Space

Robert Whittaker (1960, 1972) provided a foundation for measuring diversity at different spatial scales when he proposed that species diversity could be expressed in three ways:

1. **Alpha (α) diversity**, the number of species found at a local scale (i.e., within a habitat or local site)
2. **Beta (β) diversity**, a measure of the difference in species composition, or species turnover, between two or more habitats or local sites within a region
3. **Gamma (γ) diversity**, a measure of species richness in a region

Note that ecologists can and do measure other aspects of biodiversity besides numbers of species; they can also study genetic diversity, functional group diversity, or the relative abundance of species, among other measure-

Region X
5 species

Site 1
5 species

Region Y
6 species

Site 2
3 species

Site 3
5 species

Site 4
3 species

Figure 2.3 An illustration of alpha (α), beta (β), and gamma (γ) diversities as measured at four sites and in two regions (X and Y). Alpha diversity quantifies diversity at the local scale (i.e., within a given site or habitat). In this example, alpha diversity is greatest at sites 1 and 3 (five species each) and lower at sites 2 and 4 (three species each). Beta diversity measures the change in diversity between sites—that is, the amount of species turnover within a given region. In this example, region Y has higher beta diversity than region X (i.e., the same five species are found at both sites in Region X, whereas there is no overlap between the three species present at each of the two sites comprising Region Y). Gamma diversity measures the total diversity within a region, assessed across all sites. In this example, gamma diversity is higher in region Y (six species) than in region X (five species). (After Perlman and Anderson 1997.)

ments (Magurran and McGill 2011). For simplicity, however, we will limit our discussion here to measures of species richness; the ideas can easily be extended to other aspects of biodiversity.

The three diversity measures proposed by Whittaker are simple in principle but can be problematic to apply in the real world. The concept of alpha diversity—determining the number of species found in a specific habitat or site—is straightforward. However, the boundaries of a habitat or site are often subjective, and getting an accurate measure of species richness at a site involves important sampling considerations. Gotelli and Colwell (2001) and Magurran and McGill (2011) provide good advice on how to best measure alpha diversity in nature.

The most contentious of the diversity measures is beta diversity, which measures the difference in species composition, or species turnover, between two of more sites within a given region (Magurran 2004). Recently, there has been much discussion over the best ways to define and calculate beta diversity. Tuomisto (2010a,b) and Anderson et al. (2011) provide thoughtful summaries of how beta diversity can be measured and how the concept can be applied to different systems.

Finally, gamma diversity combines alpha and beta diversities to yield the total number of species found in a region (**Figure 2.3**). You might ask, "What do we mean by a region?" That is a very good question. Ecologists use the concept in a number of different ways, depending on the question

TABLE 2.1   Some spatial designations in community ecology[a]

Term	Definition
Biogeographical region/realm	Extremely large spatial extent, roughly corresponding to continental and subcontinental areas. Spatial limits coincide with major geological and climatic barriers that are the result of plate tectonics; these barriers (e.g., oceans, mountains, deserts) limit the dispersal of organisms over evolutionary time. Functionally defined as portions of the globe that have a shared evolutionary history and that consequently contain endemic and related taxa. There is rapid turnover in species at the boundaries of biogeographic regions.
Province	Similar to a biogeographical region/realm.
Biome or ecoregion	Very large spatial extent. Terrestrial biomes are functionally defined by climate and the vegetation and animals associated with specific climatic conditions, wherever on the globe those climatic conditions occur. Examples: desert, tropical rain forest, temperate grassland, tundra. Aquatic biomes, like terrestrial biomes, are characterized by assemblages of organisms adapted to similar conditions; they are functionally defined not by climate but by a variety of factors including salinity, water temperature and movement, light penetration, and substrate composition. Examples: River, lake, salt marsh, intertidal, open ocean.
Region	Large spatial extent, covering many square kilometers and containing a large number of habitats and communities. Sometimes used synonymously with biogeographic region or province, but more commonly used to define a somewhat smaller spatial scale. Functionally defined as an area in which the processes of speciation and extinction operate to affect biodiversity and from which species, over time, have a good probability of colonizing a local community. For example, we may refer to the "regional species pool" as the collection of species likely to colonize a local community of interest.
Local	Small spatial extent, < 1 meter to a few square kilometers. Spatial extent is determined in part by the size of organisms within a community; e.g., from microorganisms living within a tree hole or a pitcher plant leaf, up to trees in a tropical forest. Functionally, a local area or local community is defined by the likelihood of species interactions. Within a local area or local community, species have a high probability of interacting with each other and influencing each other's dynamics.

[a] The terms listed here are widely used by community ecologists to describe spatially dependent relationships and processes. Although the definitions given reflect common usage, strict definitions of these terms do not exist. It is important to note that the terms are defined both by their spatial extents (approximate area/size) and by their functional roles in community ecology (see Cornell and Lawton 1992).

of interest (**Table 2.1**). A **region** is generally considered to be large in spatial scale, extending over many square kilometers and containing a large number of habitats and communities. In addition, ecologists often use "region" to refer to an area from which species, over time, have a good probability of colonizing a local community of interest. The species likely to colonize a local community are collectively referred to as the **regional species pool**. Finally, ecologists studying global patterns of biodiversity often think about regions as parts of the globe that have distinct evolutionary histories—that is, geographic areas that are isolated by major physical boundaries. Wallace (1876) noted that Earth's biota show pronounced dissimilarities at major geographic boundaries such as continental margins, major mountain ranges, or climatic breaks. These biologically and climatically distinct regions are termed **biogeographic regions** or **biogeographic realms** (Kreft and Jetz 2010). In the study of global patterns of diversity, biogeographic regions are often used to delineate regional diversity.

Recently, McGill (2011: 482) suggested that alpha and gamma diversities are best defined by the processes that are dominant at each scale rather than by actual measurements of area. "Thus, the γ-scale is that at which evolutionary process (specifically speciation and global extinction) determine biodiversity and variation is driven by broad-scale gradients such as climate ... area ... and biogeographical, historical contingencies. ... In contrast, α-diversity is the scale at which processes traditionally studied in community ecology such as dispersal limitation, microclimate, and species interactions predominate." I agree with McGill's sentiments. Although we might wish to have more precise definitions for the three spatial components of diversity (alpha, beta, and gamma), they are inherently relative concepts. In the following section, we will examine the processes that generate diversity at regional scales (gamma diversity), focusing on the potential drivers of Earth's most pronounced biodiversity pattern: the latitudinal diversity gradient.

Explaining the Latitudinal Diversity Gradient

Dozens of hypotheses have been offered to explain the latitudinal diversity gradient (Willig et al. 2003; Mittelbach et al. 2007). These hypotheses can be grouped into four general categories:

1. "Null-model" explanations based on geometric constraints on species ranges distributed across the globe
2. Ecological hypotheses that focus on an area's carrying capacity for species
3. Historical explanations based on geologic history and the time available for diversification
4. Evolutionary hypotheses that focus on rates of diversification (speciation minus extinction)

We will consider these four classes of explanations below, recognizing that regional diversity may be determined by a combination of factors working in concert; the critical question is the relative importance of the different processes.

A null model: Geometric constraints and the "mid-domain effect"

A **null model** in ecology attempts to specify how a relationship or pattern in nature should look in the absence of a particular process or mechanism (for example, how might two species be distributed across a habitat in the absence of interspecific competition). Gotelli and Graves (1996: 3) give a more formal definition:

> [A null model can be seen as] a pattern-generating model that is based on randomization of ecological data or random sampling from a known or imagined distribution. Certain elements of the data are held constant, and others are allowed to vary stochastically to create new assemblage patterns. The randomization is designed to produce a pattern that would be expected in the absence of a particular mechanism.

Distribution
limit X   "Mid-domain"

Distribution
limit Y

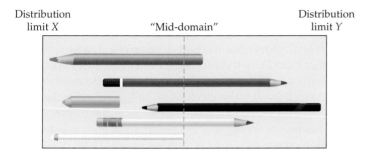

Figure 2.4   The mid-domain effect illustrated by a simple thought experiment with a pencil box. Species' ranges are represented by the pencil lengths, and limits to species distributions (by analogy, Earth's poles) are represented by the ends of the pencil box. Shaking the box so as to randomly distribute species ranges between their upper and lower limits results in ranges overlapping more in the middle of the box than at the ends; species richness is highest in the center (by analogy, near the equator).

The power of the null model approach is that we can specify the probability that an observed pattern in nature differs from the predicted "null pattern," much as we can specify how an observed experimental result may differ significantly from a random expectation in a statistical model. Null models have a rich history in community ecology. They were instrumental, for example, in focusing debate about the importance of interspecific competition during the late 1970s and early 1980s (see Chapter 1; see also Gotelli and Graves 1996).

Colwell and Hurtt (1994) proposed that the latitudinal gradient in species diversity may simply reflect the outcome of placing species ranges on a bounded domain (the globe). They reasoned as follows: Imagine a collection of different-sized pencils within a pencil box (**Figure 2.4**). Each pencil's length represents the range of a given species, and the sides of the box represent constraints placed on the distribution of these ranges (for example, by the harsh environment of Earth's poles). When the box is shaken, we find that the pencils overlap most in the center of the box. By analogy, species ranges are expected to overlap more at the equator than at the poles, and therefore species richness should show a gradient with latitude. This explanation for the latitudinal diversity gradient has been termed the **mid-domain effect**, and Colwell and colleagues suggest that it offers an appropriate null model against which other explanations for the latitudinal gradient should be measured (e.g., Colwell and Hurtt 1994; Colwell et al. 2004, 2005).

The mid-domain effect has generated considerable debate among ecologists. It has strong supporters as well as critics. Its critics have focused on two main issues: (1) whether the model is truly "neutral" to the processes that it attempts to rule out (e.g., because observed species ranges reflect the influence of climate, history, and species interactions), and (2) technical issues of model development and testing with empirical data (McCain 2007). Recent studies suggest that the mid-domain effect provides a reasonable explanation for diversity gradients within certain bounded regions, such as mountains (Watkins et al. 2006; Brehm et al. 2007; but see McCain 2007) and rivers (e.g., Dunn et al. 2006). However, by itself, the mid-domain effect appears unlikely to account for diversity gradients on a global scale. For example, using the global distribution of bird species, for which ecologists have remarkably complete data, Storch et al. (2006) predicted how bird species diversity should vary with latitude based on random dynamics of species ranges (using a generalized spreading dye model; see Jetz and Rahbek 2001). Storch et al. found that the mid-domain effect, when applied globally, explained less than 2% of the variation in avian richness. When applied within biogeographic regions, the mid-domain model did much better. However, when used in this way, the model *a priori* incorporates regional (i.e., tropical/temperate) differences in species richness and thus provides a weak test. Most interesting, however, was the observation that mid-domain type models that allowed the locations of species ranges to be positively influenced by available energy (i.e., species ranges spread preferentially into more productive areas) provided a good fit to the global bird data set. Thus, while the mid-domain effect per se appears inadequate to explain the global latitudinal diversity gradient, linking climate to species range dynamics suggests a promising avenue for future research (Gotelli et al. 2009).

Ecological hypotheses: Climate and species richness

A layperson's response to the question of why are there more species in the tropics than at the poles might be, "It's the climate, stupid." Climatic variables (e.g., mean and variance in temperature and precipitation) do explain much of the global variation in taxonomic richness (see O'Brien 1998; Kreft and Jetz 2007). Hawkins et al. (2003) found that the best single climatic variable affecting a given region explained, on average, about 64% of the variance in species richness across a broad range of taxa, and measures of energy (solar radiation, temperature), water (precipitation), or water–energy balance consistently explained variation in species richness better than other climatic and non-climatic variables. Water variables usually represented the strongest predictors of species richness in the tropics, subtropics, and warm temperate zones, whereas energy variables (for animals) and water–energy balance variables (for plants) were the strongest predictors in high latitudes (Hawkins et al. 2003; Whittaker et al. 2007; Eiserhardt et al. 2011).

Contemporary climate and species diversity are strongly correlated, especially at broad spatial scales (Field et al. 2009). But what is the mechanism behind this correlation? Of the many ecological hypotheses that have been offered to explain why there are more species where it is warm and wet, perhaps the most prominent is the **species–energy hypothesis** (also called the "more individuals" hypothesis). Climate strongly affects primary productivity, and this hypothesis postulates that the number of individuals an area can support should increase with primary productivity (energy). It also postulates that species richness varies as a function of the total number of individuals in an area because over evolutionary time a greater number of individuals can become divided into more species with viable population sizes. Therefore, there should be more species in climate zones with higher productivity, such as the humid tropics (Hutchinson 1959; Brown 1981; Wright 1983).

Currie et al. (2004) examined the predictions of the species–energy hypothesis in detail. They found that the number of individuals within a taxon (trees, birds) tended to increase with primary productivity (as measured by actual evapotranspiration, AET) and that the number of species tended to increase with the number of individuals (**Figure 2.5**). Thus, the two basic

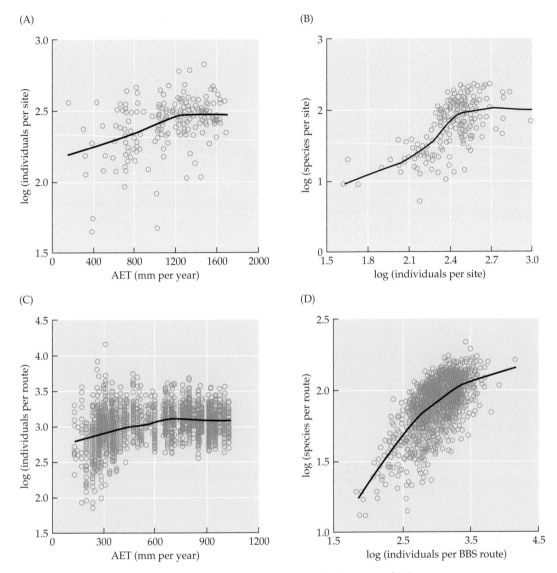

Figure 2.5 Components of the species–energy hypothesis (also known as the "more individuals hypothesis") as applied to the latitudinal diversity gradient. The premise reasons that (1) more productive environments, such as the humid tropics, can support more individuals; and (2) areas that support more individuals will contain more species. (A, B) Data for trees worldwide; Gentry's counts of individuals in 0.1-ha plots. (C, D) Data for North American birds from the North American Breeding Bird Survey (BBS routes). Areas of higher productivity (as measured by annual actual evapotranspiration, AET) contain more individuals (panels A and C), and areas with more individuals have more species (panels B and D). Thus, these data support the two assumptions of the species–energy hypothesis. (After Currie et al. 2004.)

components of the species–energy hypothesis were supported. However, the combination of the two relationships was insufficient to account for the observed increase in bird and tree diversity with decreasing latitude (**Figure 2.6**). In other words, for these two well-studied taxa, species accumulate

Figure 2.6 Data for trees world-wide and for birds in North America (see Figure 2.5) suggest that the species–energy hypothesis cannot account for the increase in species richness observed for these two taxa across broad latitudinal (productivity) gradients. Species accumulate more rapidly than would be predicted (blue dashed line) based on known species abundance relationships. (Data from Currie et al. 2004.)

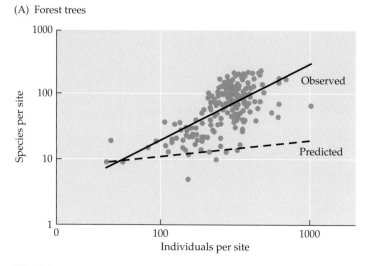

(A) Forest trees

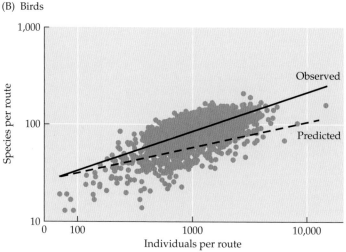

(B) Birds

more quickly as one moves from the temperate zone to the equator than would be predicted based on the increase in number of individuals (see also Evans et al. 2005, 2008). Currie et al. derived their predictions (dashed lines in Figure 2.6) from the famous species abundance distributions determined by Fisher et al. (1943) and Preston (1948, 1962), which we will discuss in more detail later in this chapter. Šímová et al. (2011) used an expanded global set of 370 "Gentry-style" (0.1 ha) forest plots to examine more closely how tree species richness varies as a function of the number of individual trees per plot. They concluded, as did Currie et al. (2004), that tree species richness does not simply follow the total number of individuals, and that variation in tree diversity worldwide is not simply the result of more productive areas supporting more individuals and therefore more species.

Why are there more species in the tropics than we would expect to find based on a fixed relationship between productivity, number of individuals, and number of species? At one level, the answer appears simple. There is growing evidence that the relationship between the number of individuals

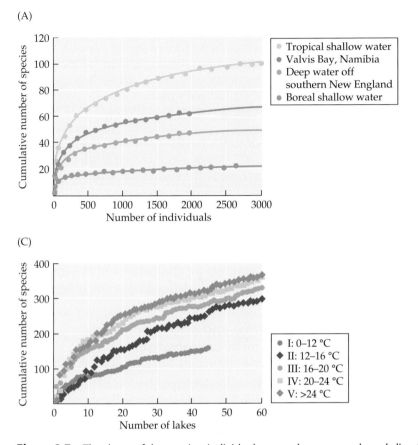

(A)

(B)

(C)

Figure 2.7   The shape of the species–individuals curve changes over broad climatic gradients. (A) The number of benthic invertebrate species found in samples collected from four marine sites ranging from boreal to tropical waters shows that many more species are encountered for the same sampling effort in warm tropical waters than in cold boreal waters. (B) Rarefaction curves for tree species richness in forest plots world-wide (constructed for sampling efforts of 100, 200, 300, 400, and 500 individual trees) show that more species are encountered for the same sampling effort at high-produc-tivity sites (darker shading) than at low productivity sites (lighter shading); productiv-ity is measured as AET. (C) The number of freshwater phytoplankton species found in North American lakes in samples from five different temperature bands shows that more species are encountered for the same sampling effort (number of lakes sampled) in regions of warm temperatures than in regions of cold temperatures. (A after Hubbell 2001, redrawn from May 1975, original data from Sanders 1969; B after Šímová et al. 2011; C after Stomp et al. 2011.)

sampled and the number of species accumulated—the **species–individuals curve**—varies predictably with climate, and that warmer regions (with higher productivity) support more species per number of individuals sampled (**Figure 2.7**). This pattern suggests that geographic variation in minimum viable population size (m) could give rise to the latitudinal diversity gradi-ent. That is, if species can persist at smaller population sizes in warmer, more equable climates, then we would expect similar sampling efforts to yield more species in tropical than in temperate climates. However, based on the relationships in Currie et al. (2004), a tenfold increase in richness would require a 6,600-fold decrease in minimum viable population size. Thus, it seems unlikely that variation in m could account for the orders-of-magnitude changes in species richness observed across latitude. Reed et al. (2003), in fact, observed no latitudinal variation in m for vertebrates, although more work on latitudinal variation in minimum viable population sizes is needed.

Latitudinal gradients in biodiversity have existed since before the time of the dinosaurs (e.g., Powell 2007, 2009), and most evidence points to a substantial increase in the magnitude of the latitudinal diversity gradient throughout the Cenozoic (the past 65 million years) (Crame 2001). Thus, evolution and the Earth's biogeographic history are likely to play important roles in the long-term development of regional differences in species diversity. Below we examine possible evolutionary and historical drivers of the latitudinal diversity gradient, focusing on two important mechanisms: (1) biogeography and the time available for speciation, and (2) latitudinal differences in **diversification rates** (simply defined as the rate of speciation minus the rate of extinction).

Historical hypotheses: The time-and-area hypothesis and the concept of tropical niche conservatism

Tropical environments are both older and more widespread in Earth's history than temperate environments. This fact alone could account for present-day differences in biodiversity between temperate and tropical regions. The idea that there has been more time for diversification in the tropics dates back to Wallace (1878), who argued that equatorial regions have suffered less from harsh climatic events (primarily "ice ages") that result in extinctions and thus set back the diversity clock. Early statements of this "time hypothesis" focused on the impacts of the Pleistocene glaciations. However, we now know that the latitudinal diversity gradient has roots far deeper than the Pleistocene, extending back through the Mesozoic (180 mya) both on land and in the oceans (Crame 2001; Powell 2007).

More recent discussions of the time hypothesis have focused on the relatively greater age and geographic extent of tropical environments (e.g., Ricklefs 1987; Wiens and Donoghue 2004; see review in Mittelbach et al. 2007). In particular, tropical environments reached their maximum extent in the early Tertiary (mid-Paleocene to early Eocene, 59–50 mya), when warm (>18°C) surface waters extended to the Arctic and the equator-to-pole temperature gradient was much reduced (Sluijs et al. 2006; Graham 2011). During this time, tropical floras existed as far north as London and southern Canada. This thermal maximum was followed by a cooling trend beginning around 45 mya and continuing to the present (**Figure 2.8**). Thus, tropical environments are older than temperate environments and therefore should have provided a longer "effective" time (i.e., time free from major extinction events as well as absolute time) for diversity to accumulate.

In addition, the greater area of the tropics during the early Cenozoic could have contributed to higher diversification rates due to area effects on rates of speciation and extinction (Terborgh 1973; Rosenzweig 1995;

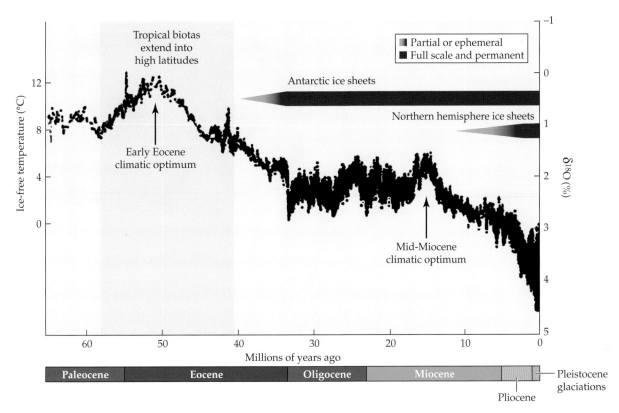

Figure 2.8 Global climate during the Cenozoic era (i.e., over the past 65 million years). A prolonged warm period (shown in yellow) persisted throughout the late Paleocene and the first half of the Eocene, during which tropical conditions extended into the high latitudes and the polar oceans were ice-free. This warm period was followed by a gradual cooling to today's climate. The measurement $\delta^{18}O$ (right-hand axis) is a proxy for temperature based on the percentage of oxygen-18 isotopes in foraminiferan fossils; the ice-free temperature scale (left axis) applies only to the warm period, which ended with the onset of large-scale glaciation in the Antarctic about 40 million years ago. (After Zachos et al. 2008.)

Chown and Gaston 2000). A number of lines of evidence support this **time-and-area hypothesis**. For example, Fine and Ree (2006) show that variation in current tree species richness among boreal, temperate, and tropical biomes is significantly correlated with the area that each biome has occupied integrated over time since the Eocene. Further, most temperate tree taxa, including such major clades as the Fagales (birches, beeches, and oaks), are embedded within ancestrally tropical groups, suggesting that tropical lineages have had more time to diversify (Ricklefs 2005). The fact that many lineages of plants and animals show tropical origins, but only a few of their clades have managed to reach the temperate zone, has led a number of ecologists to hypothesize that the transition from tropical to temperate environments presents an important physiological hurdle (e.g., Farrell et al. 1992; Latham and Ricklefs 1993; Kleidon and Mooney 2000). Although such **tropical niche conservatism** (Wiens and Donoghue 2004; Wiens et al. 2009) appears common among terrestrial organisms, it is less evident in the ocean (Roy et al. 1996; Jablonski et al. 2006), where, at least for marine mollusks, the pattern is one of tropical origin and then dispersal into temperate and polar regions (the "out of the tropics" hypothesis; see Jablonski et al. 2006; Kiessling et al. 2010).

Current latitudinal gradients in taxonomic diversity bear the clear imprint of history. However, few of the species that comprise these current diversity gradients date back 40–60 million years, to when tropical and subtropical environments dominated Earth (Antonelli and Sanmartin 2011). Thus, whereas the relative age and extent of tropical environments may

Figure 2.9 Latitudinal diversity gradient for marine brachiopod genera through late Paleozoic time, shown as a contour plot. Note that the diversity gradient weakened during the late Paleozoic ice age and then re-formed. (After Powell 2007.)

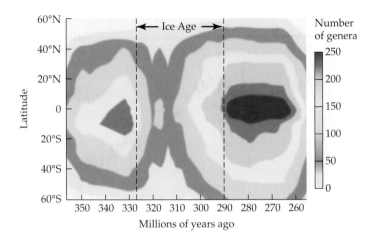

account for much of the latitudinal gradient in higher taxonomic diversity (order, family, genus), what is the link between historical climates and current species diversity? Has species richness increased steadily through time, or has it followed an irregular course of advance and retreat? The palynological (fossilized pollen) record suggests that tropical tree species richness in the Amazon lowlands was high during the Eocene tropical maximum, decreased with late Eocene and early Miocene cooling, but rebounded again in the late Miocene or Pliocene under different climatic conditions (Jaramillo et al. 2006; Mittelbach et al. 2007). Crame (2001: 182) also notes that for many taxa "the Cenozoic diversification event continued well past the prolonged late Palaeocene-middle Eocene interval of global warming." Finally, fossil data show that the latitudinal diversity gradient for marine mollusks has strengthened and weakened numerous times far back in Earth's history (**Figure 2.9**; Powell 2007). Thus, while tropical habitats are older than temperate habitats and were once very widespread, the evidence suggests that time and area alone are unlikely to provide a complete explanation for the current latitudinal gradient in species richness (Antonelli and Sanmartin 2011).

Evolutionary hypotheses: Do rates of diversification differ across latitude?

Do diversification rates (speciation minus extinction) differ between temperate and topical regions, and if so, could such variation in evolutionary rates cause the latitudinal diversity gradient? This question has puzzled evolutionary biologists for a very long time (e.g., Dobzhansky 1950; Fisher 1960), and it was G. L. Stebbins (1974) who colorfully asked whether the tropics are a "cradle" for the generation of new taxa (i.e., higher rates of speciation) or a "museum" for the preservation of existing diversity (i.e., lower rates of extinction). The answer appears to be "yes" on both counts (Jablonski et al. 2006; McKenna and Farrell 2006). Paleontologists and evolutionary biologists use two different methods to estimate diversification rates: studying fossils and phylogenies. Both methods let us look back in time, although in each case the view is imperfect.

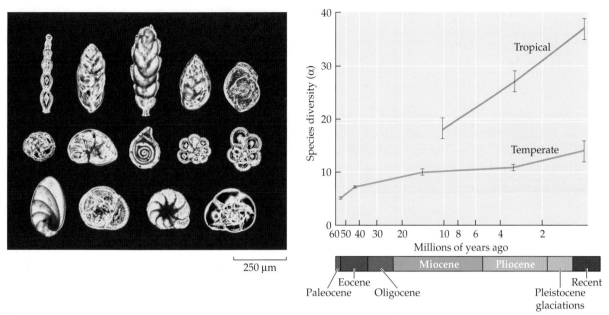

Figure 2.10   Unicellular foraminiferans secrete calcium carbonate shells that fossilize well. Over 10 million years, increase in foraminiferan diversity was 150% greater on the tropical Central American isthmus than on the temperate Atlantic coastal plain. (After Buzas et al. 2002; photograph © Science Photo Library.)

The fossil record shows that diversification rates are generally higher in the tropics than in temperate and polar regions, and the evidence suggests that this difference is due to higher rates of species origination and (probably) lower extinction rates in the tropics (reviewed in Mittelbach et al. 2007; Valentine and Jablonski 2010). But the fossil evidence is limited to a very few taxonomic groups, with most of the data coming from marine invertebrates. For example, Buzas et al. (2002) showed that planktonic foraminifera had higher gains in species richness over the last 10 million years in tropical than in temperate seas (**Figure 2.10**). Jablonski (1993) and Jablonski et al. (2006) showed that orders of marine invertebrates and genera of marine bivalves, respectively, preferentially originate in the tropics, as do species of fossil foraminifera (Allen and Gillooly 2006). Recently, Krug et al. (2009) used a different approach to compare the diversification rates of tropical and temperate/Arctic bivalves. They used a large data set of some 854 living bivalve genera and subgenera collected from around the globe, whose ages they determined from the fossil record. They found that the rate of increase in the number of bivalve genera was much higher in tropical than in temperate or polar waters (**Figure 2.11**).

Phylogenetic analyses also show higher diversification rates in the tropics than in the temperate zone for some birds, butterflies, amphibians, and plants (i.e., Cardillo 1999; Davies et al. 2004; Cardillo et al. 2005b; Ricklefs 2006; Wiens 2007; Mullen et al. 2011; Condamine et al. 2012); however, counterexamples exist as well (e.g., Farrell and Mitter 1993; Wiens et al. 2009;

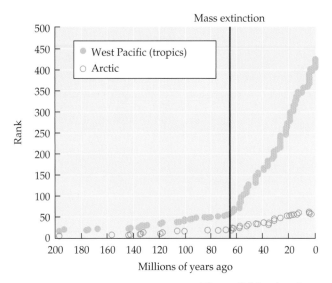

Figure 2.11 Age frequency distributions of genera of marine bivalves. The curves plot the ages of genera, ranked from youngest to oldest, in the tropical West Pacific and Arctic bioregions. The vertical line marks the mass extinction at the Cretaceous–Paleogene boundary (65 million years ago). The slopes of the two provinces differ slightly before the extinction event but diverge strongly thereafter, as many more new species evolved in the tropics than in the Arctic. (After Crame 2009, based on data from Krug et al. 2009; photograph © Shi Yali/Shutterstock.)

Weir and Schluter 2007). Technical difficulties complicate these phylogenetic estimates of diversification rates (Rabosky 2009; Gillman et al. 2010; Losos 2011; see discussion in Chapter 15), and separating the effects of speciation from those of extinction on overall diversification rates is difficult or impossible in most cases (Ricklefs 2007; Rabosky 2010). Thus, it is too early to say whether phylogenetic data will display general patterns in temperate and tropical diversification rates that will help explain the latitudinal diversity gradient, although recent studies show promise (e.g., Gonzalez-Voyer et al. 2011; Mullen et al. 2011; Condamine et al. 2012).

It is easy to understand why the tropics might be a "museum" for diversity. Extinction rates might be lower in these mild and stable environments than in the relatively harsh and less stable temperate and polar regions (although comparative data on extinction rates between temperate and tropical regions are surprisingly few). However, why might speciation rates differ between regions—that is, why should the tropics be a "cradle" for diversity? Fedorov (1966) proposed that the forces of genetic drift acting on small populations in the tropics might contribute to rapid speciation. However, this mechanism requires that tropical communities be diverse already, with many species existing in small populations, and therefore cannot explain the *origins* of diversity (Schemske 2002). Other authors have suggested that opportunities for geographic isolation may be greater in tropical regions (Janzen 1967; Rosenzweig 1995), whereas Rohde (1992) and Allen et al. (2002) have suggested that higher mutation rates and shorter generation times due to the kinetic effects of temperature could contribute to higher speciation rates in the tropics. Recently, Schemske (2002, 2009), building on earlier ideas of Dobzhansky (1950), suggested that stronger biotic interactions (e.g., predator–prey; host–parasite; symbioses) and a greater potential for coevolution in the tropics could lead to higher rates of speciation. Early evidence points to more rapid molecular evolution in some tropical taxa than in temperate taxa (e.g., Wright et al. 2003, 2006; Davies et al. 2004; Allen et al. 2006; Gillman et al. 2009), but a general link between nucleotide substitution rates and speciation rates remains to be demonstrated. Likewise, although many biotic interactions appear to be stronger and have a greater effect on species dynamics in the tropics than in higher latitudes (Schemske et al. 2009), do such interactions really facilitate speciation? More research on these questions is clearly needed.

The latitudinal diversity gradient is one of the most striking features of our natural world, and the pervasiveness of this pattern across taxa, among environments, and through time suggests that it may have a general explanation. Ecologists and evolutionary biologist are closing in on the answer. In all likelihood, multiple processes have contributed to the pattern, and sorting out their relative contributions is a major challenge. Ultimately, large-scale biodiversity patterns are the result of geographic patterns of speciation and extinction (Swenson 2011). In the end, our continuing efforts to understand large-scale gradients in biodiversity, such as the latitudinal diversity gradient, will provide valuable insights into the processes that govern the development of regional biotas and species pools.

Patterns of Biological Diversity at Different Spatial Scales

Understanding the nature of large-scale diversity patterns such as the latitudinal gradient poses a tremendous challenge to community ecologists. Equally challenging are the diversity patterns that occur at smaller spatial scales. We will consider some of these patterns in the following sections, with an eye toward the ultimate goal of being able to explain why patterns of biodiversity change across spatial scales and how these patterns are interrelated.

Productivity and species richness

As we have seen from the study of the latitudinal diversity gradient, species richness at broad spatial scales tends to increase with an increase in primary productivity (Currie 1991; Hawkins et al. 2003). Areas that have high productivity (e.g., the humid tropics) tend to have high species richness. Although the productivity–diversity relationship at regional scales is generally positive, it is often decelerating (i.e., the rate of increase in species richness declines with increasing productivity). At smaller spatial scales, the relationship between productivity and species richness is much more varied; both positive and negative relationships are common, as are hump-shaped and U-shaped relationships (**Figure 2.12**; Mittelbach et al. 2001; Gillman and Wright 2006; Partel et al. 2007). Ecologists once thought that hump-shaped productivity–diversity relationships were nearly ubiq-

Figure 2.12 Types of productivity–diversity relationships in plant communities at different spatial scales. Statistically significant productivity–diversity relationships were classified into four types: positive, negative, hump-shaped, and U-shaped. (A) Mittelbach et al.'s (2001) initial analysis of the literature showed that hump-shaped relationships were most common at smaller spatial scales, whereas positive and hump-shaped relationships predominated at the largest spatial scale. (B) Gillman and Wright's (2006) subsequent literature review concluded that positive, hump-shaped, and U-shaped relationships were equally represented at small spatial scales, and that positive relationships predominated at large spatial scales.

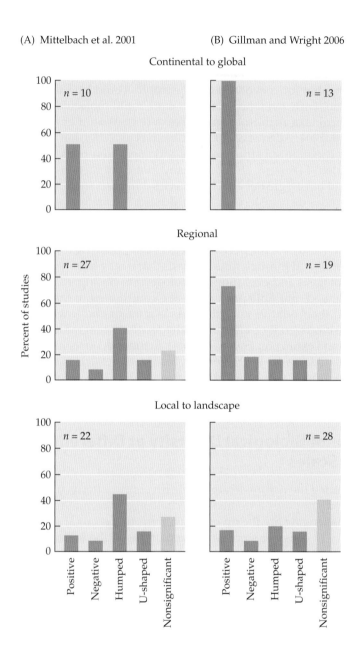

uitous at local to regional scales (e.g., Tilman and Pacala 1993; Huston and DeAngelis 1994; Rosenzweig 1995). We now know that this is not true. However, hump-shaped relationships occur often enough to demand an explanation, and this explanation is far from obvious. While the ascending limb of the hump may be the result of increased productivity allowing more species to exist at minimum viable population sizes, the descending limb is problematic. As Rosenzweig (1995: 353) notes, "to me the decrease phase presents the real puzzle: why, past a certain point, does enhanced productivity tend to reduce the number of species?"

Literally dozens of hypotheses have been proposed to explain the causes of productivity–diversity relationships, particularly the hump-shaped curve (reviewed in Abrams 1995; Rosenzweig 1995), yet there is no consensus explanation. Some of the more prominent hypotheses include:

1. There is a shift from nutrient limitation to light limitation at high productivity, resulting in a decline in plant species richness in very productive environments (Tilman and Pacala 1993).

2. Environments of intermediate productivity are more common across the landscape than are high- and low-productivity environments (especially in the temperate zone); more species have therefore evolved to utilize intermediate productivities, creating a larger species pool for those environments (Denslow 1980; Partel and Zobel 2007; Partel et al. 2007).

3. Communities are more likely to occur in alternative states at intermediate productivities; therefore, summing species richness across these alternative-state communities leads to more species at intermediate productivity (Chase and Leibold 2003b).

4. High productivity magnifies the impact of apparent competition, leading to a loss of species richness in very productive environments (Abrams 2001b).

Each of these hypotheses has some empirical and theoretical support, but none is a clear winner. There also continues to be a great deal of controversy over the shape and strength of the productivity–diversity relationship at local to regional spatial scales (e.g., Mittelbach 2010; Whittaker 2010; Adler et al. 2011). In Chapter 3 we will look at the productivity–diversity relationship in more detail, but from a different perspective—namely, how does species diversity affect productivity?

Area and species richness

Large areas contain more species than small areas. Organisms on islands provide the most prevalent examples of this **species–area relationship (SAR)**, but it exists for mainland and marine areas as well. Like the latitudinal diversity gradient, the SAR is pervasive; its discovery dates to the mid-nineteenth century, and it was one of the first diversity patterns quantified in ecology (Arrhenius 1921; Gleason 1922; Rosenzweig 1995).

The relationship between species richness and area can be described by a power function of the form

$$S = cA^z \qquad\qquad \text{Equation 2.1}$$

where S is the number of species, A is the area, and c and z are fitted constants. This relationship yields a straight line on a log–log plot (**Figure 2.13**). Other functions (e.g., exponential) also have been used to fit the SAR, and there is some recent evidence that the power function may not be its best descriptor (Kalmar and Currie 2007). However, most studies describe SARs with a power function, and the fitted constants c and z provide a useful way to compare relationships among studies (see Connor and McCoy 1979; Drakare et al. 2006).

Figure 2.13 Two species–area curves. Note that the axes are logarithmic and that the relationships are well fitted by a straight line based on Equation 2.1 ($S = cA^z$, where c and z are fitted values; the arithmetic relationship is depicted in the inset in panel A). (A) MacArthur and Wilson plotted the number of reptile and amphibian species present against the size of several islands in the Greater and Lesser Antilles. (B) One of the first species–area curves was constructed by H. C. Watson in 1859 for plant species in Great Britain. He constructed the curve based on different-sized regions within Great Britain and the island as a whole. (A after MacArthur and Wilson 1967, data from Darlington 1957; B after Rosenzweig 1995.)

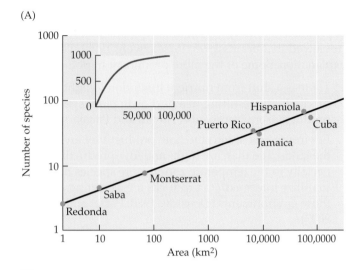

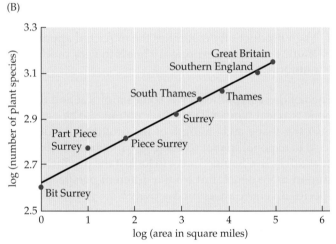

Why do larger areas contain more species? The answer has two parts. First, larger areas typically encompass a greater variety of habitats. This increase in habitat diversity with area contributes to the SAR because different species have different habitat affinities (**Figure 2.14**). However, even in areas of relatively uniform habitat, we expect larger areas to harbor more species because of demographic processes. Thus, the second part of the answer is that larger areas can support larger populations, which have a lower probability of extinction. MacArthur and Wilson (1967) developed a simple model of island biogeography to illustrate this effect (**Figure 2.15**). Simberloff (1976) tested this model experimentally by reducing the sizes of small mangrove islands off the coast of Florida with chain saws. Islands reduced in size decreased in species richness over time, whereas control islands showed no effect (Simberloff 1976). These observations and MacArthur and Wilson's (1967) theory have important implications for conservation and the design of nature preserves, as we will discuss in more detail in Chapter 12.

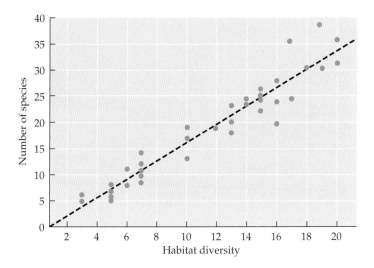

Figure 2.14   Relationship between habitat diversity and the number of terrestrial isopod species inhabiting the central Aegean Islands. Each data point represents an island; the line is the best-fit linear regression. (After Hortal et al. 2009.)

Although we understand the basic processes driving the species–area relationship, there are many aspects of SARs that continue to challenge our understanding. For example, the slope of the SAR (the fitted constant z in Equation 2.1) is of fundamental interest because it describes the rate at which new species are encountered as sampling area increases.

Figure 2.15   MacArthur and Wilson's (1967) theory of island biogeography, illustrating the effect of island size on equilibrium species richness. In MacArthur and Wilson's model, species richness on an island is a function of the rate of immigration of new species from the mainland species pool and the loss of island species due to extinction. (A) The immigration rate declines as the number of resident species on the island increases, going to zero when the island contains all the species in the mainland source pool. (B) The extinction rate increases with the number of resident species on the island because (1) there are more species to go extinct, and (2) the number of individuals per species decreases as the total number of resident species increases, and small populations are more likely to go extinct than large populations. (C) The extinction rate on large islands is lower than on small islands because large islands have more resources and should therefore support more individuals of all species. Thus, the equilibrium number of species $\hat{S}$, found at the intersection of the immigration and extinction curves) is greater on large islands than on small islands.

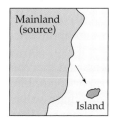

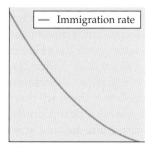

(A)

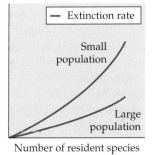

(B)

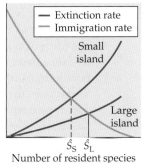

(C)

Figure 2.16 The triphasic species–area relationship (Rosenzweig 1995; Hubbell 2001). The SAR curve is steepest at the provincial scale (i.e., when diversities are compared between biogeographical regions) and is shallower at regional scales (i.e., within provinces). Allen and White (2003) and McGill (2011) showed that the slope of the SAR will approach a value of 1 when the spatial scale of study is larger than most species' ranges.

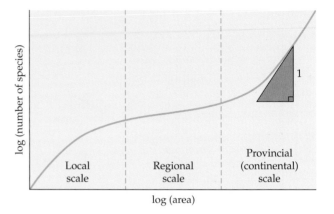

Rosenzweig (1995), following earlier work by Preston (1960), noted that z varies with spatial scale, resulting in a triphasic SAR (**Figure 2.16**; see also Shmida and Wilson 1985; Hubbell 2001; McGill 2011). The SAR is steepest (approximating unity) at the interprovincial scale (i.e., when diversities are compared among biological provinces) and is shallower within provinces (at regional scales). Provinces correspond to biogeographical regions with separate evolutionary histories (see Table 2.1). We should expect species richness to change dramatically when moving between areas with separate evolutionary histories. Moreover, Allen and White (2003) and McGill (2011) showed that the slope of the SAR will approach a value of 1 when the spatial scale of study is larger than most species' ranges. McGill (2011) provides a very general explanation for the triphasic SAR shown in Figure 2.16, based on sampling distributions and species ranges.

The majority of species–area studies have focused at the regional scale, and at this scale z-values cluster around 0.25 and generally range from 0.15 to 0.40 (Rosenzweig 1995). Early attempts to explain this apparent consistency in z-values focused on the properties of the lognormal species abundance curve (Preston 1948; May 1975; see Figure 2.17). However, recent work shows that z-values may vary considerably between study systems and that this variation is strongly linked to climate. For example, a comparison of bird species richnesses on islands and continents worldwide shows that z-values increase significantly with temperature and precipitation (Kalmar and Currie 2006, 2007). Likewise, in a meta-analysis of almost 800 SARs obtained from the literature, Drakare et al. (2006) found that the slope of the SAR increases when moving from the poles to the equator. Interestingly, neither Drakare et al. (2006) nor Kalmar and Currie (2007) found any difference in the form of the SAR between island and mainland areas, once the effect of distance-based isolation was taken into account. Drakare et al. (2006: 221) conclude from their meta-analysis that "SAR show extensive, systematic variation—they are far from being an invariant baseline … Instead, there are important differences between parameters of the SAR based on the sampling scheme (nested versus independent), across latitudes and sizes of organisms, and between different habitats."

The fact that species richness accumulates faster for a given increase in area when sampled in the tropics than when sampled in temperate zone (or in warm, wet areas compared with cool, dry areas) provides an interesting link to the latitudinal diversity gradient and other climate–diversity relationships. Moreover, it is consistent with what we know about changes in species abundance distributions across latitude. To see this, we must first delve into another prominent diversity pattern in ecology.

The distribution of species abundance

Most communities contain a few species that are common and many species that are rare. If we plot the number of species in each abundance class, we get a **species abundance distribution (Figure 2.17)**. Statistical models to describe species abundance distributions (SADs) were developed in the 1930s and 1940s, first by Motomura (1932), then by the renowned statistician Sir Ronald Fisher (working in collaboration with two lepidopterists; see Fisher et al. 1943), and also by Frank Preston, a ceramic engineer and amateur ecologist who published a classic series of papers on the topic (Preston 1948, 1960, 1962a,b).

(A)

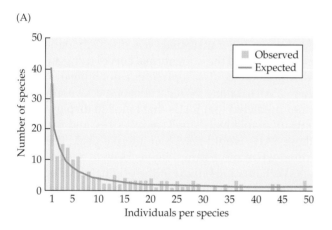

(B)

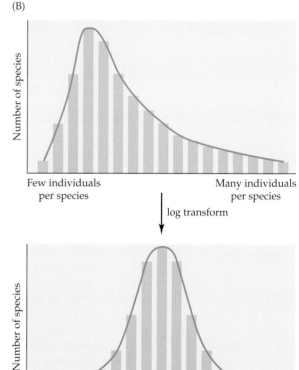

Figure 2.17 (A) An example of Fisher's log series distribution fit to data on species abundances in moths collected at light traps over a 4-year period at Rothamsted, England. (B) A hypothetical example of Preston's lognormal distribution, showing how the distribution of species abundances can be normalized when log-transformed, in this case by using an x axis where each successive abundance class represents a doubling in species abundance ($\log_2$ scale). (A after Hubbell 2001.)

Based on large samples of Lepidoptera they collected in England, Fisher and his colleagues (1943) observed that the numbers of individuals per species followed a negative binomial frequency distribution. Based on this distribution, they proposed that the number of species (F_n) having n individuals each can be expressed as

$$F_n = \alpha \frac{X^n}{n} \text{ for } n > 0 \qquad \text{Equation 2.2}$$

where X is an arbitrary scaling parameter and α is a fitted, empirical constant. Equation 2.2, referred to as Fisher's log series distribution, describes a distribution in which singleton species (i.e., those with only one individual in the sample) always make up the most abundant class (Figure 2.17A). Because α depends on the shape of the SAD, but not on the size of the sample, it is a useful descriptor of diversity and is termed Fisher's α (Magurran 2004). The total number of species S in a sample depends on the number of individuals N (the species–individuals curve) as

$$S \cong \alpha \ln\left(1 + \frac{N}{\alpha}\right) \qquad \text{Equation 2.3}$$

Preston (1948) challenged the adequacy of Fisher's log series distribution, noting that many empirical SADs display an internal mode rather than a peak in the singleton class. He further noted that these SADs were non-normal, but could be normalized when log-transformed (i.e., they are lognormal; see Figure 2.17B). Preston illustrated these lognormal curves using an x axis on which each successive abundance class represents a doubling in species abundance, which is equivalent to taking the logarithm of species abundance to base 2. Preston further predicted that SADs that failed to display an internal mode were the result of relatively small sample sizes, and that with increased sampling these SADs would reveal an internal peak. Preston's prediction was confirmed when more sampling years were added to the Rothamsted moth data (Williams 1964) used in the original development of Fisher et al.'s log series SAD (see Hubbell 2001). The lognormal SAD has since been shown to be one of the most consistent empirical patterns in community ecology (Ulrich et al. 2010).

Fisher's and Preston's models are statistical descriptors of species abundance patterns. They were developed without any real consideration of underlying biological mechanisms. What followed in the 1960s was a decade of development of ecological models of SADs based on niche theory, the most famous of which is MacArthur's "broken stick" model and its relatives (Sugihara 1980). (Excellent accounts of the development of these models can be found in Tokeshi 1999 and Magurran 2004.) Interest in these "niche-based" models and in SADs in general waned in the late 1980s and 1990s as ecologists focused their attention more on experimental studies and less on descriptive patterns. This situation changed dramatically, however, with the publication of Stephen Hubbell's book *The Unified Neutral Theory of Biodiversity and Biogeography* (2001). Hubbell proposed a radical new model that could account for both Fisher's log series and Preston's lognormal SADs and that fit the empirical data remarkably well. Hubbell's neutral theory revived interest in species abundance distributions, which currently are enjoying a lively renaissance (McGill 2006; McGill et al. 2006; Shipley et al. 2006). We will discuss Hubbell's model and what it predicts about SADs in Chapter 13.

It is important to note that SADs such as the log series and lognormal distributions specify the relationship between species richness and the number of individuals in a sample (e.g., Equation 2.3 for the log series). Both theoretically and empirically, the number of species rises as a decelerating function of the number of individuals collected (see Figure 2.7). This is because the first few individuals collected in a sample are likely to include representatives of the most common species, whereas more and more individuals must be sampled to find representatives of the rarest species. Given this continued rise in species richness with sample size, ecologists have developed methods to standardize species richness among communities sampled with different intensities (e.g., rarefaction) and to estimate the true species richness of a community based on the number of species observed in a sample of a given size. (Gotelli and Colwell 2001 provide an excellent discussion of the issues involved.)

The species–individuals curves shown in Figure 2.7 are of interest not only as examples of a general phenomenon, but also because they demonstrate an important linkage between biodiversity patterns. Note that the sampling sites for Sanders's (1969) surveys of marine benthic diversity span a broad range of latitude (see Figure 2.7A) and that the species–individuals curves become steadily steeper as one moves from the poles to the tropics. Thus, for a given number of individuals sampled, many more species will be found at low latitudes than at high latitudes. Recall that the slope of the species–area curve (z-value) also increases with a decrease in latitude (Drakare et al. 2006; Kalmar and Currie 2006, 2007). Here again, we see that more species are encountered for the same area sampled in the tropics than in the temperate zone.

The steepness of the species–individuals curve and of the species–area curve should vary similarly with latitude if the number of individuals per unit area ρ (density of individuals) remains approximately constant, such that

$$I = A \times \rho \qquad \qquad \text{Equation 2.4}$$

where A is area and I is the number of individuals. Surprisingly, we know almost nothing about how Equation 2.4 varies geographically (e.g., with latitude) for most taxonomic groups. Currie et al. (2004) show that for trees and birds, the number of individuals per sampling site (area) increases with a decrease in latitude. However, the change is relatively small (see Figure 2.5), and Enquist and Niklas (2001) found no change in tree biomass with latitude. Clearly, much more work is needed to link biodiversity patterns across different geographic regions and across different spatial scales. However, a number of generalities have begun to emerge. More importantly, by investigating these questions, ecologists have broadened their focus to consider the interrelatedness of the processes that regulate diversity across different spatial scales.

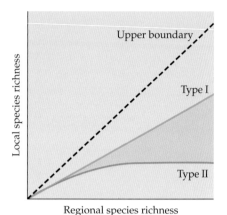

Figure 2.18 Theoretical curves for the relationship between local and regional species richness. In the type I curve, local species richness increases linearly (proportionally) with regional species richness. In the type II curve, the relationship is nonlinear and saturating because of biotic interactions and niche-filling in the local community. Real communities are likely to fall somewhere in the continuum between these two extremes (shaded area). The "boundary" represents the upper limit of the relationship, as local species richness can not exceed regional species richness. (After Cornell and Lawton 1992.)

Local–regional diversity relationships

How might we expect species richness at local and regional scales to be related, and what can that relationship tell us about the processes regulating local species diversity? Let's consider two possibilities. If the composition of local communities is determined largely by the input of species from the regional species pool, then local species richness (alpha diversity) should increase linearly with the richness of species in the region (gamma diversity); this results in the type I curve in **Figure 2.18**. On the other hand, if local communities have limited membership (Elton 1950) and species interactions such as competition and predation restrict which species are able to coexist in a community, than we would expect species richness in local communities to **saturate** with regional richness (the type II curve). In other words, there may be a limited number of niches in any community, so that no matter how high the regional species richness (and therefore no matter how large the potential pool of colonists), niches become full and the local community can support only so many species.

Cornell and Lawton (1992) conducted an initial analysis of published local–regional richness relationships. They tentatively concluded that there was little evidence for saturation and that the principal direction of control for species richness was from regional to local. Later analyses have supported Cornell and Lawton's initial conclusion that linear local–regional richness relationships are more common than nonlinear ones (e.g., Lawton 1999; Hillebrand and Blenckner 2002; Shurin and Srivastava 2005; Cornell et al. 2008; Harrison and Cornell 2008). However, the potential mechanisms driving the observed patterns are more complicated than originally envisioned.

The original predictions of linear versus saturating local–regional richness relationships were framed in terms of noninteractive versus interactive (niche-filling) communities. However, it turns out that a combination of processes can result in different patterns. At a methodological level, "pseudosaturation" and the scale at which local and regional richness are measured have been shown to affect the predicted and observed local–regional species richness relationship (Hillebrand 2005; Shurin and Srivastava 2005). However, the one study that explicitly explored the effect of measurement scale found that local–regional species richness relationships in Pacific corals were similar (and linear) across three very different scales of local communities (Cornell et al. 2008). At a more fundamental level, theoretical analyses have shown that there may be a continuum of local–regional richness relationships (from saturating to linear), depending on the relative differences between rates of local species extinction and regional dispersal (e.g., He et al. 2005; Shurin and Srivastava 2005). If local extinction is the dominant process, the relationship will be nonlinear, but a linear pattern will result if dispersal dominates. Likewise, if species are maintained in local communities via source–sink dynamics (e.g., immigration from the outside maintains species in the local community that would otherwise go extinct due to species interactions or other processes), then we would expect to see saturation of local diversity at low dispersal rates, but more linear local–regional patterns at high dispersal rates (Shurin and Srivastava

2005). Moreover, if higher regional species richness results in higher rates of species dispersal to local communities, then local richness should scale positively with regional richness, even when there is strong species sorting at the local scale (Ptacnik et al. 2010). Thus, it is risky to infer the processes regulating local species diversity only from plots of local–regional species richness, and comparisons of patterns should be accompanied by studies focused on the underlying mechanisms (see Chapter 13 and Shurin 2000; Shurin et al. 2000).

Conclusion

As we have seen in this chapter, the diversity of life varies across space and time, and there are patterns to this variation. Recent advances in technology—from satellite imagery and remote sensing to DNA sequencing and phylogenetic analysis to GIS mapping and spatial analysis—have allowed ecologists to better document these patterns at all spatial scales and to search for their interrelationships (e.g., McGill 2011). As ecologists, we are interested not only in "searching for repeated patterns" (MacArthur 1972), but also in understanding the processes that underlie these patterns. In this chapter I have highlighted some of the mechanisms thought to underlie patterns in biodiversity, focusing on those hypotheses that have (in my opinion) received the most theoretical and empirical support. My list of "most likely explanations" won't sit well with everyone, and there is still much to be learned from the study of ecological patterns. Moreover, our ability to explain spatial and temporal patterns in biodiversity is the litmus test of what we discover from detailed studies of ecological mechanisms.

Chapters 4–9 will focus on interactions between species (competition, predation, mutualism, and facilitation) and how they affect species coexistence and diversity in local communities. For a time we will put on "blinders," if you will, to the impact of regional processes on the structure of local communities as we focus on species interactions in small modules. We will return to regional processes and their effects on local communities later in the book. Before we get to the "nitty-gritty" of species interactions, however, I want to explore one more important aspect of biodiversity: How does biodiversity affect the functioning of ecosystems, and what are some of the consequences of lost biodiversity to the services that ecosystems provide?

Summary

1. Species diversity is measured at different spatial scales and can be expressed in three important ways. Alpha (α) diversity is the number of species found at a local scale. Beta (β) diversity is the difference in species composition, or species turnover, between two or more habitats or local sites within a region. Gamma (γ) diversity is a measure of regional species richness.

2. Species diversity differs between biogeographic regions. For similar-sized areas, there are many more species in the tropics than in the temperate zone. Hypotheses proposed to explain this latitudinal diversity gradient can be grouped into four general categories:

- Null-model explanations based on geometric constraints on species ranges. Null models generate patterns based on randomized data; such patterns (e.g., the mid-domain effect) would be expected in the absence of a particular mechanism, and are rejected (and mechanisms hypothesized) if observed data do not match the null model.
- Ecological hypotheses such as the species–energy hypothesis, which postulates that because the tropics tend to be highly productive, these regions can support more individuals per unit of area. A greater number of individuals presumably leads, over evolutionary time, to a greater diversity of species.
- Historical explanations such as the time-and-area hypothesis based on geologic history—that is, because tropical environments are older and have been more widespread than extratropical environments, there has been more opportunity for diversification to occur in the tropics.
- Evolutionary hypotheses focus on rates of diversification (speciation minus extinction). Speciation might be higher ("cradle" effect) in the tropics because of the effects of uniformly warm temperatures, and/or because there are more opportunities for coevolution there. Extinction rates might be lower ("museum" effect) for a number of reasons, including the more stable and equable climate.

3. At broad spatial scales, species richness generally increases with productivity. At local to regional scales, productivity–diversity relationships are more varied and tend to be either positive or hump-shaped. Several hypotheses have been proposed to explain the descending limb of humped-shaped productivity–diversity relationships (where productivity is high but diversity is low).

4. Larger areas contain more species, as they typically contain a greater variety of habitats and can support larger populations. The slope of a species–area curve varies across spatial scales and across latitude.

5. Most species are rare and a few are abundant, a relationship that can be expressed as a species–abundance distribution. The lognormal species abundance distribution is one of the most consistent empirical patterns in community ecology.

6. Relationships between local and regional species richness vary along a continuum from linear relationships, in which local species richness increases with regional species richness, to nonlinear, saturating relationships (the relationship increases to a certain point until it reaches a saturation point at which it plateaus). Linear relationships are the most common.

3 Biodiversity and Ecosystem Functioning

Species leave the ark one by one.

Thomas Lovejoy, 1986: 13

Gains and losses of species from ecosystems are both a consequence and driver of global change, and understanding the net consequences of this interchange for ecosystem functioning is key.

David A. Wardle et al., 2011: 1277

[O]ur well-being is far more intertwined with the rest of the biota than many of us would be inclined to believe.

Thomas Lovejoy, 1986: 24

We have seen that species richness varies with latitude, climate, productivity, and a host of other physical and biological variables. Community ecologists have long focused on understanding the mechanisms driving these patterns of diversity. Recently, however, ecologists have become interested in how biodiversity, in turn, may affect the functioning of ecosystems. This is no casual question. Species are being lost at a rate unprecedented since the mass extinction that claimed the dinosaurs; by the end of the current century, as many as 50% of Earth's species may disappear due to habitat destruction, overharvesting, invasive species, eutrophication, global climate change, and other effects of human activities (**Figure 3.1**; Wake and Vredenburg 2008; Pereira et al. 2010). Can we as ecologists predict the consequences of this lost biodiversity? We know that Earth's biota moves hundreds of thousands of tons of elements and compounds between the atmosphere, hydrosphere, and lithosphere every year. Living organisms obviously play a major role in the planet's dynamics, but what role does species diversity per se have in the functioning of ecosystems? Or, to put it simply, how many species do we need to ensure that ecosystems continue to function normally? Two groundbreaking papers published in 1994 opened the door to this question.

Figure 3.1 Predicted rates of species extinctions in the twenty-first century compared with extinction rates in the previous century and in the fossil record. Extinction rates are shown as the percentage of species within each group going extinct per century. Data for the fossil record and the twentieth century are from Mace et al. 2005. Projected losses in the twenty-first century are shown with upper and lower bounds and are based on several different studies (cited in Pereira et al. 2010). The different drivers of extinction are distinguished where possible. (After Pereira et al. 2010.)

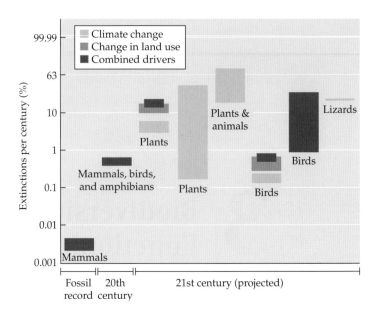

Tilman and Downing (1994), working in a grassland in east-central Minnesota, found that plots with high plant species diversity were better buffered against the effects of a severe drought; those plots maintained higher plant productivity during the drought than did low-diversity plots (**Figure 3.2**). Their results showed an apparent link between species richness, ecosystem productivity, and ecosystem stability; however, Tilman and Downing studied a set of communities in which plant species richness had been changed by adding fertilizer (diversity decreased with increased fertility). Therefore, the presence of this additional factor (added soil fertility) may have clouded their interpretation of a causal relationship between plant species richness and ecosystem productivity and stability (Huston 1997).

Figure 3.2 Tilman and Downing performed one of the first studies showing that plant species richness has a positive effect on ecosystem functioning. They showed that drought resistance was positively associated with plant species richness in four grasslands at the Cedar Creek Ecosystem Science Reserve, Minnesota. Drought resistance was measured as the loss in plant community biomass during a drought year as compared with plant biomass in the previous year (highest drought resistance is thus at the top of the y axis). Biomass ratio expresses the proportional loss in biomass due to the drought. Data points are means (±1 SE) for plots of a given species richness. Plant species richness varied among the 207 study plots as the result of previous and ongoing nutrient manipulations. (After Tilman and Downing 1994.)

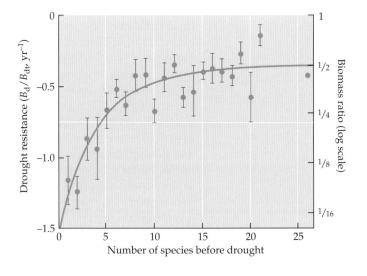

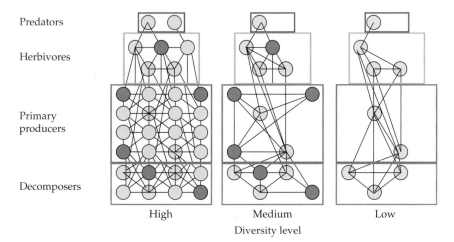

Predators

Herbivores

Primary
producers

Decomposers

High Medium Low

Diversity level

Figure 3.3 The experimental design used by Naeem and colleagues to study the effects of species losses at multiple trophic levels on ecosystem functioning. In their experiment, model communities containing four trophic levels—primary producers (annual plants), herbivores (molluscs and insects), predators (parasitoids), and decomposers (collembolids and earthworms)—were established in 14 large environmental chambers in the "Ecotron" at Silwood Park, U.K. Three diversity levels were established, such that each of the lower-diversity treatments contained a subset of the species used in the higher-diversity treatments, to mimic the uniform loss of species from high-diversity ecosystems. (After Naeem et al. 1994.)

In a second study, also published in 1994, Naeem and colleagues directly manipulated species richness in a set of small, replicated ecosystems constructed in 14 large, controlled environmental chambers at the "Ecotron" at Silwood Park, U.K. Their experiment included multiple trophic levels at three levels of diversity, such that the medium-diversity treatment contained a subset of the species in the high-diversity treatment, and the low-diversity treatment a subset of the species in the medium-diversity treatment (**Figure 3.3**). The researchers used these treatments to mimic the uniform loss of species from high-diversity ecosystems. Naeem and his colleagues found that high-diversity communities consumed more CO_2 (i.e., had higher respiration rates) than low-diversity communities and that higher plant diversity resulted in greater primary productivity (Naeem et al. 1994).

The experiments of Naeem et al. and Tilman and Downing paved the way for a veritable flood of studies that have examined relationships between biodiversity and **ecosystem functioning**. The sheer volume of this work and the rapid pace at which it is advancing make it difficult to summarize in a single chapter. Fortunately, a number of recent reviews provide a road map to this literature and summarize many of its findings (e.g., Hooper et al. 2005; Balvanera et al. 2006; Stachowicz et al. 2007; Cadotte et al. 2008; Cardinale et al. 2007, 2011). In this chapter, I highlight these findings with respect to the effects of biodiversity on four important ecosystem functions:

- Productivity
- Nutrient use and nutrient retention
- Community and ecosystem stability
- Invasibility

At the end of the chapter, I discuss some of the many questions about the biodiversity–ecosystem function relationship that remain unresolved and highlight directions for future research.

Diversity and Productivity

Does higher species diversity lead to higher ecosystem productivity? The answer, at least for low-stature plant communities, is yes. Three large-scale

Figure 3.4 An aerial view of the study plots at the Cedar Creek Ecosystem Science Reserve, located in east-central Minnesota, U.S.A. (Photo from Tilman 2001.)

studies show this effect quite clearly. In 1993, David Tilman and colleagues (Tilman et al. 1997, 2001) set up 168 plots (9 m × 9 m) at the Cedar Creek Ecosystem Science Reserve in Minnesota (**Figure 3.4**), into which they introduced seeds from 1, 2, 4, 8, or 16 grassland–savanna perennial species. The species composition for each of the different diversity treatments was randomly generated from a pool of 18 species, and species were introduced to the plots by adding 15 g of seed per m² for each species. A small army of workers weeded the plots to maintain the diversity treatments, and Tilman and colleagues followed the effects of species diversity and plant functional group diversity on ecosystem productivity, nutrient dynamics, and stability for more than 15 years. In Europe, researchers employed a similar experimental design, repeated across seven countries from Sweden in the north to Greece in the south, to look at the effects of plant diversity on ecosystem functioning (Hector et al. 1999). In this experiment, known as BIODEPTH, the regional species pool differed between sites, allowing rare insight into the potential generality of the results across different environments. A third major experiment along these lines, the Jena Biodiversity Experiment, was established in 2002 near Jena, Germany. As in the other two experiments, plant diversity was manipulated in a large number of experimental plots in order to study above- and belowground productivity and nutrient use efficiency; however, the Jena Experiment has a greater focus on examining multiple trophic level effects (e.g., Marquard et al. 2009; Hector et al. 2011).

Species richness had a strong, positive effect on primary production in each of these experiments. At Cedar Creek, productivity (measured as plant biomass) increased rapidly in response to the addition of a few plant species, but the rate of increase then slowed in an asymptotic fashion as the number of species increased (**Figure 3.5A**). The positive effect of species richness on plant biomass increased over time in the Cedar Creek experiment and in the others. Results from the pan-European BIODEPTH experiment showed variation among study sites in the strength of the relationship between biodiversity and productivity, but the general response was quite consistent: in seven of the eight study sites, productivity increased significantly with an increase in species richness (**Figure 3.5B**; note that the *x* axis in this plot is on a log scale). The Jena Experiment also showed a positive species richness–biomass relationship, as well as a posi-

(A)

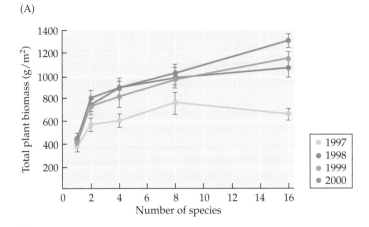

(B)

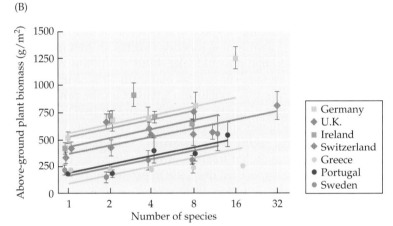

(C)

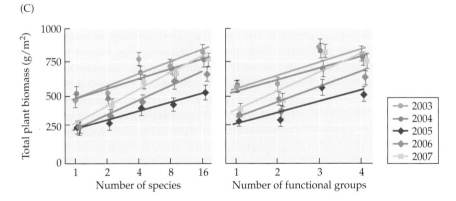

Figure 3.5 Three large-scale field experiments have shown that plant community biomass (and thus primary production) increase with increased species richness. (A) Average total plant biomass from the Cedar Creek experiment for the years 1997–2000. (B) Annual aboveground plant biomass as a function of species richness in the BIODEPTH experiment, conducted at eight sites in Europe (data from the two U.K. sites, at Sheffield and Silwood Park, have been combined). Note that the x axis displays species richness on a log scale ($\log_2$). (C) Total plant community biomass as a function of species richness (x axis, $\log_2$ scale) and plant functional group richness in the Jena Biodiversity Experiment for the years 2003–2007. Data points in all graphs are means ±1 SE. (A modified from Loreau 2010, original data from Tilman et al. 2001; B modified from Loreau 2010, original data from Hector et al. 2002; C modified from Marquard et al. 2009.)

tive functional group richness–biomass relationship (**Figure 3.5C**). Hector et al. (2011) recently re-analyzed the results of these three experiments to address whether biomass production is more strongly affected by species richness or by species composition (i.e., the overall variation between the different species compositions used). Counter to some earlier predictions (e.g., Hooper et al. 2005), Hector et al. found that the number and types of species present in experimental grassland communities were of similar importance in determining aboveground productivity.

The positive effects of species richness on biomass production observed in grasslands can now be extended to a more general conclusion: Increasing species richness increases biomass production in primary producers. A recent meta-analysis of 368 independent experiments that manipulated plant or algal species richness in a variety of terrestrial, freshwater, and marine ecosystems found that the most diverse polycultures attain, on average, 1.4 times more biomass than the average monoculture (Cardinale et al. 2011). This result was consistent for both aquatic and terrestrial systems. Cardinale and colleagues went on to fit a variety of mathematical functions to the observed diversity–productivity relationships and found that 79% of the relationships were best-fit by a function that was positive and decelerating. These results suggest that some fraction of species can be lost from communities with minimal loss in productivity, but that beyond a certain level of species loss, productivity declines rapidly. The researchers then attempted to calculate the fraction of species needed to maintain maximum producer biomass by fitting a saturating function (Michaelis–Menten curve) to the data. They concluded that it would take about 90% of the maximum number of species used in experiments to maintain 50% of maximum producer biomass. However, this conclusion is based on the problematic assumption that a fitted saturating function can be extrapolated to estimate maximum biomass production as species richness goes to infinity. Cardinale et al. (2011) caution against taking their estimates too literally, and at this point we simply don't know what fraction of species may be lost from a community before we see a specified decrease in productivity.

Mechanisms underlying the diversity–productivity relationship

Why might productivity increase with species richness? Ecologists have focused on two likely mechanisms: niche complementarity and species selection (Huston 1997; Loreau and Hector 2001). **Niche complementarity** may occur if species differ in the way they use limiting resources. Plant species, for example, may differ in their phenology, physiology, rooting depths, or nutrient requirements. If a community consists of species that differ in their niches, the overall efficiency of resource use by the community may increase with an increase in species richness, leading to higher overall productivity. Complementarity effects may also result from facilitation (e.g., the presence of nitrogen-fixing legumes may increase the growth rates of non-nitrogen-fixing species). On the other hand, **species selection** (also called the **sampling effect**) may lead to increased productivity with increased species richness if diverse communities are more likely to contain more productive species that come to dominate the community (Aarssen 1997; Huston 1997; Wardle 1999; Loreau and Hector 2001).

Two types of positive evidence for niche complementarity have been documented. The first is **transgressive overyielding**: plots containing species mixtures yield more biomass than any monoculture plot (Loreau and Hector 2001). The second is a positive relationship between species richness and ecosystem functioning, combined with a negative relationship between the extent of ecosystem functioning and the degree of niche overlap between species (**Figure 3.6**) (Wojdak and Mittelbach 2007). It is also possible for niche complementarity and species selection to act jointly

(A)

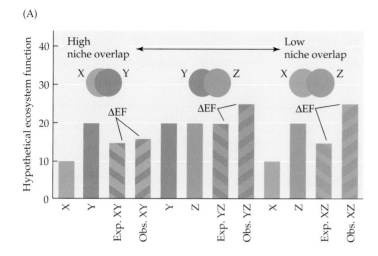

(B)

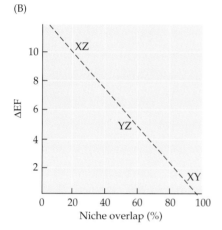

Figure 3.6 According to the niche complementarity hypothesis, combinations of species that differ in their resource use should have a greater impact on ecosystem functioning than combinations of species that are similar in resource use. The figure here presents a hypothetical example illustrating this effect for three species (X, Y, and Z) grown in monoculture and in two-species combinations. The null expectation for the two-species combinations is the average of each species' performance in monoculture, whereas the prediction of the niche complementarity hypothesis is that the difference between observed and expected ecosystem functioning will depend on species resource use. (A) When species are very similar in resource use (i.e., their niches), the difference between observed and expected ecosystem functioning (ΔEF) will be small. When species' niches are very different, ΔEF will be large. Thus we expect a negative relationship (B) between the degree of niche overlap and the difference between observed and expected ecosystem functioning. (After Wojdak and Mittelbach 2007.)

to increase ecosystem functioning. The relative contributions of these two mechanisms may be partitioned statistically using the methods outlined by Loreau and Hector (2001). Such statistical partitioning is possible when we can identify the contributions of individual species to ecosystem functioning within a mixture of species. In terrestrial plant communities, it is possible to measure individual species contributions to total primary production (by measuring plant standing biomass). For many other ecosystem attributes, however, it is impossible to identify how individual species contribute to the functioning of polycultures (Wojdak and Mittelbach 2007). Therefore, most of what we know about the relative effects of niche complementarity and species selection on ecosystem functioning comes from studies of productivity in plants.

Studies of terrestrial plant communities suggest that species selection (sampling) effects are the dominant drivers of ecosystem functioning early in an experiment, whereas niche complementarity effects may become stronger as communities mature (Cardinale et al. 2007). In the Cedar Creek experiment, the percentage of plots exhibiting transgressive overyielding increased with

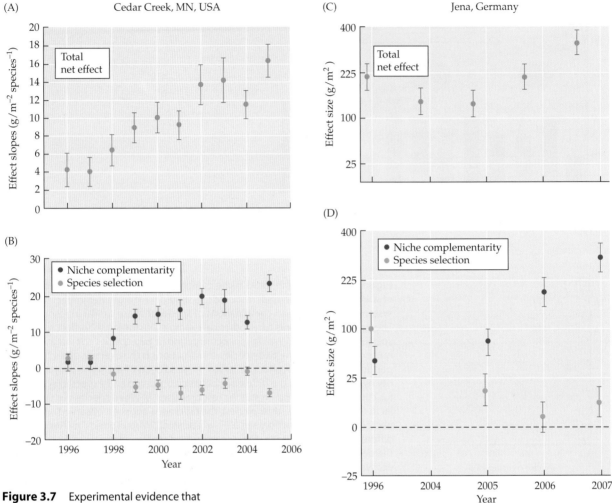

Figure 3.7 Experimental evidence that the contributions of niche complementarity and species selection may change over time. Data points in all graphs are means ±1 SE. (A, B) Species richness/biomass relationship in the Cedar Creek experiment, 1996–2006. (A) The net effect of species richness on community biomass production (i.e., the slope of the relationship) increased over the 10-year period. (B) Contributions of niche complementarity and species selection to the overall slope of the species richness/ productivity relationship. The complementarity effect increases over time, whereas the selection effect decreases. (C, D) Contributions of niche complementarity and species selection to the total net effect of biodiversity on production in the Jena Experiment, 2003–2007. The y axis is plotted on a square root scale. (A, B modified from Fargione et al. 2007; C, D after Marquard et al. 2009.)

increasing species richness in later years (Tilman et al. 2001). Thus, there is evidence of niche complementarity late, but not early, in the experiment. Fargione et al. (2007) and Marquard et al. (2009) applied the methods of Loreau and Hector (2001) to partition the effects of niche complementarity and species selection in this same experiment. They found that niche complementarity effects increased over time, whereas species selection effects decreased over time (**Figure 3.7**). Selection effects decreased through time because those species that came to dominate the mixtures (the competitive dominants) were not the same species that exhibited the highest production in monoculture. The overall (net) effect of species richness on community production increased over time because niche complementarity effects increased more strongly than species selection effects decreased. Fargione et al. (2007) hypothesize that the mechanism for increasing niche complementarity through time is an increase in nitrogen input (by legumes) and increased retention of nitrogen in the system.

A meta-analysis of 44 plant community studies (Cardinale et al. 2007) showed that niche complementarity may be as important as species selection (based on Loreau and Hector's 2001 methods for statistical partitioning of effects) and that the importance of niche complementarity increases over time (see also Cardinale et al. 2011). In experiments lasting 2–5 plant generations, species mixtures outyielded the most productive monocultures in about 12% of cases. A more extensive meta-analysis covering primary producers in aquatic and terrestrial ecosystems found transgressive overyielding in 37% of the studies, which means that 63% of the time diverse polycultures yielded less biomass than the single highest-yielding monoculture (Cardinale et al. 2011). On average, the most diverse polyculture in each study yielded just 0.87 times the biomass of the highest-yielding monoculture. Thus, Cardinale and colleagues (2011: 581) concluded that "there is presently little evidence to support the hypothesis that diverse polycultures out-perform their most efficient or productive (constituent) species."

To date, niche complementarity effects have been demonstrated almost exclusively in terrestrial plant communities, most often when focal species differ qualitatively in their functional traits (e.g., nitrogen-fixing versus non-nitrogen-fixing plants; Spehn et al. 2002; Heemsbergen et al. 2004; Fargione et al. 2007; Marquard et al. 2009). Few studies have attempted to use quantitative differences among species (measured along some niche axis) to predict the impact of niche complementarity on ecosystem functioning (e.g., Norberg 2000; Wojdak and Mittelbach 2007). In fact, as Ives et al. (2005: 112) state, "We know of no conclusive evidence from empirical studies that resource partitioning [niche complementarity] is the mechanism underlying an effect of consumer diversity on resource or consumer density."

The limited evidence for such a clear and overriding effect of niche complementarity on ecosystem functioning is surprising, given its central place in the theory (e.g., Tilman et al. 1997). In the next section, however, we will discuss two novel experiments that better elucidate the role of niche complementarity in driving the positive effect of species richness on nutrient retention. Moreover, there is recent evidence from plant communities that soil pathogens (e.g., microbes, fungi) with species-specific negative effects on plant growth may act in concert with niche complementarity to determine the diversity–productivity relationship. In experiments that varied plant species richness growing in natural soil versus soil that was sterilized to kill all microbes (via gamma radiation) or fungi (via fungicides), Schnitzer et al. (2011) and Maron et al. (2011) showed that a pronounced relationship between plant species richness and plant productivity was dependent on having living microorganisms present in the soil. The authors propose that soil-borne herbivores and pathogens have strong negative density-dependent effects on plant productivity at low species richness because the plants are likely to be growing near (infected) conspecific neighbors. In more diverse plant communities, however, species-specific soil-borne diseases or herbivores will have less of a negative effect on productivity because a given plant is less likely to be in close contact with an infected individual of the same species. The authors conducted additional experiments to document the species-specific, negative density-dependent effects of soil pathogens and conclude that "niche-based/competitive process and soil pathogen effects … may act in concert" to drive the diversity–productivity relationship (Maron et al. 2011: 40).

Diversity, Nutrient Cycling, and Nutrient Retention

If species exhibit resource partitioning and niche complementarity, then more diverse communities should use resources more efficiently, and the amount of unused resources in an ecosystem should decline as species richness increases. These predictions have been supported in a number of studies that involved manipulating plant diversity in terrestrial ecosystems, where nitrogen availability limits primary productivity (e.g., Tilman et al. 1996; Hooper and Vitousek 1997, 1998; Niklaus et al. 2001; Scherer-Lorenzen et al. 2003; Oelmann et al. 2007). Oelmann et al. (2007) present a particularly complete analysis of nitrogen availability in plant communities of varying species richness. They showed that an increase in plant species richness (which varied from 1 to 16 species) reduced available nitrogen (NO_3^-) concentrations in the soil and decreased dissolved organic nitrogen (DON) and total dissolved nitrogen (TDN). Nitrogen contained in the aboveground vegetation also correlated positively with species diversity, indicating that total N uptake increased with increasing diversity. A similar study by Dijkstra et al. (2007), again varying species richness from 1 to 16 species, found that an increase in plant species richness reduced leaching loss of dissolved inorganic nitrogen (DIN) as a consequence of more efficient nutrient uptake by the plant community. Dijkstra and colleagues found that DON, which does not dissolve easily and may not be directly available for plant uptake (Neff et al. 2003), was lost at a higher rate in more species-rich plots.

The weight of the evidence shows a positive effect of plant species richness on nitrogen use efficiency in terrestrial plant communities. Cardinale et al. (2011) found in their review of 59 studies (56 on grasslands and 3 on freshwater algae) that species-rich polycultures reduced nutrient concentrations, on average, 48% more than the mean monoculture. However, they found little evidence that the polycultures used resources more efficiently than the most efficient monocultures. Standing nutrient concentrations in soil or water were higher, rather than lower, in polycultures than in the most efficient monocultures (Cardinale et al. 2011).

Two recent studies (Northfield et al. 2010; Cardinale 2011) employed novel experimental designs to test whether species-rich polycultures use resources more efficiently than the most efficient monocultures, and if so, whether this more efficient resource use was due to niche complementarity. Most studies of the effects of species richness on ecosystem functioning use a "substitutive" experimental design, where the number of species is varied among treatments but the total number of individuals in held constant. Northfield and colleagues (2010) crossed this design with another that varied the number of individuals within each species to generate a full "response surface design." Their new design allowed them to test for complementarity in resource use by different species of predatory insects (i.e., bugs, parasitic wasps) feeding on aphids that infect crop plants (see Chapter 8 for a discussion of substitutive and response surface experimental designs in relation to interspecific competition). They reasoned that if different predator species use the prey resource differently (e.g., by feeding in different ways or in different microhabitats) then the total resource use of any single predator species should plateau (saturate) with increasing predator density, but the total resource use of more diverse predator communities should not plateau, or should plateau at higher levels (Northfield et al. 2010). They found that a diverse predator community exhibited better aphid control than any single predator species. Moreover, they showed that "no single consumer species, even at high density, was capable of driving resources to as low a level as could be achieved by a diverse mix of consumers—the hallmark of resource partitioning" (Northfield et al. 2010: 346–347).

Bradley Cardinale used a very different experimental design to show that niche complementarity plays a significant role in the way biodiversity in stream algae improves water quality (Cardinale 2011). Nitrogen runoff into streams and rivers is a significant source of pollution in many areas. In laboratory streams with high habitat heterogeneity, an increase in algal species richness (up to 8 algal species) dramatically reduced the amount of dissolved nitrogen (NO_3^-) in the stream water. However, when the stream environment was made more homogeneous, Cardinale found that the loss of niche opportunities for the different algal species led to reduced algal diversity and a reduction in nitrogen uptake. In streams with diverse habitats, most of the positive effect of algal species richness on water quality (by reducing nitrogen levels) was due to niche complementarity (**Figure 3.8**).

Figure 3.8 Effects of algal species richness on nitrogen uptake and algal biomass in streams differ depending on whether or not the different species can partition resources. (A) Streams with heterogenous habitats allow for resource partitioning by the 8 algal species in the study. (B) Where stream habitats are homogeneous, opportunities for resource partitioning are limited and a single species becomes dominant. Data points are means ±1 SE, with best-fit functions plotted as solid lines. Dashed horizontal lines and shaded areas show means ±1 SE for *Stigeocloneum*, the species that achieved the highest values in monoculture. The rightmost panels of each sequence show the proportion of the increases in algal density driven by niche complementarity (NC) or by selection effects (SE, the effect of the dominant species, in this case *Stigeocloneum*). (After Cardinale 2011.)

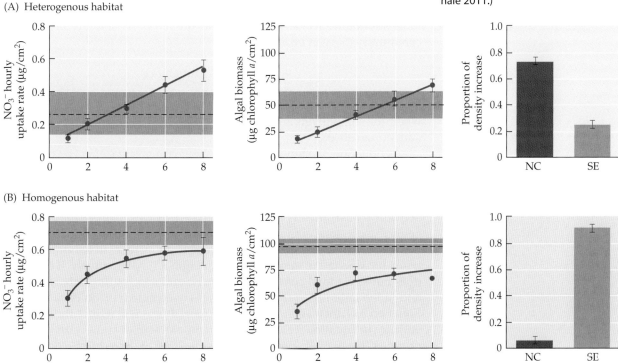

Diversity and Stability

The notion that diversity begets stability in ecosystems has a long and venerable history in ecology. In the 1950s, MacArthur (1955), Elton (1958), Odum (1959) and others suggested that communities containing more species should be better buffered against the effects of species extinctions, species invasions, or environmental perturbations. Theoretical development of these ideas followed shortly thereafter (e.g., Levins 1970; May 1971, 1972; MacArthur 1972), culminating at the time with May's 1973 monograph, *Stability and Complexity in Model Ecosystems*. In this book, May reached the surprising conclusion that diversity does not promote ecosystem stability; instead, the more species a community of competitors contains, the less likely it is to return to equilibrium after a perturbation. The topic of diversity and ecosystem stability has been controversial ever since (McNaughton 1977, 1993; Pimm 1984; Ives and Carpenter 2007). In recent years, experimental explorations of the effect of species diversity on ecosystem functioning have shed new light on the question of whether diversity begets stability. The answer, it turns out, depends in part on how we view stability.

Temporal stability—the consistency of a quantity (e.g., species abundance) over time—has been the primary focus of empirical ecologists studying diversity–stability relationships (McNaughton 1977; Pimm 1984). Operationally, temporal stability is calculated as the variance in species abundance (usually biomass) measured over time and scaled to the mean abundance. This measure of stability can be applied to the entire community or to its constituent species. The temporal stability of an individual species (S_i) depends on the species' mean abundance (μ_i) and the standard deviation in its abundance (σ_i) as μ_i/σ_i, whereas the temporal stability of the total community (S_T) can be expressed as

$$S_T = \frac{\mu_T}{\sigma_T} = \frac{\Sigma\,\text{Abundance}}{\sqrt{\Sigma\,\text{Variance} + \Sigma\,\text{Covariance}}} \qquad \text{Equation 3.1}$$

where the summations are taken over all species in the community (Lehman and Tilman 2000).

As Equation 3.1 shows, diversity can have a positive effect on temporal community stability either by increasing total species abundance, by decreasing the summed variance in species' abundances, by decreasing the summed covariance in species' abundances, or by any combination of the above. We have already seen that increased species richness may result in an increase in total community biomass (overyielding). Therefore, all else being equal, overyielding will contribute to an increase in temporal stability by increasing the numerator of Equation 3.1. Alternatively, if diversity does not affect total community biomass and species fluctuate randomly and independently in their abundances over time (i.e., no covariance in Equation 3.1), then an increase in species richness will reduce the summed variance in Equation 3.1 and increase temporal stability purely on statistical grounds (Doak et al. 1998). Tilman et al. (1998) referred to this phenomenon as the "portfolio effect" because the same principle applies to a fixed financial investment that is diversified across a portfolio of stocks. Finally, if species show a correlated response to environmental fluctuations, then any negative covariance in response will lead to a positive effect of diversity on community temporal stability, whereas positive covariance will have the opposite effect. Interspecific competition would be expected to result in negative covariances among species, whereas facilitation or mutualism should result in positive covariances. Thus, we can visualize the portfolio effect as operating through a reduction in summed variances among species, whereas the covariance effect results from a reduction in the summed covariances (Lehman and Tilman 2000).

Recently, Loreau (2010; see also Loreau and de Mazancourt 2008) have taken issue with the above approach of partitioning the positive effect of species richness on temporal stability into components resulting from species' variances and covariances in abundance, arguing that this approach is too dependent on specific assumptions of how species interact. Instead, they suggest, a more general conclusion is that the main mechanism driving the stabilizing effect of species diversity on community abundance (numbers or biomass) is simply the asynchrony of species responses to environmental fluctuations.

Does diversity affect temporal stability and, if so, can we interpret its effects in light of asynchronous responses of species to the environment? In the Cedar Creek experiment, temporal stability in community biomass increased dramatically with an increase in species richness (Tilman et al. 2006). On average, the treatment plots with the highest diversity were about 70% more stable in their biomass production over time than were monocultures (**Figure 3.9A**). On the other hand, the stability of individual plant species declined significantly with an increase in the number of species in the community (**Figure 3.9B**). Therefore, diversity had a strong, positive effect on the temporal stability of the total plant community, but a negative effect on the stability of individual species abundances. Other studies have shown a similar positive effect of species richness on the temporal stability of total community productivity (McNaughton 1977) and other ecosystem properties (McGrady-Steed et al. 1997), although there are some significant exceptions (Pfisterer and Schmid 2002; Bezemer and van der Putten 2007). In contrast, Proulx et al. (2010) found that "bottom-up" effects (e.g., nutrient supply) tended to promote temporal stability in plant biomass at both population and community levels in the Jena Experiment.

Diversity may also serve to stabilize yields of economically important species when those species are composed of populations (or stocks)

Figure 3.9 Temporal stability in ecosystem functioning (total community biomass in this case) increases with species richness. (A) Temporal stability, measured as the ratio of mean total plant biomass per plot (μ) to its temporal standard deviation (σ), as a function of species richness in the study plots at Cedar Creek from 1996 to 2005. (B) Unlike community temporal stability, the temporal stability of individual plant species was a negative function of species richness for the period 2001–2005. (After Tilman et al. 2006.)

(A)

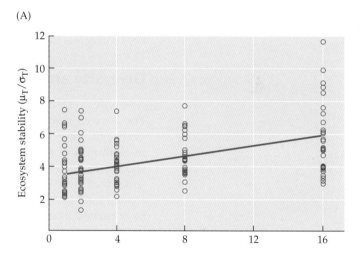

(B)

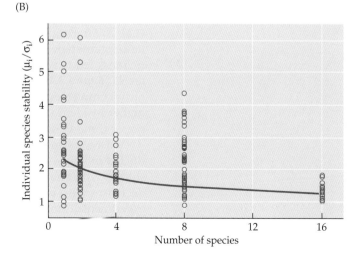

Number of species

that differ in their ecologies and life histories (i.e., within-species diversity). Schindler et al. (2010) provide a dramatic example of the stabilizing consequences of stock diversity to the yield of sockeye salmon (*Oncorhynchus nerka*) harvested from Bristol Bay, Alaska (a multibillion dollar natural resource). The salmon returning to Bristol Bay each year comprise multiple stocks that spawn in the nine major rivers flowing into the bay, and each river stock contains tens to hundreds of locally adapted populations distributed among tributaries and lakes. Variation in fish life histories means that Bristol Bay sockeye spend anywhere from 1 to 2 years in fresh water and 1 to 3 years in the ocean as they complete their life cycles. As a consequence of this local adaptation and life history diversity, the different sockeye populations of the Bristol Bay fishery tend to fluctuate asynchronously in abundance. Schindler et al. (2010) use 50 years of population data to show that variability in annual Bristol Bay sockeye returns is twofold less than it would be if the system consisted of a single homogeneous population. Furthermore, a homogeneous sockeye population would likely see 10 times more frequent fisheries closures than the existing diverse salmon population. These results provide a dramatic demonstration of the importance of maintaining population diversity for stabilizing ecosystem services and the economies that depend on them.

Diversity and Invasibility

Are more diverse communities more resistant to invasion by exotic species? The answer, it seems, is yes, but some important questions remain. Fifty years ago, Elton (1958) proposed that successful invaders must overcome resistance from the resident community, and that this "biotic resistance" should increase with the number of resident species. Elton (1958: 117) reasoned that more diverse communities are more likely to repel invaders due to increased interspecific competition: "This resistance to newcomers can be observed in established kinds of vegetation, [due to] competition for light and soil chemicals and space." Indeed, Fargione and Tilman (2005), using the Cedar Creek study plots, found that both the biomass and the number of invading plant species declined as the richness of the resident plant community increased (**Figure 3.10A, B**; see also Levine 2000; Seabloom et al. 2003). Fargione and Tilman then went on to demonstrate the importance of interspecific competition, showing that increased resident species richness led to higher root biomass (**Figure 3.10C**) and lower soil nitrate levels (**Figure 3.10D**). This higher nutrient use efficiency was due to both niche complementarity and a species selection effect, in that treatments with higher resident species richness tended to contain more warm-season, C_4 grasses, and C_4 grasses caused the largest reduction in soil nitrogen levels.

Most experimental studies on diversity and invasibility have been conducted in low-stature plant communities. However, there are a number of studies with animals that support the general conclusion that invasibility decreases with increased diversity. For example, Stachowicz et al. (1999) experimentally manipulated the species richness of sessile, suspension-feeding marine invertebrate communities ("marine fouling communities") raised on tiles and looked at the establishment success of an invasive species, the

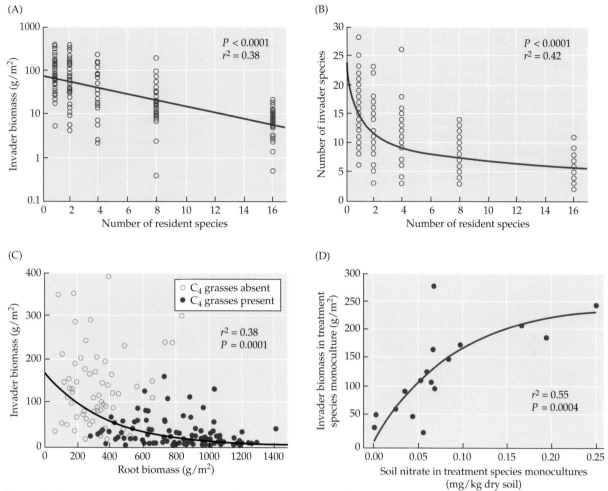

Figure 3.10   Invasion resistance increases with species richness (1-16 resident species) in study plots at the Cedar Creek Ecosystem Science Reserve. Invasion resistance is measured as (A) invader biomass per plot and (B) number of invading species per plot. The mechanisms behind this effect appear to be (C) an increase in root biomass and (D) a decrease in available nitrate in plots with greater species richness. (After Fargione and Tilman 2005.)

sea squirt (ascidian) *Botrylloides diegensis*. They found that the survival of *Botrylloides* recruits decreased with an increase in the native community's species richness, and that this result was consistent with a reduction in the space available for colonization as community species richness increased (**Figure 3.11**).

Open space was consistently lower in communities with more species due to natural cycles in the abundances of individual species (i.e., a kind of portfolio effect). Shurin (2000) also found that the plankton communities of freshwater ponds in Michigan were less invasible by zooplankton species the higher the number of resident species. By experimentally reducing the

Figure 3.11 Species-rich marine fouling communities occupy more space and are less susceptible to invasion than less diverse communities. (A) Survival of recruits of an exotic invader, the sea squirt *Botrylloides diegensis*, as a function of native species richness. (B) Availability of free space in the community as a function of native species richness. Lines are best-fit regressions; larger dots indicate two coincident data points. (After Stachowicz et al. 1999.)

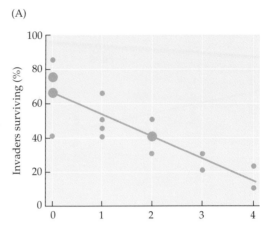

(A)

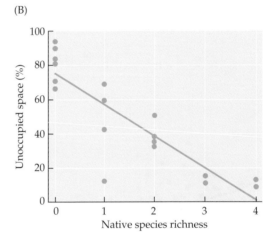

(B)

density of the resident species, Shurin showed that interspecific competition from the resident community was a significant factor limiting colonization by introduced species. (Additional examples of reduced invasibility in diverse communities can be found in Stachowicz et al. 2002; France and Duffy 2006; and Maron and Marler 2007.)

The studies discussed above support the hypothesis of Elton (1958) and others (e.g., MacArthur 1970; Case 1990) that resistance to invasion increases as resident species richness increases, and that the mechanism underlying this response is a reduction in resource availability as species richness increases. But what exactly do we mean by "resistance to invasion"? Are species-rich communities non-invasible, or do invasions of these communities simply take longer or require greater propagule pressure compared with species-poor communities? Levine and colleagues addressed this question with a meta-analysis of resistance to exotic plant invasions (Levine et al. 2004). Reviewing 13 experiments from 7 studies, they found that "although biotic resistance significantly reduces the seedling performance and establishment fraction for individual invaders, it is unlikely to completely repel them" (p. 976). They based their conclusion in part on the observation that the most diverse natural plant communities are also those with the highest number of invaders (Levine et al. 2004; see also Robinson et al. 1995; Wiser et al. 1998; Lonsdale 1999; Stohlgren et al. 1999; Levine 2000), a paradox we return to in the next section.

Unanswered Questions

Our understanding of how biodiversity affects ecosystem functioning has increased tremendously in recent years. Yet many important questions remain. I touch on a few of these unanswered questions below.

MULTIPLE TROPHIC LEVELS   Most ecological communities contain a complex web of interacting species at multiple trophic levels—detritivores, herbivores, carnivores, and primary producers. Yet most biodiversity experiments have focused on studying the effects of diversity at a single trophic level (usually primary producers). How representative of other systems are the results of these experiments? We do not know the answer to this question; what we *do* know is that trophic-level interactions are important in the functioning of ecosystems, and that a change in the number and/or kind of top-level consumers can have significant consequences that cascade down the food chain (see Chapter 11). Moreover, we know that the number of species per trophic level tends to decline from the bottom to the top of the food web, and that species are more likely to be lost from higher than from lower trophic levels (due to smaller population sizes, longer generation times, and greater risk of overexploitation; Petchey et al. 2004). If biodiversity is lost preferentially from higher trophic levels, this loss may result in cascading species extinctions and large effects on ecosystem functioning (Petchey et al. 2004; Thebault et al. 2007). Addressing these issues will require studies conducted at multiple trophic levels (e.g., Downing 2005; Duffy et al. 2005; Steiner et al. 2006; Vogt et al. 2006; Long et al. 2007; Downing and Leibold 2010; Northfield et al. 2010), along with better estimates of the probability of species loss at different points in the food web.

COMMUNITY ASSEMBLY OR SPECIES LOSS?   Our interest in understanding how species diversity contributes to ecosystem functioning is driven in part by the unprecedented rate at which species are being lost from the biosphere. This rapid loss of diversity has led some ecologists to argue for more experiments that examine the effects of species removals from ecosystems, instead of (or in addition to) manipulating biodiversity by constructing assemblages of varying species richness drawn randomly from a regional species pool (Diaz et al. 2003; Wootton and Downing 2003). Removing species from natural ecosystems, however, raises many ethical and practical concerns. Thus, there have been relatively few experiments using species removals to probe the link between biodiversity and ecosystem functioning. In general, such studies (e.g., Symstad and Tilman 2001; Smith and Knapp 2003; Wardle and Zackrisson 2005; Suding et al. 2006; Valone and Schutzenhofer 2007; Walker and Thompson 2010) have shown natural ecosystems to be rather resilient to declines in diversity. However, we know from decades of previous work that the loss of particular species can result in dramatic ecosystem changes (see discussion in Wootton and Downing 2003). Thus, removal studies raise the perplexing question of which species we should remove, given that most natural communities are diverse and that it is impossible to remove all species combinations.

If we knew which species were most likely to be lost from a community, however, we could more realistically simulate the effects of species extinctions in nature. Some traits, such as habitat specialization, small population size, limited geographic range, large body size, high trophic level, and low fecundity, make species more vulnerable to extinction (Pimm et al. 1988; Tracy and George 1992; McKinney 1997; Purvis et al. 2000; Cardillo et al. 2005). Some of these traits may be associated with particular ecological functions within a community. For example, Solan et al. (2004) showed that body size in marine invertebrates affects both the probability of species extinction and the extent of bioturbation of marine sediments by invertebrate communities. Thus, future research may be able to better predict the consequences of species loss if we knew how particular traits affect both a species' probability of extinction and the role that species plays within a community (Gross and Cardinale 2005; Srinivasan et al. 2007).

GLOBAL EXTINCTION AND LOCAL ECOSYSTEM FUNCTIONING: IS THERE A MISMATCH OF SPATIAL SCALES?   Although the current unprecedented loss of species globally is often cited as the reason we need to better understand how biodiversity affects ecosystem functioning, there is something of a mismatch of spatial scales in this argument (I thank Dov Sax for bring this to my attention and for raising the following points).

The global loss of species is a tragedy of unknown consequences. What matters to ecosystem functioning, however, isn't how many or what percentage of species are lost globally. Rather, what matters to ecosystem functioning is how biodiversity is changing at local or regional scales—these are the scales at which we expect species diversity to impact the functioning of ecosystems. And at local and regional scales, there is little evidence that species richness is generally decreasing. Although some communities and ecosystems have shown dramatic declines in species richness in recent decades, in just as many (probably more) cases, species richness has actually *increased* due to species invasions (Sax and Gaines 2003; Jackson and Sax 2010). Species invasions, habitat modification, and global climate change are affecting the species composition of local and regional communities worldwide, introducing new species and new functional traits into communities, modifying trophic structure and species interactions, and creating "novel" communities. Therefore, one might argue that what we really need is more research on the consequences to ecosystem functioning of changing biodiversity (writ large), not simply the consequences of species loss, and that our focus should be on the functional traits of species gained and lost from communities and on the consequences of such "trait changes" for ecosystem processes (Wardle et al. 2011).

THE INVASION PARADOX   In nature, areas of high native plant diversity also contain the greatest number of invading plant species (Herben et al.

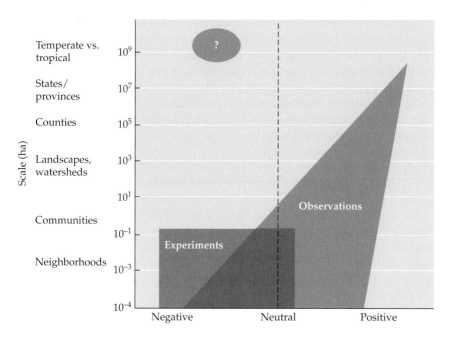

Figure 3.12 Conceptual diagram of the "invasion paradox." Most broad-scale observations in nature find positive correlations between native and exotic species richness (the tropics may be an exception to this pattern, indicated by the circle at upper left). On the other hand, local-scale studies (most of which are experimental) tend to find negative correlations between native species richness and the number of species of exotic invaders. (After Fridley et al. 2007.)

2004; Fridley et al. 2007) (**Figure 3.12**). This positive correlation between native and exotic species richness runs opposite to that found in most experimental manipulations of biodiversity, in which the most species-rich communities show the lowest rate of invasion by exotics (e.g., Fargione and Tilman 2005). Fridley et al. (2007) refer to this mismatch as the "invasion paradox." How do we reconcile the observational and experimental patterns?

The most likely possibility is that in nature, the same environmental factors that support high native plant diversity (e.g., habitat heterogeneity, resource availability) also support high exotic plant diversity, leading to positive covariation in native and exotic species richness. Small-scale experimental studies, on the other hand, seek to control this environmental variation and focus only on the effects of native diversity on invasion success. While this explanation is plausible, the reason for the "invasion paradox" is far from settled, and there is considerable debate over the observed patterns (Rejmanek 2003; Taylor and Irwin 2004; Harrison 2008; Stohlgren et al. 2008). Nevertheless, as Levine et al. (2004: 982) note, if environmental factors "override any negative effect of diversity [on invasibility], that diversity effect cannot be strong, let alone absolute."

HOW IMPORTANT ARE DIVERSITY EFFECTS IN NATURE? As we have seen in this chapter, studies of synthetic communities constructed in the field and in the laboratory show that increasing species diversity generally leads to greater and more stable biomass production, more efficient resource use, and reduced invasibility. Species removal experiments are more equivocal in their findings and tend to show smaller impacts of diversity change (e.g., Smith and Knapp 2003; Wardle and Zackrisson 2005; Suding et al.

2006). Nevertheless, the overwhelming conclusion from almost 20 years of experimental work following the pioneering papers of Naeem et al. (1994) and Tilman and Downing (1994) is that biodiversity does have an impact on the functioning of ecosystems (Cardinale et al. 2011). What is still unknown is how strong this impact is relative to the effects of other ecological interactions, such as competition, predation, or disturbance.

Only one experimental study that I am aware of (Wojdak 2005) has compared the relative impacts of varying species richness with those of what might be considered the classic "top-down" and "bottom-up" perturbations—changes in predator abundance and changes in nutrient input. In a study of simulated freshwater ponds, Wojdak varied grazer (gastropod) species richness in combination with the presence or absence of the grazer's predator; he also varied resource abundance by adding two limiting nutrients (phosphorous and nitrogen) to some treatments. He found that the effects of species richness on ecosystem functioning were strong, and comparable in magnitude to the effects of varying predation pressure or resource abundance.

Despite the evidence from experimental studies, a number of ecologists question whether species diversity per se has a significant effect on the functioning of natural ecosystems (e.g., Grime 1997; Huston 1997; Huston et al. 2000; Grace et al. 2007). They argue that experimental manipulations of species diversity tend to have their strongest effects at relatively low species richness, often at much lower numbers of species than are found in natural communities. They also argue that there is a big difference between demonstrating that a process operates in a (controlled) experiment and showing that the same process is a potent force in nature. It is possible that the effects of species diversity on ecosystem functioning in nature are simply overwhelmed by other interactions with the abiotic and biotic environment. How do we address this question? One possibility is to compare the patterns observed in experimental manipulations of biodiversity with those observed across natural diversity gradients and ask if there is any congruence. In the case of community invasibility, as we have seen, there is a distinct mismatch between observational patterns and experimental results.

THE TWO SIDES OF DIVERSITY AND PRODUCTIVITY   Ecologists have struggled to reconcile the positive effect of species richness on productivity (biomass) found in many biodiversity experiments with observations from natural communities showing that richness and productivity often are not positively related (see Chapter 2). For example, Grace et al. (2007) examined diversity–biomass relationships in 12 natural grassland ecosystems using structural equation models to determine significant pathways of interaction between four variables: species richness, plant biomass, abiotic environment, and disturbance. They found that abiotic environment and disturbance had significant effects on species richness and biomass in a number of ecosystems, and also that biomass had significant effects on species richness in 7 of the 12 ecosystems studied. However, in no case did they find that species richness had a significant effect on plant biomass. Grace et al. (2007: 680) conclude that "these results suggest that the influence of small-scale diversity on productivity in mature systems is a weak force, both in absolute terms and relative to the effects of other controls on productivity."

Ecologists are also attempting to understand how diversity and productivity are functionally related across different spatial scales. As we saw in Chapter 2, there is a long history of looking at patterns of productivity and diversity. In these studies, the two most commonly observed patterns are a hump-shaped relationship, in which species richness peaks at intermediate productivities, and a monotonic (but often decelerating) increase in species richness with productivity (Mittelbach et al. 2001; Gillman and Wright 2006; see Figure 2.12). Traditionally, these observational studies have assumed that environmental productivity is the driver of species richness, whereas more recent experimental studies manipulating biodiversity have found an impact of species richness on productivity. Current research is attempting to reconcile these patterns. For example, Loreau (2000, 2010) and Schmid (2002) hypothesized that resource supply and species traits determine the maximum number of competing species that can coexist at a site, whereas the number of species at a site determines biomass production and the efficiency of resource use. This combination of drivers can generate different forms of the productivity–diversity relationship across sites (as seen in the empirical data; see Figure 2.12). Yet, for all those forms, controlled experiments would show that species richness has a positive effect on productivity within a site.

Recently, Gross and Cardinale (2007) developed a mathematical model in which species compete for resources locally, and in which their regional abundance is determined by their dynamics across all local communities (i.e., in a metacommunity). In this model, biodiversity is a hump-shaped function of resource supply in the local community, and both total biomass production and the efficiency of resource use in the metacommunity increase as diversity increases. Thus, Gross and Cardinale's model is able to account for some of the joint patterns seen in nature and in experimental manipulations of diversity. However, it remains to be seen whether these simple models of competition for resources can account for the broad-scale variation in biodiversity seen across large spatial scales in the natural world.

Conclusion

We started this chapter with a simple question: How many species are needed to ensure the successful functioning of ecosystems? Clearly, we do not yet know the answer to this question. However, ecologists have made great strides toward an answer in a very short time. I imagine that in another 10 or 15 years, we may well have an answer—at least, a qualified one. This much we do know: (1) species play a fundamental role in the functioning of ecosystems, and (2) the species composition of communities worldwide is changing at an unprecedented rate. Taken together, these two facts make the question of how biodiversity affects ecosystem functioning one of vital importance. In addition, the study of biodiversity and ecosystem functioning has played a vital role in helping to bring together the disparate fields of community and ecosystem ecology (Loreau 2010).

Summary

1. Higher species diversity is generally associated with higher primary production. It is still unknown what fraction of species could be lost from a community before we would see a major decrease in primary production.

2. Niche complementarity and species selection are two mechanisms that may tie species richness to production:

   • Niche complementarity may occur if species differ in the way they use limiting resources, in which case the overall efficiency of resource use by the community may increase with an increase in species richness, leading to higher production. Complementarity effects may also result from facilitation.

   • Species selection (also called the sampling effect) may lead to increased productivity if species-rich communities are more likely to contain more-productive species that come to dominate the community.

3. A number of studies have shown that species-rich communities use resources more efficiently than less diverse communities, provided the species exhibit resource partitioning.

4. Temporal stability of communities (total biomass or total species abundance) tends to be greater in more diverse communities due to asynchrony in species responses to environmental fluctuations.

5. Species-rich experimental communities are more resistant to invasion by exotic species than are less diverse communities, but this "biotic resistance" does not allow them to repel invaders indefinitely.

6. More research into the effects of species diversity on ecosystem functioning is needed in several areas, including effects at multiple trophic levels; how particular traits affect both a species' probability of extinction and its functional role within a community; how biodiversity is changing at local and regional scales; the factors underlying the "invasion paradox"; the strength of species richness effects relative to those of other ecological interactions; and how diversity and productivity are functionally related across different spatial scales.

The Nitty-Gritty

Species Interactions in Simple Modules

4 Population Growth and Density Dependence

It is becoming increasingly understood by population ecologists that the control of populations, i.e., ultimate upper and lower limits set to increase, is brought about by density-dependent factors.

Charles Elton, 1949: 19

[Elton's] statement was somewhat premature.

Peter Turchin, 1995: 19

The reproductive capacity of biological organisms is amazing. A single bacterium dividing every 20 minutes under optimal conditions would cover the Earth knee-deep in less than 48 hours. The world population of our own species (*Homo sapiens*) has grown from fewer than 10,000 breeding couples in Africa some 70,000 years ago (Ambrose 1998), to about 300 million people in the first century CE, to 1 billion sometime around 1880, to 7 billion by late 2011. The current growth rate adds roughly 80 million people per year to the planet, according to the U.S. Census Bureau. Thomas Malthus (1789) recognized that this tremendous capacity for population increase could not be maintained in an environment with limited resources. His writings had a profound influence on Charles Darwin, who appreciated the far-reaching consequences of the fact that many more individuals are born than can possibly survive. Since the time of Malthus and Darwin, the study of population growth and population regulation has occupied a central place in ecology. As William Murdoch (1994: 272) put it, "Population growth underlies most … ecological problems of interest, such as the dynamics of diseases, competition, and the structure and dynamics of communities … . [Population] regulation is essential to long-term species persistence and … is still the central dynamical question in ecology."

The first part of this chapter reviews the basic mathematics of population growth under conditions where resources are unlimited (the exponential growth model) and where resources are limited (the logistic growth model). For most of you, this material will be familiar. However, it is useful to see how these simple models of population growth are derived from first

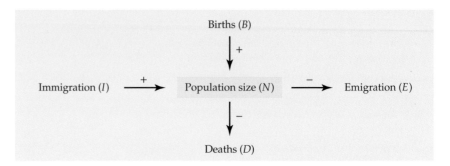

Figure 4.1 Factors affecting the size of a local population. (After Case 2000.)

principles, as they form the foundation for more complex models of species interactions that will be covered in the following chapters on predation, competition, and mutualism. In deriving the exponential and logistic growth equations, I have borrowed rather shamelessly from the excellent presentation in Gottelli (2008).

In the second part of this chapter, we will examine the role of density dependence in population regulation. Ecologists have debated the importance of density dependence in natural populations for almost 100 years, and it is important to know something of this debate. In addition, the study of density dependence in single-species populations leads naturally to the question of **compensatory dynamics** in communities—the idea that in a resource-limited community, an increase in the density of some species should be balanced by a decrease in the density of other, interacting species—is the basis for most theories of community stability and ecosystem resilience (Holling 1973; Gonzalez and Loreau 2009). After we have examined the evidence for density dependence and compensatory dynamics in nature, we will be ready to delve more deeply into the study of species interactions, including a mechanistic look at some of the processes underlying population regulation, such as competition for limited resources (Chapters 7 and 8) and density-dependent mortality due to predators (Chapters 5 and 6).

Exponential Population Growth

Consider a population growing in an unlimited environment, such as a bacterium introduced to a culture flask or a pregnant rat arriving on an oceanic island free of predators. If we let N_t represent the number of individuals in a population at time t, then the size of the population at the end of the next time interval N_{t+1} will depend on the number of individuals born B, the number that die D, the number that immigrate into the population I, and the number that emigrate from the population E during that time interval (**Figure 4.1**). Thus, the population size at time $t + 1$ can be written as

$$N_{t+1} = N_t + B + I - D - E$$

and the change in population size from time t to time $t + 1$ as

$$N_{t+1} - N_t = B + I - D - E$$

For simplicity, we will assume a closed population (i.e., no immigration or emigration), in which case

$$N_{t+1} - N_t = B - D$$

We will also assume that population growth is continuous—that is, that time steps are infinitely small and that births and deaths happen continuously. This assumption allows us to use the mathematics of differential equations (i.e., calculus). If you have forgotten most of your calculus or never understood it all that well in the first place, don't worry. All you really need to recognize here is that dN/dt is the rate of change in the number of individuals N with regard to time (**Figure 4.2**).

The rate of population growth dN/dt can now be written as

$$dN/dt = B - D$$

It is useful at this point to express birth and death rates on a per capita (i.e., per individual) basis. Therefore, we will define a constant b as the per capita birth rate and a constant d as the per capita death rate, such that

$$dN/dt = bN - dN$$

or

$$dN/dt = (b - d)N$$

Letting $r = (b - d)$,

$$dN/dt = rN \qquad\qquad \text{Equation 4.1}$$

Equation 4.1 thus describes how a population will grow in an unlimited environment (see Figure 4.2), where birth and death rates are unaffected by population density. The rate of population growth is governed by the constant r, which is the **instantaneous rate of increase**. When resources are unlimited, r is at its maximum value for a species, referred to as a species' **intrinsic rate of increase**. However, as we will see below, birth and death rates may vary with population density.

To predict the size of the population at any given time t, we need the integral form of Equation 4.1, which is

$$N_t = N_0 e^{rt} \qquad\qquad \text{Equation 4.2}$$

where N_0 is the initial population size at time zero and e is the base of the natural logarithms (≈ 2.718).

Equations 4.1 and 4.2 describe a population undergoing **exponential growth**, where numbers increase (or decrease if r is negative) without bound (see Figure 4.2). Populations are expected to grow exponentially when resources are unlimited, such as may occur when a species colonizes a new environment. Perhaps the most famous example of exponential growth began with the introduction to Australia of about 20 European rabbits (*Oryctolagus cuniculus*) by English settlers in 1859. In their new environment, the rabbits multiplied like, well, like rabbits. Within less than a decade, there were literally millions of rabbits in Australia; in 1907, the government constructed a 1800-km "rabbit-proof" fence in an attempt to limit their spread, but to little avail. There are many other examples of

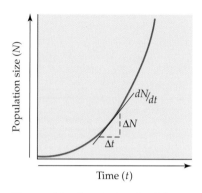

Figure 4.2 Population size (*N*) plotted as a function of time (*t*) for a population growing exponentially. The rate of population change (ΔN) during a time interval (Δt) is equal to $\Delta N/\Delta t$. If we apply the methods of calculus and assume an infinitely small time step, then the rate of population growth can be written as dN/dt, which is equal to the slope of the line tangent to the curve.

introduced or recolonizing species showing exponential population growth (e.g., Van Bael and Pruett-Jones 1996; Madenjian et al. 1998); however, we do not expect such growth to be sustained in a limited environment. For example, the current estimated mean value of the net growth rate for Australia's rabbit population is zero (Hone 1999).

Logistic Population Growth

As a population increases in size, its growth rate will at some point begin to decrease due to the depletion of resources or due to the direct negative effects of a high population density, such as the accumulation of toxic wastes, or aggression between individuals. This negative feedback between population size and per capita growth rate is the essence of what ecologists call **density dependence** (Abrams 2009). A good working definition is found in Turchin (2003: 155): "density dependence is some (non-constant) functional relationship between the per capita rate of population change and population density, possibly involving lags." Note the emphasis here on the *per capita* rate of population change.

In a closed population, the negative feedback between population density and population growth must occur through changes in per capita birth rate, changes in per capita death rate, or both. The simplest way to model these changes is to assume that per capita birth rates or death rates change linearly with population size. Following Gottelli (2008: 26), we can modify the equation for exponential population growth (Equation 4.2) to

$$dN/dt = (b' - d') N \qquad \text{Equation 4.3}$$

where b' and d' now represent the per capita density-dependent birth rate and per capita density-dependent death rate, respectively. We can express the linear change in the birth rate b' with population size N as

$$b' = b - aN$$

and the linear change in the death rate d' with population size N as

$$d' = d + cN$$

where a and c are constants (**Figure 4.3**).

We can now modify the equation for population growth in an unlimited environment (Equation 4.1) to include the effects of density dependence on per capita birth and death rates by changing

$$dN/dt = (b - d)N$$

to

$$dN/dt = (b' - d')N$$

or by substitution,

$$dN/dt = [(b - aN) - (d + cN)]N \qquad \text{Equation 4.4}$$

(Note that in above formulation, birth rates could become negative at large values of N, a nonsensical result.)

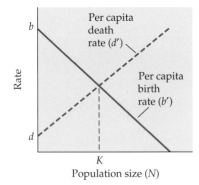

Figure 4.3 Density-dependent per capita birth (b') and death (d') rates in the logistic model of population growth. The per capita birth rate declines linearly with population size ($b' = b - aN$), and the per capita death rate increases linearly with population size ($d' = d + cN$). Where the two curves intersect (i.e., when birth rate equals death rate), population size is unchanging and the population is at its carrying capacity (K). (After Gotelli 2008.)

Gotelli (2008: 27) showed how the terms in Equation 4.4 may be rearranged and simplified to

$$dN/dt = rN[1 - \frac{(a+c)}{(b-d)}N]$$
Equation 4.5

where $r = (b - d)$. Because a, b, c, and d are all constants, we can define a new constant K as

$$K = \frac{(b-d)}{(a+c)}$$

Thus Equation 4.5—the equation for density-dependent population growth—can now be written as

$$dN/dt = rN(1 - \frac{N}{K})$$
Equation 4.6

Alternatively, it is sometimes written as

$$dN/dt = rN(\frac{K-N}{K})$$
Equation 4.7

Equation 4.6/4.7 is the well-known **logistic equation** for population growth in a finite environment. First introduced by Pierre François Verhulst (1838), the logistic equation has played an important role in both the study of population dynamics and the development of models of species interactions (Lotka 1925; Volterra 1926). Because of its importance to ecology, we will explore the properties of the logistic growth equation in more detail here.

The logistic equation predicts a sigmoidal pattern of population growth through time (**Figure 4.4A**), with population size reaching a stable maximum at K, the **carrying capacity**. We can determine the number of individuals N in the population at any time t from the integral form of Equation 4.6, which is

$$N_t = N_0 \frac{K}{(K-N_0)e^{-rt} + N_0}$$
Equation 4.8

It is useful to visualize how the population growth rate dN/dt and the *per capita* population growth rate dN/Ndt vary as functions of population size. As **Figure 4.4B** shows, the population growth rate dN/dt is a peaked function of N, reaching its maximum when $N = \frac{1}{2}K$. This maximum growth rate corresponds to the inflection point in the curve shown in Figure 4.4A. The per capita growth rate dN/Ndt behaves differently, however: it declines linearly with N, starting at a maximum value equal to r and declining to zero when $N = K$ (**Figure 4.4C**).

(A)

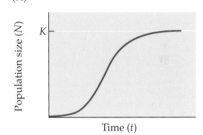

(B)

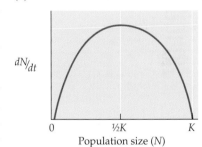

(C)

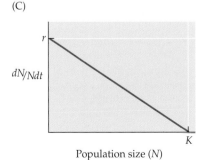

Figure 4.4 Properties of a population exhibiting logistic growth. (A) Change in population size N as a function of time. (B) The population growth rate, dN/dt, peaks at an intermediate population size (equal to half the carrying capacity, or $\frac{1}{2}K$). (C) The per capita population growth rate, dN/Ndt, decreases linearly with an increase in population size, declining from a maximum value equal to the species' intrinsic growth rate, r, and reaching zero at the species' carrying capacity K.

The above analysis shows how the logistic growth model may be derived from the exponential growth model by assuming linear density dependence in per capita birth and death rates and a maximum population size of K. In fact, we get the same result if only the birth rate or the death rate is linearly density-dependent.

The logistic equation is a very simple model for population growth and, as such, it makes four important assumptions that are rarely met in nature (Turchin 2003; Gotelli 2008):

1. It is assumed that the per capita growth rate is a linear function of N (see Figure 4.4C). [Sibly and Hone (2002) provide empirical examples showing both linear and nonlinear density dependence.]
2. It is assumed that population growth rate responds instantly to changes in N (that is, that there are no time lags).
3. It is assumed that the external environment has no influence on the rate of population growth.
4. It is assumed that all individuals in the population are equal (that is, that there are no effects of individual age or size).

Despite these unrealistic assumptions, the logistic equation continues to be taught and used in ecology because, as Turchin (2003: 49) notes, "it provides a simple and powerful metaphor for a regulated population, and a reasonable starting point for modeling single-population dynamics, since it can be modified to address all four criticisms listed above."

This short review of population growth has (hopefully) served to refresh your memory of the basic tenets underlying exponential and logistic growth. We will build on this foundation and on the insights gained from the properties of the logistic equation (see Figure 4.4) in the remainder of the book. There is a wealth of knowledge on the dynamics of single-species populations that we will not discuss in this book, including the effects of time lags and age or size structure on population dynamics as well as the complex dynamics that may be exhibited by populations in models and in nature (e.g., limit cycles and chaos). Interested readers should consult Case (2000), Turchin (2003), Vandermeer and Goldberg (2003), and Gotelli (2008) for an introduction to these more complex dynamics.

The logistic equation can be readily modified to incorporate the effects of additional species. We will make use of a modified logistic model (e.g., the Lotka–Volterra equations for predator–prey dynamics and for interspecific competition) when we dig into the "nitty-gritty" of species interactions in the next few chapters. However, before moving on to the study of interspecific interactions, let's take a closer look at the evidence for density dependence in nature and at some of its consequences.

The Debate over Density Dependence

Although the existence of density-dependent population regulation is widely accepted today, it was once the topic of one of the most acrimonious and long-running debates in ecology. As noted in Chapter 1, during the early twentieth century, there were two opposing schools of thought on the matter. On the one side, Nicholson, Lack, and Elton believed strongly that for populations and species to persist without growing to unimaginably large numbers, they must be subject to density-dependent regulation. On the other side, Andrewartha and Birch saw no evidence for density dependence in natural populations, and they questioned whether density dependence was even necessary to explain the persistence of species. This debate over the factors regulating populations continued for decades. In the 1980s, Dempster (1983) and Strong (1986) argued that populations spend most of their time fluctuating in a density-independent manner, and that it is only when their numbers become very large or very small that density-dependent factors come into play, effectively setting floors and ceilings to population size.

At the same time, some of the debate over density dependence and population regulation shifted subtly to include the question of whether populations exhibit equilibrium or nonequilibrium dynamics. The observation that species abundances in many ecosystems seemed to fluctuate caused ecologists to question whether population models that predict a stable equilibrium density, such as the logistic growth model developed above, had any relevance to the real world. For many, there was broad acceptance that most communities were in a nonequilibrium state. One of my favorite commentaries from this time comes from a review of a symposium entitled "The Shift from an Equilibrium to a Non-equilibrium Paradigm in Ecology," held during the 1990 annual meeting of the Ecological Society of America. In this review, Murdoch (1991: 50) gives a flavor for the dispute over equilibrium and nonequilibrium dynamics that existed at the time.

> Certainly the early proponents of this idea [nonequilibrium dynamics] were viewed as radicals. The ultimate punishment awaiting iconoclasts, however, is canonization, and the title and contents of this symposium, together with talks in several others, suggest that a similar fate is now befalling those few hardy souls who long ago threw in their lot with Andrewartha. For a significant segment of the ESA, it seems, the Reformation is accomplished and, to confuse some history, there is a new pope in Avignon.

Murdoch goes on to express his doubts, however, about the wisdom of embracing this paradigm shift from equilibrium to nonequilibrium ecology:

> Perhaps it is the certitude with which the equilibrium point of view was discarded that impels me to its defense, even though my Society of Ecological Anarchists membership card bears a low number. Is it really true that there is no evidence for stable equilibria in Nature? That an equilibrium view can never be fruitful? Surely not. Convergence in density following experimental perturbation has been demonstrated in at least three systems (perhaps the only ones tested). Is it easy to explain the amazing constancy of the Thames heron population over 50 years without some reference to equilibrium? Regular cycles are surely not examples of stable point equilibria, but an equilibrium-centered view certainly leads to the closely related idea of a stable-limit cycle, which offers at least a plausible explanation for many apparently oscillating populations. It's messy, but I'm afraid there will need to be more than one pope.

Murdoch's comments, besides displaying a keen wit, bring up an important issue that I want to address before going further into our examination of density dependence. The question relates to the usefulness of simple models for understanding the real-world dynamics of populations and species. Personally, I believe that simple models, despite their "unrealistic assumptions," are very useful for developing our insight into the factors that govern population growth and determine the outcome of species interactions. As I mentioned earlier, we will build on the logistic equation to develop models for predator–prey interactions and for interspecific competition. In addition, we will look for equilibrium solutions to these models when we examine the factors that promote (or destabilize) species coexistence. This isn't to say that we believe these models to be literal descriptions of nature, or that we expect the world to be at equilibrium. All populations are affected by exogenous factors, and their densities fluctuate to various degrees. However, as Turchin (1995: 25–26) notes (and it is worth quoting him at length):

> [T]he notion of equilibrium in ecological theory [can be] replaced with a more general concept of *attractor* (e.g., Shaffer and Kot 1985). In addition to point equilibria, there are periodic [limit cycles], quasiperiodic, and chaotic attractors (e.g., Turchin and Taylor 1992). The notion of attractor can be generalized even further, so that it is applicable to mixed deterministic/stochastic systems. … In the presence of noise, a deterministic attractor becomes a stationary probability distribution of population density (May 1973), which is the defining characteristic of a regulated dynamical system. Put in intuitive terms, the equilibrium is not a point, but a cloud of points (Wolda 1989; Dennis and Taper 1994). Thus, if we define equilibrium broadly as a stationary probability distribution, then *being regulated* and *having an equilibrium* are one and the same thing. The whole issue of equilibrium versus nonequilibrium dynamics becomes a semantic argument (see also Berryman 1987).

Evidence for density dependence in nature

Although the early ecologists debated whether density dependence was necessary for population regulation, it is now widely accepted that regulation cannot occur in the absence of density dependence (Murdoch and Walde 1989; Hanski 1990; Godfray and Hassell 1992; Turchin 1995). Thus, the way to detect population regulation is to test for density dependence. Furthermore, density dependence leading to population regulation must take a specific form (Turchin 1995, 2003). First, it must operate in the right direction (there must be a tendency for the population to return to its previous size after a disturbance). Second, this return tendency must be strong enough and quick enough to overcome the effects of disturbance. Most tests for density dependence have examined a time series of estimated population densities or manipulated density experimentally. Detecting density dependence in nature, however, is surprisingly difficult.

The most common approach to looking for density dependence is to examine how the growth rate of a population (r) varies with population density in a long time series of data. Turchin (1995: 28) proposed a simple population model for use in such a test:

$$r_t \equiv \ln N_t - \ln N_{t-1} = f(N_{t-1}) + \varepsilon_t$$

where f is typically a linear function of either N_{t-1} (e.g., Dennis and Taper 1994) or log N_{t-1} (e.g., Pollard et al. 1987) and ε_t is a term representing random density-independent factors. The above model may be criticized on many grounds, including the fact that the time interval of the censuses used often does not correspond to the time scale at which density dependence works (it is often continuous). Still, as Turchin (1995) notes, the above model is usually adequate to address the simple question of whether there is evidence for regulation in the population. Pollard and colleagues proposed looking for regulation by testing the correlation between r_t and log N_{t-1}, whereas Dennis and Taper's test regresses r_t on the untransformed population density N_{t-1}.

Using the methods outlined above, Woiwod and Hanski (1992) analyzed almost 6000 time series of aphid (94 species) and moth (263 species) populations recorded at various sites in Britain. They found density dependence in 69% of aphid and 29% of moth time series. But if short time series (<20 years) are removed from the analysis, the percentage of studies showing density dependence jumps to 84% and 57%, respectively. Holyoak (1993) and Wolda and Dennis (1993) also found that the percentage of studies showing significant density dependence increased with the length of the time series. Further, Turchin and Taylor (1992) found density dependence in 18 of 19 forest insect populations and in 19 of 22 vertebrate populations using tests of delayed as well as direct density dependence, and they observed cyclic and chaotic as well as stable dynamics (Murdoch 1994). Finally, Brook and Bradshaw (2006) conducted a meta-analysis of long-term abundance data from 1198 species spanning a broad range of taxa (vertebrates, invertebrates, and plants). Using a variety of analytical methods, including multi-model inference as well as the traditional methods outlined above, they found "convincing evidence for pervasive density dependence in the time series data" and "little discernible difference across major taxonomic groups" (Brook and Bradshaw 2006: 1448). Moreover, they found a convincing trend toward a zero net growth rate (implying equilibrium) as the number of generations in the time series increased (**Figure 4.5**). These results lead to an inescapable conclusion: when examined with sufficiently long time series, the evidence for density dependence is consistent and overwhelming for many different types of species (Turchin 1995).

An alternative approach to testing for density dependence is to experimentally alter the densities of natural populations and look for density-dependent changes in birth rate, death rate, and population size (Nicholson 1957; Murdoch 1970). Harrison and Cappuccino (1995) reviewed 60 experimental studies that tested for density dependence in this way in a wide range of animal taxa. They found that of the 84 effects examined in the 60 studies, almost 80% showed negative density dependence (positive density dependence was detected twice). Thus, they concluded that experimental manipulations of population density corroborate the results from time-series analysis. In other words, density-dependent population regulation is common and widespread among animal taxa, although Harrison and Cappuccino caution that their conclusions are based on a relatively small number of experimental studies. Interestingly, Harrison and Cappuccino (1995) found in their survey that negative density dependence was more

(A)

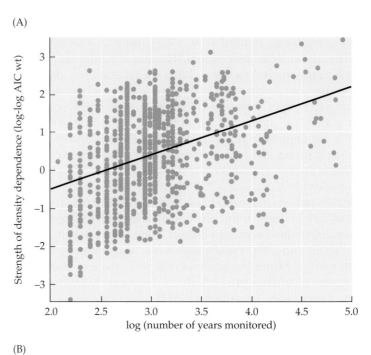

(B)

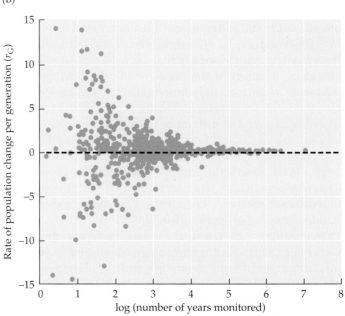

Figure 4.5 The results of a meta-analysis examining the evidence for density dependence in population time-series data for 1198 species in nature. (A) The strength of the evidence for density-dependent regulation, measured as log–log transformation of summed model fits Akaike's Information Criterion, or AIC weights, of three density-dependent models, increases as the length of the population time series increases. (B) The average rate of per-generation population change as a function of the number of generations monitored, showing that the observed net rate of population change tends toward zero (dashed line) as the number of generations monitored increases. (After Brook and Bradshaw 2006.)

commonly the result of competition than the result of control by predators and other natural enemies.

This last result highlights an important and timely question in the study of density-dependent regulation: What factors contribute most to population regulation ("bottom-up" forces such as competition for limiting resources, or "top-down" forces such as death by predators and other natural enemies)? In addition, we would like to know which demographic parameters (e.g., reproduction, survival) respond most strongly to changes in population size, as well as how the strength of density dependence varies with organism age, body mass, and ecological context (Bonenfant et al. 2009). For many, the answers to these questions are far more meaningful than wondering whether density dependence does or does not occur in a population (Krebs 1995). Sibly and Hone (2002) provide a good review of the different approaches ecologists have taken to studying density dependence (i.e., testing for the existence of density dependence versus testing for the mechanisms causing density dependence). Other important foci in the study of population regulation are the relative impacts of endogenous versus exogenous forces on population change, and whether regulation is most often the result of local processes or the outcome of metapopulation dynamics (Murdoch 1994; Turchin 1999).

Positive density dependence and Allee effects

Up to this point, we have focused our discussion of density dependence on those factors that cause the per capita population growth rate to decline with density, because negative density dependence is crucial for population regulation. However, a number of factors may result in *positive* density dependence (i.e., a positive correlation between density and per capita population growth), particularly at very low population densities (Case 2000). These factors are commonly referred to as **Allee effects**, named after the zoologist W. C. Allee (Allee et al. 1949). For example, at low population densities, the inability to find a mate may limit the ability of a species to colonize new environments (Gascoigne et al. 2009). A nice example of an Allee effect comes from the work of Sarnelle and Knapp (2004) and Kramer et al. (2008), who studied the recolonization success of zooplankton in high alpine lakes in the Sierra Nevada of California. Two large-bodied zooplankton species, the copepod *Hesperodiaptomus shoshone* and the daphnid cladoceran *Daphnia melanica*, were native to these lakes prior to their being stocked with trout in the early to mid-1900s. Large-bodied zooplankton, however, are a preferred prey for trout, and following the introductions, *Hesperodiaptomus* and *Daphnia* disappeared. When trout were removed from some study lakes in the late 1990s, *Daphnia* rapidly recolonized. *Hesperodiaptomus*, however, failed to reappear in many of the study lakes where they were once native. Sarnelle and Knapp (2004) hypothesized that the copepods were unable to recover from low densities because of a strong Allee effect. Copepods and daphnids both produce resting eggs that may remain dormant in lake sediments for many years, and these resting eggs are the most likely source of colonists in these isolated alpine lakes. *Daphnia* can reestablish a population from only a few colonists (potentially a single, hatched resting egg) because they can reproduce asexually (by parthenogenesis) as well as sexually. Copepods, however, are obligately sexual and must find a mate in order to reproduce. Kramer et al. (2008) hypothesized that the inability of *Hesperodiaptomus* to find mates and reproduce at very low densities was the reason they were unable to recolonize lakes following trout removal. Experimental introductions of copepods at varying densities supported their hypothesis and provided one of the few experimental tests of Allee effects in nature.

Other factors that may lead to positive effects of population density on per capita growth rates include group protection from predators, amelioration of a harsh environment by neighbors (see examples in Chapter 9), and increased foraging efficiency in groups (Dugatkin 2009). Kramer et al. (2009) provide a recent review of the evidence for Allee effects, in which they conclude that density often has positive effects on various components of individual fitness, particularly at low densities. However, the total impact of density on per capita population growth must include the combined effects of both positive and negative density-dependent factors. This recognition has led ecologists to distinguish "component Allee effects," in which density has a positive impact on some component of fitness, from "demographic Allee effects," in which the total impact of density on per capita population growth rate is positive (Stephens et al. 1999). Kramer et al. (2009: 341) concluded that "although we find conclusive evidence for Allee effects due to a variety of mechanisms in natural populations of 59 animal species, we also find that existing data addressing the strength and commonness of Allee effects across species and population is limited." In the most recent analysis of demographic Allee effects, Gregory et al. (2010) found only limited evidence for such Allee effects in 1198 natural populations. Using a best-fit model approach, they concluded that positive density feedbacks might influence the growth rate of about 1 in 10 natural populations. In their review of mate-finding Allee effects, Gascoigne et al. (2009) found only a few examples of demographic Allee effects. Thus, while examples of component Allee effects are common in nature, especially at low population densities, the evidence suggests that the demographic consequences of density on population growth are largely negative; or, simply put, demographic Allee effects are rare. Therefore, it appears that we are on safe ground in focusing on negative density dependence when constructing simple models of population growth (such as the logistic equation). We will build on the logistic model as we begin our study of species interactions in resource-limited environments in Chapters 5, 7, and 9.

The functional form of density dependence

Before we conclude our discussion of density dependence in single-species populations, it is worth noting that most models of density dependence (including the logistic model) do not explicitly incorporate resources, even though competition for resources is assumed to be the major factor driving negative density dependence in most systems. In the logistic model, density dependence acts in a linear fashion—that is, per capita population growth declines linearly with population density (see Figure 4.4C) due to (unspecified) linear effects of density on per capita birth rates and/or per capita death rates (see Figure 4.3). How might we expect the functional form of density dependence to look in more realistic population models that explicitly incorporate consumer–resource interactions? Recent studies suggest that the form of density dependence in most consumer–resource interactions will be highly nonlinear because consumer density has little impact on per capita population growth rate when small population sizes are small and consumers are nearly saturated with resources (Abrams 2009a,b). As Abrams has shown, we can use data from harvested populations to probe the form of density dependence, since the impact of population density on resource-dependent growth rates is equivalent to that of density changes resulting from per capita removal rate in a harvested population (because removal must equal growth in order for the population to persist at some constant or average level). This approach allows us to better characterize the form of density dependence in consumer–resource models, and perhaps in some natural populations as well.

Compensatory Dynamics in Communities

The concept of density dependence in a population may be extended to the community level under the rubric of **density compensation** (MacArthur et al. 1972). If we assume that the environment can support some fixed number (or biomass) of individuals that use the same set of resources or occupy the same trophic level (i.e., that there is a community-level carrying capacity), then any increase in the abundance of one species should be mirrored by a decrease in the abundance of other species. Or, in the case of islands (the situation examined by MacArthur et al. 1972), a decrease in species richness on islands compared with the mainland should be compensated by an increase in the density of the island species. This notion that the density (or biomass) of all individuals within a functional group should be a conserved quantity (perhaps fluctuating around some mean value) has been termed "compensatory dynamics," "community-level density dependence" (Houlahan et al. 2007; Gonzalez and Loreau 2009; Tanner et al. 2009), or "zero-sum dynamics" (Hubbell 2001; Ernest et al. 2008). Just as in the debate over density dependence within single-species populations, the evidence for compensatory dynamics in communities has been hotly contested.

Both experimental manipulations and time-series analyses have been used to test for density compensation. An excellent example of compensatory dynamics in a natural community comes from the work of James Brown and his colleagues, who conducted a long-term experiment on rodents, plants, and ants in the Chihuahuan Desert near the town of Portal, Arizona. In 1977, Brown and colleagues established 24 fenced plots (50 m × 50 m) to which various rodent exclusion treatments were applied (including control plots to which all rodents had access). Gates in the plot fences excluded or permitted access by different rodent species. Some of these gates excluded the largest rodent species (members of the genus *Dipodomys*, the kangaroo rats). In their study of compensatory dynamics, Brown and colleagues focused primarily on comparing the natural dynamics of the rodent communities in the control plots with those in the kangaroo rat removal plots. This system is an especially good one for studying compensatory dynamics because the rodent species are all granivores and are therefore supported by the same resource base.

Importantly, this long-term experiment has been conducted against a background of environmental change—specifically in precipitation levels—that has resulted in a vegetation shift from grasses to woody shrubs and a concomitant shift in rodent species composition. Ernest et al. (2008) found evidence for compensatory dynamics at the study site. Over time, the rodent community shifted from dominance by grassland species to dominance by shrubland species (**Figure 4.6A**). However, species richness at the site remained quite constant despite the considerable species turnover (**Figure 4.6B**). Moreover, the shift in species composition of the rodent community resulted in a decrease in average rodent body size and an increase in total rodent abundance (**Figure 4.6C, D**). In combination, these changes in body size and abundance resulted in no change in the total energy flux (resource consumption) by the rodent community as a whole (**Figure 4.6E**). Thus, Ernest et al. (2008) concluded that the consistency in total energy use through time by the rodent community showed evidence of density compensation (i.e., the resource base could support a constant total energy use, regardless of the species composition). Tanner et al. (2009) provide another example of community-level density dependence based on 27 years of data on coral community dynamics from Australia's Great Barrier Reef. Hawkins et al. (2000) also found clear evidence of density compensation in stream invertebrate communities in comparisons between disturbed and reference sites for more than 200 taxonomic groups and 200 study sites.

Other approaches to looking at density compensation have examined variation and covariation in species densities over time. Houlahan et al. (2007) examined 41 different plant and animal data sets to assess patterns of covariation among population abundances. They predicted that compensatory dynamics within communities should result in a preponderance of negative covariation in species densities (i.e., that some species' populations would grow larger as others decreased). Instead, they found that 31 of the 41 data sets had more positive covariances than would be expected by chance; in other words, populations in the community were growing larger or smaller together. Based on these results, they concluded "that variability in community abundance appears to be driven more by processes that cause positive covariation among species (e.g., similar responses to the environment) than processes that cause negative covariation among species (e.g., direct effects of competition for scarce resources)" (Houlahan et al. 2007: 3276). Mutshinda et al. (2010) also concluded that environmental variation plays a much larger role than intra- or interspecific interactions in determining temporal variation in species abundances in an analysis of time-series data from a range of taxa including moths, fishes, macrocrustaceans, birds, and rodents. Most of the observed environmentally driven covariation between species was positive, and Mutshinda and colleagues (2010: 2923) concluded that "compensatory dynamics are in fact largely outweighed by environmental forces, and the latter tend to synchronize the population dynamics." However, Ranta et al. (2008) cautioned that the lack of negative covariation in species densities over time does not imply an absence of competition. Furthermore, Peter Abrams (personal communica-

(A)

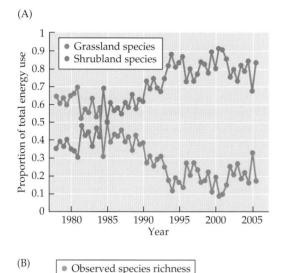

(C)

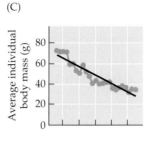

(D)

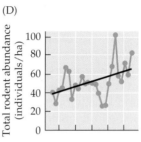

(B)

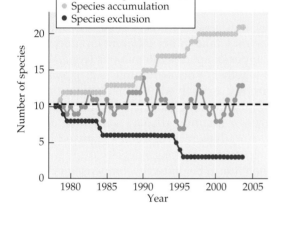

(E)

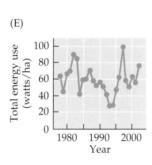

Figure 4.6   Evidence for density compensation and compensatory dynamics in a long-term study of a desert rodent community in the Chihuahuan Desert near Portal, Arizona. (A) Temporal trends in the proportion of total energy flux through the rodent community demonstrated a shift in dominance from grassland species to shrubland species. (B) Trends in rodent species richness demonstrated that the long-term average species richness (dashed line) was essentially constant despite considerable species turnover. (C–E) Trends in mean individual rodent biomass (C), total rodent abundance (D), and summed rodent metabolic rate, an index of community-level energy use (E). Regression lines are plotted for significant relationships. (After Ernest et al. 2008.)

tion) notes that the predictions of most multispecies models of competition are not strongly affected by the addition of environmental variation whose effects are positively correlated among species. If environmental variation affects different species in a similar manner, then time series will often show positive cross-species correlations in density, but mean densities of species will follow the underlying predictions of competition models.

What can we conclude about the strength of compensatory dynamics in communities? For now, the evidence is mixed. Strong compensatory dynamics have been detected in a number of field experiments. However, analyses of time-series data suggest that environmental effects tend to trump impacts of intra- or interspecific density dependence, or both. Clearly, the matter is far from settled and there is much work yet to do. This work is important because compensatory dynamics figure into many aspects of community and ecosystem theory, including the relationship between biodiversity and ecosystem functioning (Doak et al. 1998; see Chapter 3), species abundances in neutral models (Hubbell 2001; see Chapter 13), species coexistence in variable environments (Chesson 2000; see Chapter 14), and predator–prey interactions in multi-prey communities (Abrams et al. 2003; see Chapter 6). Gonzalez and Loreau (2009) provide an excellent overview of some of the causes and consequences of compensatory dynamics in communities.

Conclusion

Understanding the factors that regulate population growth is fundamental to understanding all aspects of ecology. In this chapter, we examined how simple mathematical models of population growth (the exponential and logistic equations) may be derived from basic principles, and we discussed the significance of density dependence to population regulation and the persistence of species. We will build on this foundation as we move on to study species interactions and the mechanisms that underlie predator–prey dynamics, interspecific competition, and mutualism/facilitation. The debates over whether density dependence is necessary for population regulation and whether density dependence is common in nature have been largely settled: affirmative on both accounts. However, important questions remain regarding the factors that contribute most to density dependence (i.e., top-down versus bottom-up forces), how and when these factors operate, and what form density dependence may take under different environmental conditions and with different types of species interactions (e.g., Abrams 2009a,b, 2010). The question of whether or not communities display strong compensatory dynamics is currently unresolved. As noted above, the idea that species assemblages exhibit something akin to a carrying capacity established by limited resources or that they display something akin to density compensation (or zero-sum dynamics) underlies many ecological theories based on resource competition. Although density compensation is unlikely to operate everywhere (or to be perfectly compensatory when it does occur), we should expect to find evidence for it in most natural systems where interspecific competition is important.

Summary

1. In exponential growth, a population's size increases without bound at an instantaneous rate of increase, r. Exponential growth can occur when resources are unlimited. As a population increases, however, its per capita growth rate will begin to decrease due to the depletion of resources or other negative effects of high population density. This negative feedback is referred to as density dependence.

2. The logistic growth equation models the linear effects of density dependence on population size. It predicts a sigmoidal pattern of population growth, with population size reaching a stable maximum at K, the carrying capacity.

3. The most common approach to detecting density dependence in natural populations is to examine how r varies with population density in a long time series of data. Experimental studies of density-dependence alter a population's density and then look at changes in birth rate (b) and/or death rate (d).

4. A number of factors, such as the inability to find a mate at low population densities, may result in positive density dependence. These factors, whose results are commonly referred to as Allee effects, are particularly important at very low densities.

5. "Component Allee effects," in which higher density has a positive impact on some component of fitness, are relatively common in nature, but the overall demographic consequences of high density for population growth are largely negative.

6. Density dependence at the community level is referred to as density compensation. If we assume that the environment can support some fixed number (or biomass) of individuals that use the same resources, then any increase in the abundance of one species should be mirrored by a decrease in the abundance of others. This phenomenon, called compensatory dynamics, has been demonstrated in some populations but remains a topic of controversy.

5 The Fundamentals of Predator–Prey Interactions

Models are tools to understand the real world and … they can sharpen our intuition about ecological mechanisms. Even the simplest model, which may match no living system, can be useful.

William Murdoch et al., 2003: 3

Tigers and deer do not stroll together.

Chinese proverb

In any community, species exist within a network of other potentially interacting species. Each species in this network consumes resources and is itself consumed by other species. The **consumer–resource link** is the fundamental building block from which increasingly complex food chains and food webs may be constructed (**Figure 5.1**; Murdoch et al. 2003). In this and the following four chapters, we will focus on the consumer–resource link to develop a mechanistic understanding of species interactions, including predation, interspecific competition, and mutualism/facilitation. Here we start with what is arguably the simplest consumer–resource interaction: one predator species feeding on one species of prey (**Figure 5.2A**). From this simple beginning, we will move on (in Chapter 6) to consider predators that feed on a variety of prey species (**Figure 5.2B**). Next we will examine consumer–resource interactions from the perspective of competition (Chapters 7 and 8) and mutualism and facilitation (Chapter 9), after which we will put the pieces together by exploring interactions between multiple consumers and multiple resources using the constructs of ecological networks and food webs (Chapters 10 and 11). Each increase in complexity allows us to consider additional properties of multispecies consumer–resource interactions (Holt 1996).

In this chapter, we encounter two important concepts that should be part of every ecologist's toolkit for the study of species interactions. The first is the Lotka–Volterra model (here formulated for predator and prey); the second is isocline analysis. The Lotka–Volterra model for two interacting species is the simplest possible consumer–resource model, and ecologists

Figure 5.1 Simple consumer–resource interaction networks: a food chain (A) and a food web (B). Consumer–resource links (represented by arrows) are the building blocks of these networks. Plus signs next to the arrows indicate positive effects on consumer species from eating a resource species; minus signs indicate negative effects on the resource species being consumed.

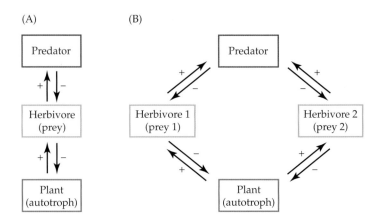

have had a love–hate relationship with it for a long time. Its critics would toss it in the dustbin of ecology as overly simplistic, outdated, and just plain wrong, but its supporters see it as a useful heuristic tool for gaining insight into the processes that govern the outcome of species interactions. (I am in the latter group.) Isocline analysis is introduced here as a means to visually interpret the outcomes of the Lotka–Volterra model. The usefulness of isocline analysis, however, extends well beyond the analysis of Lotka–Volterra type models, and we will make use of the technique again in considering consumer–resource models of interspecific competition and in the study of food webs.

This chapter is different from most others in this book in that it focuses almost exclusively on developing and analyzing simple mathematical models, with very few examples from nature. I have de-emphasized empirical examples here because I want to move quickly beyond the study of a single predator species feeding on a single species of prey (which is a relatively rare interaction in nature). However, the study of such a simple predator–prey system is a useful place to start our exploration of the nitty-gritty of species interactions. In the next chapter, we will examine more complex and realistic

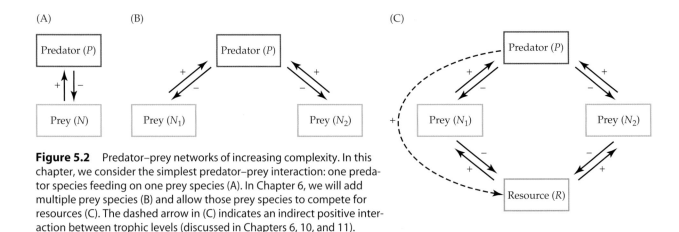

Figure 5.2 Predator–prey networks of increasing complexity. In this chapter, we consider the simplest predator–prey interaction: one predator species feeding on one prey species (A). In Chapter 6, we will add multiple prey species (B) and allow those prey species to compete for resources (C). The dashed arrow in (C) indicates an indirect positive interaction between trophic levels (discussed in Chapters 6, 10, and 11).

interactions between predators and their prey, including many empirical examples that tie theoretical concepts of predation to the natural world.

Predator Functional Responses

A fundamental property of any predator–prey interaction is the predator's **feeding rate** (the number of prey consumed per predator per unit time). We expect this rate to increase as prey density increases; however, the relationship may take different forms. For example, at low prey densities, a predator's feeding rate should be directly proportional to prey density (Turchin 2003), such that

$$\text{feeding rate} = aN \qquad \text{Equation 5.1}$$

where N is the population density of the prey and a is a constant of proportionality known as the predator's per capita attack rate. However, at high prey densities, a predator's feeding rate should reach some maximum set by the time required to capture and consume an individual prey item—referred to as that prey's **handling time**. Holling (1959) termed the relationship between prey density and predator feeding rate the predator's **functional response**, and he classified predator functional responses into three basic types (**Figure 5.3**).

The **type I functional response** presumes that a predator's feeding rate (measured in units of number of prey consumed per individual predator per unit time) increases linearly with prey density (Equation 5.1; Figure 5.3A). Although mathematically simple, the type I functional response is

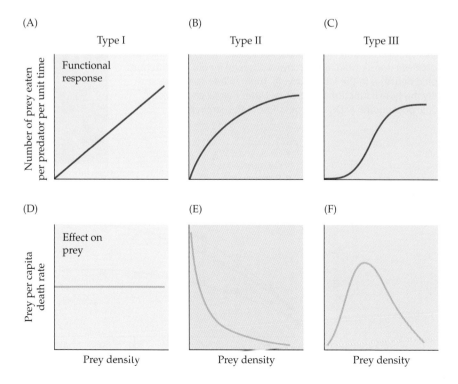

Figure 5.3 The predator's functional response and its impact on the prey's per capita death rate. The top row (A–C) shows the three types of functional responses described by Holling (1959). In each case, the number of prey eaten per predator per unit time increases with increasing prey density. The increase in predation rate with prey density is linear for the type I response, saturating for the type II response, and sigmoidal for the type III response. The bottom row (D–F) shows how the per capita death rate of the prey population changes with prey density under each type of predator functional response.

unrealistic for most predators, as it assumes that a predator can consume a constant fraction of the prey population per unit time regardless of prey density. Consequently, the per capita death rate of the prey does not vary with its own density (Figure 5.3D). Some filter-feeding organisms such as zooplankton may fit this relationship over a range of prey densities (Jeschke et al. 2004), but the consumption rate of most predators will approach a maximum set by the prey's handling time. We could incorporate a maximum feeding rate into the type I functional response by allowing feeding rate to increase linearly with prey density up to a point where it abruptly becomes constant (Holling 1959). However, a more realistic scheme is to model the predator's functional response as a curve that smoothly approaches an asymptote.

Holling's **type II functional response** (Figure 5.3B) describes such a relationship, where

$$\text{feeding rate} = \frac{aN}{1+ahN} \qquad \text{Equation 5.2}$$

and where h is the handling time (see Gotelli 2008 pp. 136–137 for a simple derivation of Equation 5.2). The type II functional response has been shown to fit the feeding rates of a wide variety of predators (**Figure 5.4**; Hassell 1978; Jeschke et al. 2004), and it is to be expected whenever a single predator is foraging without interference for a single type of prey (Turchin 2003). Note that at very low prey densities, the ahN term in the denominator is small and the feeding rate approximates a type I functional response. At very high prey densities, the feeding rate approaches a maximum rate equal

Figure 5.4 The number of aphids consumed in 24 hours by three species of adult ladybird beetles (*Cheilomenes sexmaculata*, *Coccinella transversalis*, and *Propylea dissecta*) feeding at different aphid densities in the laboratory. All three beetle species exemplify the type II functional response. (After Pervez and Omkar 2005; photograph © Dr. J. Burgess/Photo Researchers, Inc.)

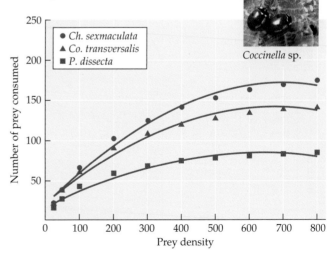

Coccinella sp.

to the inverse of the prey's handling time ($1/h$). Note too that under a type II functional response, an individual prey's probability of death declines as prey density increases (Figure 5.3E). It is not important that the type II functional response fit Equation 5.2 exactly; rather, the defining property of the type II functional response is that it increases at a decreasing rate (see Abrams 1990 for examples).

Holling characterized the **type III functional response** as a sigmoidal relationship (Figure 5.3C), in which a predator's feeding rate accelerates over an initial increase in prey density but then decelerates at higher prey densities. A useful formulation for the type III functional response is

$$\text{feeding rate} = \frac{cN^2}{d^2 + N^2} \qquad\qquad \text{Equation 5.3}$$

where $c = h^{-1}$, the maximum feeding rate, and $d = (ah)^{-1}$, the half-saturation constant (the prey density at which the feeding rate is half the maximum; see Turchin 2003 for alternative formulations and further discussion of the functional response). With a type III functional response, the prey's per capita death rate peaks at an intermediate prey density (Figure 5.3F).

Holling (1959) suggested that the type III functional response might best apply to vertebrate predators (which might develop a search image for abundant prey), whereas invertebrates might best fit type II. However, it is now thought that the tendency of predators to fit either the type II or type III functional response is more a function of foraging mode. Predators feeding on a single prey type (specialists) commonly exhibit a type II functional response, whereas predators that switch between feeding on different prey types (generalists) tend to exhibit a type III functional response. Predators may also have a type III functional response if they adjust their foraging effort adaptively because it is costly to look for prey when there are very few around (see Abrams 1982 for theory and Sarnelle and Wilson 2008 for a likely example). Because our current focus is on the interaction between a predator and a single prey species, we will defer discussion of prey switching and the type III functional response to Chapter 6, where we consider the interactions between selective predators and multiple prey.

In all the formulations of the functional response mentioned above, a predator's feeding rate is determined only by the density of its prey, and feeding rates are unaffected by the number of other predators present (except as mediated through changes in prey density). In other words, these formulations are **prey-dependent**. However, in some situations, predators may interfere with one another's ability to capture prey, or they may facilitate one another's success by group foraging. In such cases, the formulation of the functional response should include the density of predators as well as the density of prey. A simple way to handle this requirement is to divide the number of prey available by the number of predators (P) and substitute the quantity N/P for N in the above formulations of the functional response (Equations 5.1–5.3). There is a sizable literature on this form of **ratio-dependent predation** (e.g., Arditi and Ginzburg 1989; Akçakaya et al. 1995), and there has been considerable debate as well about which form of the functional response (prey-dependent or ratio-dependent) provides the best starting point to construct simple predator–prey models. Most predator–prey models incorporate prey-dependent processes, perhaps because prey density must affect the functional response, whereas predator dependence may or may not be important. Here, we will use the traditional prey-dependent formulations (Equations 5.1–5.3) to examine the factors affecting the abundance and stability of predator–prey populations.

The Lotka–Volterra Model

The simplest model describing the population dynamics of a single predator species feeding on a single prey species was developed independently in the 1920s by Alfred Lotka (1925) and Vito Volterra (1926). In the **Lotka–Volterra model**, the prey population (N) is assumed to grow exponentially in the absence of predators:

$$dN/dt = rN \qquad \text{Equation 5.4}$$

where N is the number of individuals in the prey population and r is the prey's per capita population growth rate. If the predator has a linear (type I) functional response (i.e., the number of prey killed is directly proportional to prey density), and if predation is the only source of mortality for the prey population, then the net rate of change in prey numbers is

$$dN/dt = rN - aNP \qquad \text{Equation 5.5}$$

Note that the units for a, the predator's per capita attack rate, are the number of prey eaten per prey per predator per unit time, and that a corresponds to the slope of the predator's linear (type I) functional response (see Figure 5.3A).

In the absence of prey, the predator population is assumed to decline exponentially as

$$dP/dt = -qP \qquad \text{Equation 5.6}$$

where q is the predator's per capita mortality rate (which is density-independent); the units for q are the number of deaths per predator per unit time. Growth of the predator population depends on the number of prey consumed per unit time, which is equal to aNP, and on the predator's efficiency at turning prey eaten into new predators, which we assume to be a simple constant (f). Thus, the birth rate of the predator population is equal to $faNP$, and the dynamics of the predator population is described by

$$dP/dt = faNP - qP \qquad \text{Equation 5.7}$$

Together, Equations 5.5 and 5.7 make up the Lotka–Volterra predator–prey model.

Isocline analysis

The properties of the Lotka–Volterra model can be investigated by determining the zero growth isocline for each population. **Zero growth isoclines** define conditions in which a population is neither growing nor declining. For a prey population whose dynamics are governed by Equation 5.5,

$$dN/dt = 0 \text{ when } rN = aNP$$

Thus, the prey population will be at equilibrium (neither growing nor declining) when the density of predators P equals the quantity r/a. Likewise, for the predator population,

$$dP/dt = 0 \text{ when } faNP = qP$$

and the predator population will be at equilibrium when the density of prey N equals q/fa.

It is useful to plot these zero growth isoclines on a graph where the x axis is prey density (N) and the y axis is predator density (P). Because q, f, a, and r are constants in the Lotka–Volterra model, the zero growth isoclines for the predator and the prey are straight lines. Looking first at the prey isocline (**Figure 5.5A**), we see that at predator densities less than r/a, the prey population increases, and at predator densities greater than r/a, the prey population decreases. Thus, a specific number of predators (equal to r/a) will maintain the prey population at equilibrium (zero growth). Looking next at the predator isocline (**Figure 5.5B**), we see that at prey densities less than q/fa, the predator population decreases, and at prey densities greater than q/fa, the predator population increases. Thus, a specific number of prey (equal to q/fa) will maintain the predator population at equilibrium. Figure 5.5 illustrates an important but potentially confusing point: it is the number of predators present that determines whether the prey population is at equilibrium, $dN/dt = 0$. Likewise, it is the number of prey present that determines whether or not the predator population is at equilibrium, $dP/dt = 0$.

Figure 5.5 Zero growth isoclines for the Lotka–Volterra predator–prey model. (A) The prey's zero growth isocline is a horizontal line. The prey population will increase at predator densities below the line and decrease at predator densities above the line. At a predator density of r/a, the prey population will not change ($dN/dt = 0$). (B) The predator's zero growth isocline is a vertical line. The predator population will increase at prey densities to the right of the line and decrease at prey densities to the left of the line. At a prey density q/fa, the predator population will not change ($dP/dt = 0$). (C) Predator and prey zero growth isoclines plotted on the same graph. The intersection of the two isoclines defines four regions in predator–prey state space, and the vectors (narrow-lined arrows) show the response of the predator and prey populations in each region. The heavy diagonal arrows show the combined trajectories of the predator and prey populations in each region (i.e., the resultant vectors). Over time, the predator and prey populations will cycle continuously, as illustrated by the ellipse.

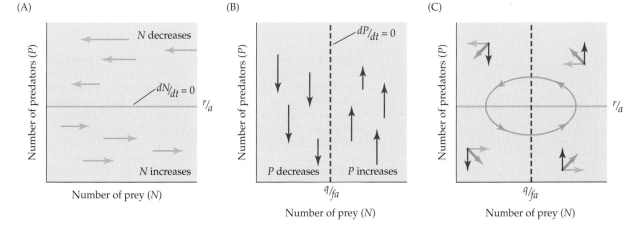

(A)

(B)

(C)

Next let's plot the predator and the prey isoclines on the same graph (**Figure 5.5C**). Note that the two isoclines intersect at a single point; only at this equilibrium point will prey and predator populations remain constant. Away from the intersection of the predator and prey isoclines, the numbers of predator and prey will change over time. How they change is illustrated by the arrows (vectors of population change) in each of the four quadrants in Figure 5.5C. For example, in the upper right-hand quadrant, both prey and predators are abundant. The predator population increases in this region because prey numbers are above those needed to support zero growth in the predator population (r/a). In this same region, however, the prey population is declining because predator abundance is above that required to hold the prey population in check (q/fa). As a result, the joint trajectory of the two populations (illustrated by the heavy arrow) will move the system (over time) into the upper left-hand quadrant. In the upper left-hand quadrant, both predator and prey numbers will decline. A little reflection on the projected dynamics of predator and prey in each quadrant of Figure 5.5C should convince you that the population sizes will track counterclockwise in state space (i.e., all possible abundances of predators and prey) along an approximate ellipse.

How does this tracking in state space in Figure 5.5C correspond to changes in the population growth curves for predator and prey over time? We can see the answer in **Figure 5.6A**. Both predator and prey populations cycle smoothly in abundance over time, with the amplitude and period (time required for a full oscillation) remaining unchanged. The predator population always lags behind the prey population by about one-fourth of a cycle, whereas the amplitude of the cycles is determined by the initial population densities (densities that start farther from the equilibrium point result in cycles of larger amplitude).

The Lotka–Volterra model is neutrally stable. That is, its predator and prey populations will cycle indefinitely, with the same amplitude and period; however, if the system is perturbed in any way (e.g., something comes along to change predator or prey numbers), that perturbation will cause the system to cycle with a similar period but a new amplitude determined

(A)

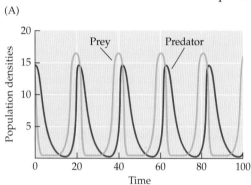

(B)

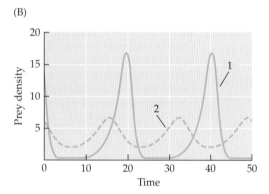

(C)

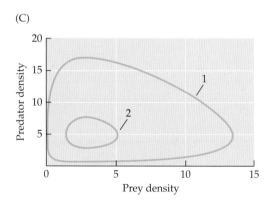

Figure 5.6 (A) In the Lotka–Volterra model, predator–prey population dynamics show continuous cycles over time. The predator population lags one-fourth of a cycle behind the prey population. (B) An external perturbation to the system (here, a change in the initial densities of predator and prey, but no change in model parameter values) results in a change in the amplitude of the cycles (only the prey cycles are shown here for clarity). The prey densities from Figure 5.6A are shown (1) along with the densities resulting from a change in initial densities (2). (C) Prey densities plotted in state space (i.e., the density of prey at time t plotted against the density of predators at time t) as predicted by the Lotka–Volterra model, with different starting densities (1, 2) but the same parameter values. (After Murdoch et al. 2003.)

by the size of the perturbation (**Figure 5.6B,C**). A model displaying such initial condition-dependent oscillations cannot be a good representation of the real world, where perturbations are common. Moreover, as Murdoch et al. (2003) note, the Lotka–Volterra model is "structurally unstable," and almost any change in the model's structure changes the predicted dynamics. Such sensitivity is often leveled as a criticism at the Lotka–Volterra model. However, sensitivity can also be a useful property because it allows us to explore how adding different ecological processes to the simple model will change the dynamics of predator and prey (Murdoch et al. 2003).

Adding more realism to the Lotka–Volterra model

Notably absent from the simple Lotka–Volterra model are (1) self-limitation in the prey or predator population and (2) a realistic predator functional response. No prey population can grow without bounds, even in the absence of predation. Therefore, a more realistic predator–prey model needs to include a density-dependent term specifying how the prey's growth rate declines with prey density. One possibility [originally proposed by Volterra (1931)] is to model prey growth in the absence of predation using the logistic equation (see Chapter 4). In this case, the prey's equation becomes

$$dN/dt = rN\left(1 - \frac{N}{K}\right) - aNP \qquad \text{Equation 5.8}$$

In Equation 5.8, we now have two factors regulating the growth of the prey population: predator density and prey density. The effect of adding prey self-limitation to the model is to rotate the prey zero growth isocline in a clockwise direction (**Figure 5.7A**) so that it has a negative slope everywhere. Intuitively, this makes sense: as prey density increases, it takes fewer predators to keep the prey population in check (zero predators are required when the prey population is at K).

Adding self-limitation to the prey population dramatically changes the stability of the predator–prey interaction. The intersection of the predator and prey isoclines in Figure 5.7A defines an equilibrium point (where $dN/dt = 0$ and $dP/dt = 0$), just as it did in Figure 5.5C. However, if predator or prey densities are perturbed, the populations will recover and return to the equilibrium point (population oscillations will damp out over time; **Figure 5.7B**). Thus, self-limitation in the prey population acts to stabilize the predator–prey interaction.

The stability of the predator–prey equilibrium can be determined by formal stability analysis (see Murdoch et al. 2003 for a good introduction to local stability analysis). However, when the predator's per capita growth rate depends on prey density (and not on its own density), we can apply a simple rule of thumb to judge the stability of the equilibrium point marked by the intersection of predator and prey isoclines. If the predator isocline intersects the prey isocline on its negative slope (i.e., Figure 5.7A), then the predator–prey equilibrium is stable. Intersections on a positively sloped prey isocline are unstable. This rule of thumb glosses over some important complexities, but it will suffice for our heuristic treatment of the factors that affect predator and prey dynamics in simple models.

(A)

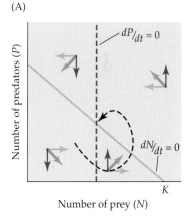

(B)

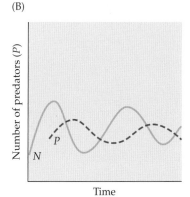

Figure 5.7 Predator–prey isoclines (A) and predator–prey dynamics (B) for a model of the Lotka–Volterra type when prey self-limitation (logistic growth) is included.

The Rosenzweig–MacArthur Model

The analyses presented above have assumed a linear (type I) predator functional response, where each predator consumes a constant fraction of the prey population per unit time. A type I functional response might be a good approximation at low prey densities, but it cannot be true at high prey densities. When prey are abundant, the predation rate will reach a maximum determined by handling time. Clearly, in order to build a more realistic predator–prey model, we need a more realistic functional response. Rosenzweig and MacArthur (1963) were the first to incorporate the more realistic type II functional response into a predator–prey model that also includes self-limitation in the prey, where

$$dN/dt = rN\left(1 - \frac{N}{K}\right) - \frac{aNP}{1+ahN} \qquad \text{Equation 5.9}$$

$$dP/dt = \frac{faNP}{1+ahN} - qP \qquad \text{Equation 5.10}$$

This model may well be the simplest representation of an actual predator–prey interaction, and it is often used in theoretical ecology (Turchin 2003).

In Rosenzweig and MacArthur's model, the prey zero growth isocline has a hump (**Figure 5.8**). It is important to understand why this is so. Recall that with a type II functional response, the prey's per capita death rate declines with each increase in prey density (see Figure 5.3). Therefore, as prey density goes up, the number of predators required to keep the prey population in check must also increase. This relationship generates the initial ascending limb of the humped prey isocline. However, self-limitation in the prey population will eventually take over, and at the prey's carrying capacity (*K*), no predators are required to control the prey population—hence, the descending limb of the humped prey isocline. Thus, adding a bit more biological realism to the Lotka–Volterra model (i.e., a maximum predator feeding rate and self-limitation in the prey population) changes the shape of the prey isocline and (as we will see below) allows us to better represent the observed dynamics of real predator–prey populations. The Rosenzweig–MacArthur model exhibits a wider and more realistic range of dynamics than does the Lotka–Volterra model, including conditions in which predator and prey coexist at a stable equilibrium point, or in a stable limit cycle, or one or both populations go extinct. A formal analysis of the conditions leading to these outcomes is beyond the scope of this book (see Murdoch et al. 2003 for a good introduction to this topic). Luckily, we can understand much about the behavior of the system simply by noting whether the predator isocline intersects the prey isocline to the right or the left of the hump. Intersections to the right of the hump result in a stable equilibrium point, to which predator and prey densities return if perturbed away from it (Figure 5.8A, C). Intersections to the left of the hump do not have a stable equilibrium point. Rather, predator and prey populations end up in a **limit cycle**, in which predator and prey densities cycle endlessly (Figure 5.8B, D). It is important to note that limit cycles are

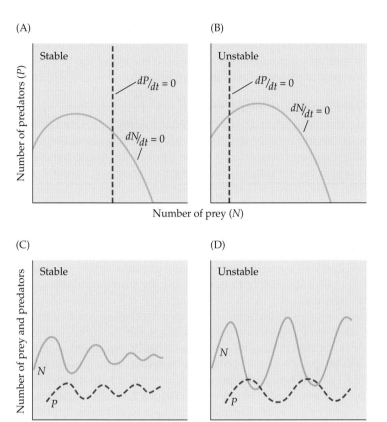

Figure 5.8 Predator and prey isoclines for the Rosenzweig–MacArthur model. Note that under this model the prey zero growth isocline is humped rather than the straight line of the Lotka–Volterra model seen in Figure 5.5A (see text for explanation). Left-hand panels illustrate stable predator–prey interactions; right-hand panels illustrate unstable interactions (limit cycles).

a common feature of predator–prey models of this type (May 1973) and that they are fundamentally different from the predator–prey cycles generated by neutral stability in the Lotka–Volterra model (Murdoch et al. 2003). A predator–prey system in a limit cycle responds to perturbation by returning (over time) to its original cycle (**Figure 5.9**), whereas a predator–prey system displaying neutral stability responds to perturbation with a change in the amplitude of the cycle (see Figure 5.6). If, however, we assume some minimum population density for the persistence of predator or prey (due to finite population size), then systems displaying limit cycles may not recover

Figure 5.9 An example of a stable limit cycle for predator and prey. (A) In a stable limit cycle, the densities of predator and prey cycle perpetually in a counterclockwise direction (marked by the heavy closed loop), tracing out the densities over time shown in (B). If we perturb the populations, pushing them to a point either above or below the limit cycle (black dots in A), they return to the cycle. (After Murdoch et al. 2003.)

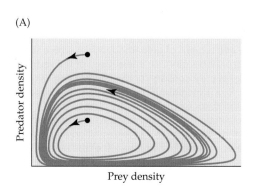

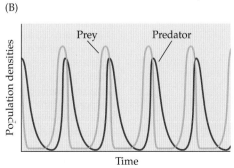

from perturbation, which may lead to the extinction of the predator, or to the extinction of both predator and prey. Such outcomes are possible when predator and prey isoclines intersect well to the left of the hump (see May 1973 for a general discussion of limit cycles).

The suppression–stability trade-off

We can understand what drives the predator–prey interaction toward stability or instability by considering the relative efficiency of the predator at controlling its prey. Intersections to the right of the hump describe a situation in which the predator is relatively inefficient at consuming prey or turning prey into new predators. In this case, the prey population exists at a high density (relative to its carrying capacity K_1) and is predominantly self-limited. Intersections to the left of the hump describe an efficient predator that holds its prey population at a low density (relative to its carrying capacity K_3). Thus, stability is more likely the closer the prey population is to its own carrying capacity and when the predator is relatively inefficient (and is therefore less likely to overexploit the prey population and drive it to low densities).

Rosenzweig (1971) recognized a surprising consequence of the fact that the predator–prey interaction becomes less stable the more responsible the predator is for controlling its prey. Presumably, we might expect that an increase in the prey's food supply (and therefore an increase in its carrying capacity, K) would benefit the prey population and help it better withstand the impact of predation. However, as **Figure 5.10** shows, increasing the prey's carrying capacity may have exactly the opposite result. Increasing carrying capacity (K) may cause the predator isocline to intersect the prey isocline to the left of the hump, destabilizing the equilibrium and potentially causing one or both species to go extinct. Rosenzweig (1971) termed this result the **paradox of enrichment**, and he suggested that it could explain the loss of species in ecosystems experiencing increased nutrient input (e.g., eutrophication in aquatic ecosystems). Although the term "paradox of enrichment" is entrenched in the literature, Murdoch et al. (2003) suggest that this phenomenon might better be termed the **suppression–stability trade-off**, to emphasize the fact that

(A)

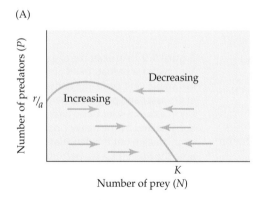

(B)

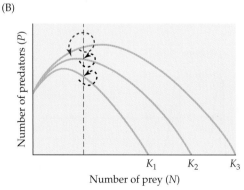

Figure 5.10   An illustration of the "paradox of enrichment" (also known as the "suppression-stability tradeoff"; see Murdoch et al. 2003) using the Rosenzweig–MacArthur predator–prey model. (A) The prey's hump-shaped zero growth isocline and regions of predator densities where the prey population is increasing or decreasing. (B) Intersection of the vertical predator isocline with a series of prey isoclines for three increasing values of prey carrying capacity (K_1, K_2, K_3). Increasing the prey's carrying capacity shifts the intersection of the predator isocline from the right to the left of the prey isocline's hump. Intersections to the right of the hump are stable, and perturbations to the system will return to a stable equilibrium point over time (black dashed arrows). Intersections to the left of the hump are unstable (red dashed arrow) and may result in the extinction of the predator or of both predator and prey.

the further a predator suppresses the prey equilibrium below the prey's carrying capacity, the more likely the system is to be unstable.

Density-dependent predators

Up to this point, we have modeled a predator population whose growth rate is dependent only on the density of its prey. In the Lotka–Volterra model (with its type I functional response),

$$dP/dt = faNP - qP$$

In the Rosenzweig–MacArthur model (with its type II functional response),

$$dP/dt = \frac{faNP}{1+ahN} - qP$$

The predator isocline is a straight, vertical line in both models. We can see this by setting $dP/dt = 0$ and solving for N, which yields $N = q/fa$ and $N = q/fa - ahq$, respectively.

As Rosenzweig (1977: 373) notes, the concept of a vertical predator isocline can be confusing. "If the victim [prey] isocline has provided difficulties, the vertical predator isocline has been positively mystifying not only to students, but also to some theoretical ecologists." The key to understanding the vertical predator isocline comes from recognizing that

> slope in isoclines is generated only if the density of the species in question directly and instantaneously affects its own birth or death rates. The victim [prey] isocline slopes because more victims mean less resources per victim per instant, and less loss to the same number of predators per victim per instant. Are such things true about every predator? No. Some predators just collect food and ignore each other. At any instant, their rate of success depends only on the amount of food that exists. Of course, in the next instant, far less food will be present if there are 10^6 predators catching it, compared to 10^4. But that is taken care of by a change in *food density*. The larger population finds itself in the next instant with less food and may well decline. In order for the predator isocline to slope positively, some predators must actually encounter the same particle of resource at the same instant, so that at most but one gets fed. Or they must both need the same den or territory.

Predators that interfere with one another are expected to show direct density dependence and have a "bent" predator isocline that curves to the right (**Figure 5.11**). Predator interference can be defined broadly to include the energetic costs of defending a territory or chasing away other predators, as well as physically fighting with another predator over a food item. The energetic costs of interference result in a rightward bend in the predator isocline. A simple way to think about this is to recognize that as predator density increases, each predator will expend more energy interacting with other predators. Therefore, each predator must encounter a higher density of prey to obtain enough energy to replace itself. Predator interference may also arise if the presence of more predators scares prey, so that prey spend

(A)

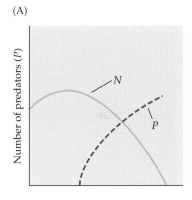

(B)

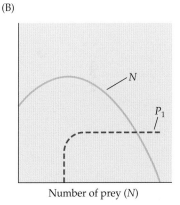

Number of prey (N)

Figure 5.11 Self-limitation in the predator population results in a rightward bend in the predator isocline. (A) The type of bent predator isocline expected if the predator population exhibits interference competition. (B) A steeper bend in the predator isocline is expected if the predator population becomes limited by a second resource (e.g., nesting sites).

more time in a refuge when more predators are present, thus effectively reducing prey abundance (Abrams 1982). An important outcome of including direct density dependence in the predator population is that it tends to stabilize the predator–prey interaction (i.e., the predator isocline is more likely to intersect the prey isocline to the right of the hump; see Figure 5.11).

Resource Harvesting

Ecologists have learned a lot from the study of simple predator–prey models. However, it is important to recognize that few predators in nature are so specialized that they feed on only one species of prey. In the next chapter, we will consider what happens when predators feed on multiple species of prey. Before moving on, however, we will take the opportunity to consider one more, highly specialized, example of a one predator–one prey system.

Humans consume a plethora of living resources (e.g., fish and shellfish; trees), and their interaction with these resources can be viewed from the perspective of predator (humans) and prey (resources). The unique aspect of this interaction is that humans have the potential to regulate harvest in such a way as to increase the resource's long-term yield; that is, they may act as "prudent predators." The idea that other animals might also behave as "prudent predators" was raised a number of years ago (e.g., Slobodkin 1968, 1974). However, most researchers now agree that this is unlikely, as it would require either group selection (see Pels et al. 2002) or other special circumstances (see Munger 1984). Here we use the simple predator–prey models developed in this chapter to explore some basic questions of harvest management. A detailed discussion of population harvesting, however, is beyond the scope of this chapter; interested readers should consult Clark (1990), Hart and Reynolds (2002), Walters and Martell (2004), and others for more advanced treatments.

A population growing logistically will have its highest growth rate (maximum dN/dt) at a density equal to one-half its carrying capacity (see Figure 4.4B). Other models of population growth will show somewhat different relationships between dN/dt and N, but in general, the highest rate of production of new individuals (the **net recruitment rate**) will occur at a density of N that is less than K (Roughgarden 1997). This net recruitment rate can be considered the surplus production of the population (**yield**) that is available for exploitation (harvest). If we were to harvest the resource population in **Figure 5.12A** at a rate equal to h_2, which just touches the peak of the resource population growth curve, we would achieve the highest possible rate of sustained harvest. In the terminology of resource management, this is the **maximum sustained yield** (**MSY**). Harvesting at a higher rate (h_3) will drive the resource population to extinction. Harvesting at a lower rate (h_1) will result in two potential population equilibria. The equilibrium point at the right is stable and results in a long-term harvest that is less than MSY. The equilibrium point at the left is unstable: if the resource population moves below this point, it will decline to extinction; if the resource population moves above this point, it will increase until it hits the stable equilibrium point at the right.

The harvesting strategy shown in Figure 5.12A is termed **fixed quota harvesting** (or **constant quota harvesting**) because a fixed amount of the resource is harvested per unit time (e.g., number of fish per fishing season, or number of prey per year or per reproductive cycle) regardless of the size of the resource population. A second harvesting strategy would be to harvest at some fixed level of effort. We can think of **fixed effort harvesting** as being equivalent to a predator with a type I functional response, where the number of prey harvested in a season (H) is expressed as

$$H = aEN \qquad \text{Equation 5.11}$$

where a is a measure of harvesting efficiency (similar to the predator's attack rate in the functional response sense), E is the harvesting effort (e.g., the number of days of harvest), and N is prey density. If we plot the har-

Figure 5.12 Fixed quota harvesting (A) and fixed effort harvesting (B, C). In each case, the resource (prey) population is assumed to grow in logistic fashion, such that the resource growth rate (dN/dt) is greatest at intermediate prey densities (i.e., between zero and K). (A) Prey are harvested at a fixed rate, or "quota"; three different rates are illustrated. The intersection of the harvesting rate (number of prey harvested per unit time) with the prey population growth rate (number of prey produced per unit time) defines an equilibrium point, at which population size does not change. (B) Under fixed effort harvesting, the number of prey harvested per unit time increases with effort and with prey density. Prey mortality rates are shown for four different levels of harvesting effort. The intersection of each effort with the growth curve represents an equilibrium point. (C) The relationship between harvesting effort and yield is also a hump-shaped curve, peaking at the optimal harvesting effort (i.e., the harvesting effort that supplies the MSY). Thus, if we have sufficient data on the observed harvesting rate for a given harvesting effort, we can specify the harvesting effort that supplies the MSY even without knowing the absolute size of the prey population.

(A) Fixed quota

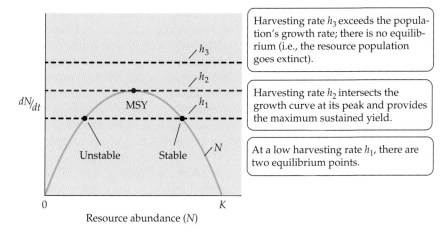

(B) Fixed effort

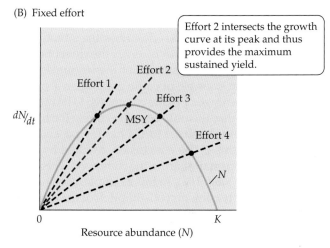

(C)

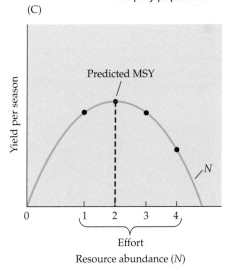

vest rate given by Equation 5.11 and the prey's net recruitment rate on the same graph we get the expected harvest from different levels of harvesting effort (**Figure 5.12B**). It turns out that the relationship between harvesting effort and yield is also a hump-shaped curve (**Figure 5.12C**), which peaks at the optimal harvesting effort (i.e., the harvesting effort that supplies the maximum sustained yield, MSY). Figure 5.12C demonstrates an important result: even without knowing the absolute size of the prey population, we can specify the harvesting effort that supplies the MSY if we have sufficient data on the observed harvesting rate for a given harvesting effort. These data are often available for marine fisheries and have been used to set harvest quotas.

How well do harvesting strategies based on simple predator–prey models work in the real world? Unfortunately, not very well. The fixed quota harvesting strategy is particularly vulnerable to overexploitation. Managers often have a poor idea of the true abundance of the resource in nature, as well as a poor picture of the resource recruitment curve (which may vary from year to year). Therefore, attempting to set a harvest quota to obtain MSY often results in overexploitation of the resource, leading to population collapse (May et al. 1979). Fixed effort harvesting is less likely to result in overexploitation because the total yield varies with the resource population size. Therefore, if the resource density declines, yield also declines. If yield declines below MSY, then the appropriate action by regulators is to reduce effort and allow the resource population to recover. However, political pressure may result in just the opposite response, as harvesters lobby for an increase in effort to restore their harvest (e.g., a lengthening of the harvest season). Furthermore, a decline in harvest may drive up the price for the resource, resulting in additional market pressure for an increase in harvesting effort. As such, economic pressures interacting with biological processes can produce a feedback loop that results in the overexploitation of resources even under fixed effort regulation.

Even if we could accurately estimate population densities and MSY for a resource, our ability to harvest a shared resource in a sustainable fashion would often be compromised by what Hardin (1968) termed the **tragedy of the commons**. The tragedy of the commons describes the situation in which each entity harvesting an open resource receives the direct benefit of harvesting that resource, whereas all consumers of the resource share in the costs. Hardin (1968) used the example of cattle grazing on a shared pasture to illustrate the tragedy of the commons: "The rational herdsman concludes that the only sensible course for him to pursue is to add another animal to his herd. And another; and another. … But this is the conclusion reached by each and every rational herdsman sharing the commons. Therein is the tragedy" (Hardin 1968: 162). In this example, each herder benefits himself by increasing his consumption of the resource (i.e., by adding another animal to his herd), whereas all the herders pay the cost when the pasture becomes overgrazed. Reynolds and Peres (2006) nicely illustrate how the tragedy of the commons can be incorporated into a simple cost–benefit (yield) model (**Figure 5.13**) to reveal how economic pressures

Figure 5.13 The tragedy of the commons illustrated for a fixed effort fishery model, in which costs are proportional to effort and benefits are proportional to yield. In this example, profits are maximized at E_P (where the difference between the benefit and cost curves is greatest); the maximum sustained yield (MSY) is obtained at E_{MSY}. Fishing effort, however, will be pushed beyond these points due to the tragedy of the commons, eventually reaching the point where costs equal benefits (E_C) and every fisherman is just breaking even. At this point, the fishery is overharvested. (After Reynolds and Peres 2006.)

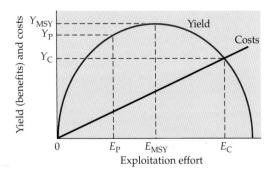

will inevitably push harvesting of an open-access resource to the point of overexploitation.

In summary, the concept of maximum sustained yield, although soundly grounded in the theory of predator–prey interactions, and widely applied in fisheries management up until the 1980s, has proven to be ineffective as a management tool for both biological and economic reasons. More modern and sophisticated approaches to managing for sustainable yields include surplus production models, dynamic pool models, population matrix models, and multispecies models (see Reynolds and Peres 2006 for a short summary of these approaches, and Hart and Reynolds 2002 and Abrams and Matsuda 2005 for more detailed explanations).

Figure 5.14 shows the results from a size-structured, multispecies fisheries model parameterized for 21 fish species in the Georges Bank ecosystem (Worm et al. 2010). The predictions of this more complex and realistic ecosystem model show how the rate of resource exploitation (harvest) affects many parameters of interest. In particular, note that the multispecies maximum sustained yield (MMSY; total fish biomass harvested) is achieved at an intermediate rate of exploitation, as we might expect based on our knowledge of predator–prey models. Moreover, the authors found that a wide range of exploitation rates ($0.25 < \mu < 0.60$) yielded catch rates within 90% of MMSY. The ecosystem consequences of these different exploitation rates, however, are significant. An exploitation rate of $\mu = 0.60$ leads to population collapse in almost half of the fish species in the ecosystem, whereas a rate of $\mu = 0.25$ is expected to rebuild total fish biomass, increase average fish body size (length), and strongly reduce the risk of species collapses with little loss in long-term yield.

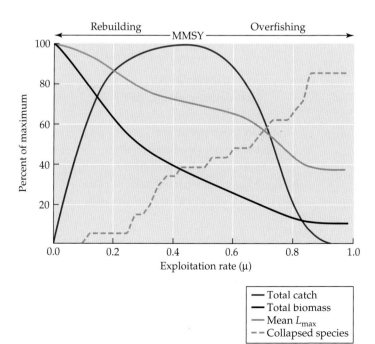

Figure 5.14   Predictions from a size-structured, multispecies fisheries model show how exploitation (harvesting) rates affect the fish community. The exploitation rate μ is the proportion of available fish biomass harvested in each year. "Mean L_{max}" is the average maximum length for the 21 fish species in the community (a measure of maturation, since most fish grow continually throughout their lives). "Collapsed species" refers to the percentage of species whose population sizes will be reduced to below 10% of their unexploited biomass. The model was parameterized for 19 target and 2 nontarget species in the Georges Bank fish community and includes size-dependent growth, maturation, predation, and fishing. (After Worm et al. 2010.)

These results suggest that it is possible to achieve high sustainable yields while maintaining most fish species in an exploited ecosystem if harvesting rates are reduced well below MMSY. Using available data from 10 large marine ecosystems, Worm et al. (2010) found that up to the 1990s, 6 of the 10 ecosystems studied had exploitation rates substantially higher than predicted MMSY. However, since the 1990s, harvesting rates in most of the ecosystems have shown substantial declines, such that they are now at or below the modeled MMSY. Thus, there is some cause for optimism about the future of fisheries, and as Worm et al. (2010: 584) note,

> In fisheries science, there is a growing consensus that the exploitation rate that achieves maximum sustainable yield should be reinterpreted as an upper limit rather than a management tool. This requires overall reductions in exploitation rates, which can be achieved through a range of management tools … [including] a combination of traditional approaches (catch quotas, community management) coupled with strategically placed fishing closures, more selective fishing gear, ocean zoning, and economic incentives.

However, Worm et al. are quick to point out that their analysis is based on data from a small number of well-studied ecosystems, accounting for less than one-fourth of the world's fisheries area and catch. As Reynolds and Peres (2006: 274) note, "parameter-hungry models are useless for the vast majority of the world's exploited species, because we usually lack even the most fundamental information on the biology of the organism and the activities of the people who exploit them." Thus, the question of how to harvest the world's fisheries sustainably remains a challenge, especially in many poorer regions and in areas outside national jurisdictions (Ludwig et al. 1993; Pauly et al. 2002; Worm et al. 2010).

In the hope of ending on a somewhat lighter note, I'll close this chapter with "An epitaph for the concept of maximum sustained yield," penned by the fisheries biologist Peter Larkin (1977):

M.S.Y.

1930s–1970s

Here lies the concept, MSY.
It advocated yields too high,
And didn't spell out how to slice the pie.
We bury it with the best of wishes,
Especially on behalf of fishes.
We don't know yet what will take its place,
But hope it's as good for the human race.

Summary

1. The relationship between prey density and predator feeding rate is referred to as the predator's functional response. Predator functional responses can be classified into three basic types:

   - In a type I functional response, a predator's feeding rate increases linearly with prey density (unrealistic for most predators).
   - In a type II functional response, a predator's feeding rate increases with prey density, but at a steadily decreasing rate. This relationship can be modeled as a curve that approaches an asymptote.
   - A type III functional response is a sigmoidal relationship in which a predator's feeding rate accelerates over an initial increase in prey density, then decelerates at higher prey densities.

2. The Lotka–Volterra model is a simple model of the dynamics of a single predator species feeding on a single prey species. It assumes that the prey population will grow exponentially in the absence of predators; that the feeding rate of predators follows a type I functional response; and that growth of the predator population depends on the number of prey consumed per unit time.

3. The Lotka–Volterra model produces coupled oscillations in predator and prey populations. Adding self-limitation to the prey population (e.g., incorporating a carrying capacity K) acts to stabilize the system, which returns to a stable equilibrium point after being perturbed.

4. The Rosenzweig–MacArthur model of predator–prey dynamics incorporates the more realistic type II functional response, which causes the prey isocline to be humped. If the predator isocline intersects the prey isocline on the right of the hump, the intersection defines a stable equilibrium point, and the system will return to this equilibrium point following a perturbation. If the predator isocline intersects the prey isocline on the left of the hump, the system is unstable (predator and prey populations oscillate in a limit cycle); perturbations will cause the amplitude of the cycle to change, and one or both species may go extinct.

5. Predator–prey systems are more likely to be stable when the predator is relatively inefficient; this phenomenon is called the suppression–stability trade-off. Increasing the carrying capacity for prey may decrease the stability of the system, a phenomenon known as the paradox of enrichment.

6. Adding predator self-limitation (e.g., interference from other predators or limiting resources for predators) will cause the predator isocline to bend to the right and tends to stabilize the predator–prey system.

7. The net recruitment rate of a prey population can be considered the surplus production (yield) that is available for exploitation (harvest). In most cases, net recruitment rate peaks at a population density well below the carrying capacity (e.g., in the Lotka–Volterra model, recruitment rate is greatest at one-half of K).

8. Harvesting a prey population such that its density is maintained at the maximum net recruitment rate should achieve the highest possible rate of sustained harvest, referred to as the maximum sustained yield (MSY).

9. Under a fixed quota harvesting strategy, a fixed amount of the resource is harvested per unit time regardless of the size of the resource population. Under a fixed effort harvesting strategy, the resource is harvested at some fixed level of effort. But human economic pressures interacting with biological processes result in overexploitation, as illustrated by the tragedy of the commons.

6 Selective Predators and Responsive Prey

[M]ost carnivores do not confine themselves rigidly to one kind of prey; so that when their food of the moment becomes scarcer than a certain amount, the enemy no longer finds it worth while to purse this particular one and turns its attention to some other species instead.

Charles Elton, 1927: 122

The food of every carnivorous animal lies therefore between certain size limits, which depend partly on its own size and partly on other factors. There is an optimum size of food which is the one usually eaten and the limits actually possible are not usually realized in practice.

Charles Elton, 1927: 60

In the previous chapter, we focused on the simplest possible predator–prey interaction: one predator species feeding on a single species of prey. Although this is a logical place to start, such extreme specialization is rare in nature (Williamson 1972; Schoener 1989). In reality, most predators feed on a variety of prey types. The fact that a predator may consume different prey species has important consequences, both from the perspective of the predator and from that of its prey. From the predator's point of view, there are several interesting questions. Are predators selective? Do they prefer some prey types while ignoring others? Does their preference change with prey density or prey behavior? How should we expect a predator's diet to change under different ecological conditions? From the prey's vantage point, selective predation has important consequences for mortality rates, thereby affecting prey behaviors, adaptations, and population dynamics as well as the coexistence and diversity of prey species. In this chapter, we explore some of these issues, starting from the predator's point of view and asking why predators prefer some prey types over others.

Predator Preference

A predator's preference for a given prey type can be defined as the difference between the proportion of that prey type in a predator's diet compared with the proportion of that prey type present in the environment. Thus, preference can be positive (a prey is selected for) or negative (a prey is selected against). It is useful to express predator preference in a way that allows comparisons between prey types, or between predators under different environmental conditions. Therefore, ecologists have developed a number of indices to measure preference (Manly 1985). The most widely used of these is the index first proposed by Manly (1974) and expanded on by Chesson (1978, 1983). The Manly–Chesson index calculates predator preference for prey type i as

$$\alpha_i = \frac{d_i/N_i}{\sum\limits_{j=1}^{k}(d_j/N_j)}$$

Equation 6.1

where $i = 1, 2, \ldots, k$, and where k is the number of prey categories, d_i is the number (or proportion) of prey of type i in the predator's diet, and N_i is the number (or proportion) of prey of type i in the environment. The index α_i ranges from 0 to 1. Prey types that are consumed in proportion to their abundance in the environment (i.e., no preference) have $\alpha_i = 1/k$; $\alpha_i > 1/k$ indicates positive preference for a prey type, and $\alpha_i < 1/k$ indicates negative preference for a prey type.

Why should predators prefer to eat some prey types and not others? We can approach this question by breaking down the act of predation into its component parts (**Figure 6.1**). Three main factors influence predator preference:

1. The probability that a prey item will be encountered (e.g., its "visibility" in the broadest sense)

2. The probability that an encountered prey item will be attacked (the predator's choice to go after a prey item or not)

3. The probability that an attacked prey item will be successfully captured and eaten

If a prey item is cryptic and hard to see, then we would expect it to be relatively rare in the predator's diet and to be selected against simply because it is "encountered" at a low rate. In general, factors that increase a prey's encounter rate or increase a predator's capture success should have a positive effect on predator preference.

What about active choice by the predator (see Figure 6.1)? Why do predators choose to go after some prey types and ignore others? Since the mid-1960s, ecologists have sought a general answer to this question by recognizing that natural selection should favor individuals that harvest food efficiently (Krebs 1978). Therefore, just as a functional morphologist might ask how natural selection has modified the jaw structure of an organism to maximize its crushing force, we can ask what set of foraging decisions should result in the most efficient energy capture. This approach

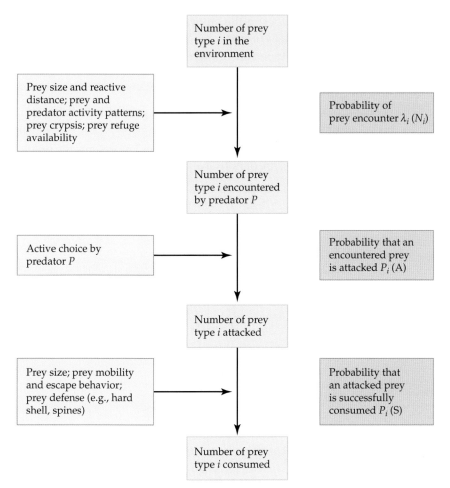

Figure 6.1 Principal elements of the predation process (center column) and factors affecting the process. The right-hand column lists the three main factors that determine the probability that a prey of type i will be consumed. Listed to the left are attributes of predators and prey that influence these probabilities. (After Mittelbach 2002.)

to understanding diet choice has become known as **optimal foraging theory (OFT)**.

Optimal foraging theory leads to a model of predator diet choice

Consider a predator actively moving through its environment and encountering potential prey items (e.g., a bird hopping from branch to branch searching for insects). If prey are randomly distributed and encountered sequentially, and if the predator's decision to pursue and eat a prey item results in time spent handling prey that is unavailable for searching, then we can model a predator's rate of energy gain as follows.

Let E be the energy gained during a feeding period of length T, which is composed of time spent searching for prey (T_s) and time spent handling prey items (T_h). If the predator expends the same amount of energy during searching and during handling, then the predator's net rate of energy gain (E_n/T) is

$$E_n/T = \frac{E}{T_h + T_s}$$

Now let there be k prey types in the environment, with each prey type i having the following characteristics:

λ_i = number of prey of type i encountered in one unit of search time

E_i = net energy gain from each prey of type i

h_i = handling time for each prey of type i

P_i = probability that the predator will pursue a prey of type i after it is encountered.

The variable P_i is under the control of the predator and represents the predator's choice to attack a prey type or not. Given these properties, it follows that

$$E = \sum_i^k \lambda_i E_i T_s P_i \text{, and}$$

$$T_h = \sum_i^k \lambda_i h_i T_s P_i \text{, then}$$

$$E_n / T = \frac{\sum_i^k \lambda_i E_i T_s P_i}{T_s + \sum_i^k \lambda_i h_i T_s P_i} \text{, or} \qquad \text{Equation 6.2}$$

$$E_n / T = \frac{\sum_i^k \lambda_i E_i P_i}{1 + \sum_i^k \lambda_i h_i P_i}$$

Equation 6.2 may look familiar—it is of the same general form as Holling's equation for a type II predator functional response (Equation 5.2), except that Equation 6.2 includes multiple prey types, the energy gained from each prey of type i (E_i), and a parameter P_i representing the probability that a predator will pursue a prey of type i.

Equation 6.2 is the standard **optimal diet model**. It was derived by several ecologists during the early 1970s (e.g., Schoener 1971; Emlen 1973; Maynard Smith 1974; Pulliam 1974; Werner and Hall 1974; Charnov 1976) as they wrestled with the question of what rules of prey choice would yield the greatest energy gain per unit time spent foraging (i.e., the optimal diet). The model makes two general predictions and two more specific predictions about optimal diet choice. Its general predictions are that (1) foragers should prefer the most profitable prey (those that yield the most energy per unit handling time, E_i/h_i) and that (2) an efficient forager should broaden its diet to include more low-value prey as the abundance of higher-value prey decreases (Stephens and Krebs 1986; Sih and Christensen 2001). There is considerable evidence that predators prefer to feed on the most profitable prey when prey are abundant (**Figure 6.2**) and that predator diets narrow

(A) Crabs feeding on mussels

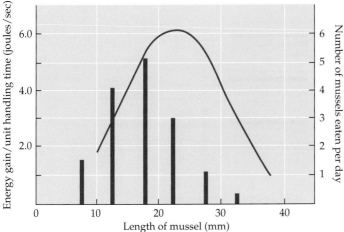

(B) 15-Spined stickleback and prey

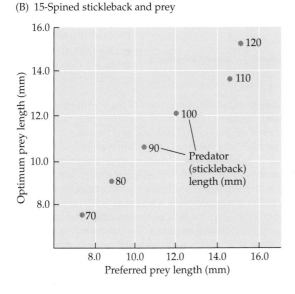

Figure 6.2 Predators prefer to eat more profitable prey. (A) In a laboratory study of crabs feeding on mussels (Elner and Hughes 1978), prey profitability (energy gain versus handling time) peaked at intermediate mussel sizes (solid curved line), and crabs preferred to eat mussels of intermediate sizes when offered a choice (histograms). (B) The preferred size of prey consumed in the wild by a small fish, the 15-spined stickleback (*Spinachia spinachia*) plotted against the stickleback's optimal prey size, based on prey mass/handling time (see Kislaliogu and Gibson 1976). The number next to each data point denotes stickleback length; both optimal and preferred prey sizes increase with the predator's size. (After Krebs 1978.)

as the abundance of more profitable prey increases (**Figure 6.3**). Thus, the general predictions of optimal foraging theory are well supported (Sih and Christensen 2001).

The standard optimal diet model also makes two quite specific predictions about the nature of prey selection and how it should change with prey abundance. The first of these predictions is that prey types are either always eaten upon encounter ($P_i = 1$) or never eaten upon encounter ($P_i = 0$), which is known as the **zero one rule** (Stephens and Krebs 1986). The second is that the inclusion of a prey type in the diet depends only on its profitability and on the characteristics of prey types of higher profitability (i.e., the inclusion of a prey type does not depend on its own encounter rate; Stephens and Krebs 1986). To understand where these predictions come from, we can use a simple algorithm to determine the predator's optimal

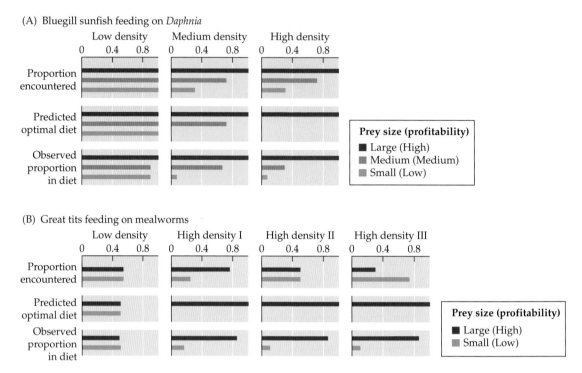

(A) Bluegill sunfish feeding on *Daphnia*

(B) Great tits feeding on mealworms

Figure 6.3 Two classic experiments tested the predictions of the standard optimal diet model. (A) Small bluegill sunfish (*Lepomis macrochirus*) in wading pools were offered a mixture of three size classes of zooplankton (*Daphnia*), and their diet choice was observed at low, medium, and high prey densities (data from Werner and Hall 1974). Ranking of prey profitability (*Daphnia* biomass/handling time) was large prey > medium prey > small prey. Histograms show the ratios of encounter rates with each *Daphnia* size class at the three prey densities, along with the predicted optimal diets and the observed bluegill diets. (B) For this experiment, Krebs et al. (1977) trained great tits (*Parus major*) to feed from a moving conveyor belt, onto which the researchers placed different-sized mealworms at different densities. The conveyor belt was then manipulated to adjust the rate at which birds encountered prey. Histograms show the proportion of large and small mealworm prey encountered; predicted to comprise the optimal diet; and observed to be eaten by the birds. (After Krebs 1978.)

diet. Following Charnov (1976), we rank prey items from most profitable (highest E_i/h_i) to least profitable (**Figure 6.4**). We then add prey types to the diet in rank order (starting with the most profitable prey) until the net rate of energy gain (E_n/T) is maximized (written as E_n^*/T^*). The optimal diet is the set of all prey types (i) whose value satisfies

$$E_i/h_i \geq E_n^*/T^*$$

Equation 6.3

In words, Equation 6.3 says that a predator should choose to pursue a prey item only if the rate of energy it gains from consuming that prey item (E_i/h_i) is greater than or equal to the rate of energy it gains by eating all

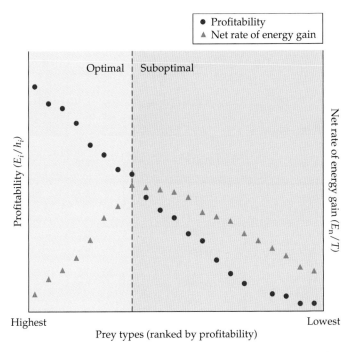

Figure 6.4   Charnov's graphic illustration of how to determine the predator's optimal diet from the standard optimal diet model (Equation 6.2). Prey types are first ranked by their profitability (E_i/h_i); then the net rate of energy gain by the predator is calculated by adding prey types to the diet in rank order. The optimal diet is that set of prey types that maximizes the predator's total net rate of energy gain (E_n/T). This optimal prey set contains all prey types (red dots) whose profitabilities rank above the prey type at which E_n/T first becomes greater than E_i/h_i, denoted by the dashed line. (After Charnov 1976.)

prey items of higher rank. Note that the decision to include a prey type in the diet depends on the prey's profitability but not on its rate of encounter, and that a prey item is either always eaten or always ignored—the two specific predictions of the optimal diet model. How well do foragers in nature match these two predictions?

Looking again at Figure 6.3, which summarizes the results of two classic tests of the optimal diet model, we see that the foragers exhibited close, but not perfect, correspondence to the optimal diet model's predictions. In both studies, less profitable prey were dropped from the diet as the abundance of more profitable prey increased. And, in the case of great tits feeding on different-sized mealworms (see Figure 6.3B), increasing a bird's encounter rate with prey that were outside the optimal diet had only a minor impact on the bird's observed diet. However, in both studies, foragers included some "suboptimal" prey in their diets.

Only a few studies have attempted to test the optimal diet model in the field (e.g., Meire and Ervynck 1986; Richardson and Verbeek 1986; Cayford and Goss-Custard 1990) because of the difficulty of estimating rates of prey

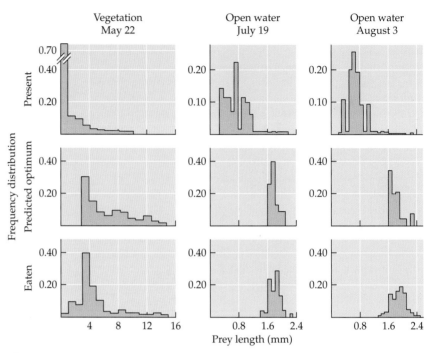

Figure 6.5 A field test of optimal diet predictions. The top row graphs the size and frequency distributions of prey found in samples from two habitats in a small Michigan lake (vegetation and open water); the predicted diets for bluegill sunfish (with an average body length of 125 mm) are in the center graphs; and the actual diets of bluegills feeding in the two habitats on the dates indicated are in the bottom row. The data support the bluegill preference for larger prey, as predicted by the optimal diet model. Note that in the vegetation habitat, the fish were able to find some larger-sized prey items than were sampled by the researcher. (After Mittelbach 1981.)

encounter. As part of my graduate work, I attempted such a test with bluegills in a small Michigan lake (Mittelbach 1981). I found that bluegills preferred to eat large prey, and that their diets were similar to those predicted by the standard optimal diet model (**Figure 6.5**). Meire and Ervynck (1986) found a similar result for oystercatchers feeding on mussels on tidal flats in the Netherlands.

The examples given above—and nearly all other tests of the optimal diet model—have shown that feeding efficiency and maximization of energy gain play important roles in determining predator diet choice. In their review of optimal foraging theory, Sih and Christensen (2001) concluded that 87% of studies (31 of 35) that included quantitative tests supported the predictions of the optimal diet model. Thus, optimal foraging theory has significantly increased our understanding of why some prey are selected and some ignored. However, in every study reviewed by Sih and Christensen, predators included some prey types in their diets that were not in the predicted optimal set (see also Pyke 1984; Stephens and Krebs 1986). Does this mean that optimal foraging theory is fundamentally flawed? Some would say yes, arguing that animals can't possibility make the "optimal" choice (Gray 1986; Pierce and Ollason 1987). However, we know that some of the assumptions of the standard optimal diet model are unlikely to be true (i.e., the assumption of perfect predator knowledge of prey quality and prey density). Therefore, we would not expect the theory to make totally accurate predictions. What we need to know is how "imperfect knowledge" affects prey selection and what more realistic models of prey selection imply for the population dynamics of predator and prey.

I share Rosenzweig's (1995: xvii) puzzlement over the relative neglect of optimal foraging studies today:

> Here is a fundamental research program (Mitchell and Valone 1990) originally suggested by no less than Charles Elton (1927). It is linking up behavior, natural selection and ecology. Yet, its critics, taking no time to understand its mathematical substance, and confusing 'optimal' with 'perfect,' consider it dead because … Well, I really don't know why they think it is dead. Students of optimality know that 'constraints' prevent perfect outcomes in the real world. But, 'constraints' are not foreign to optimality theory. They form one of its central features. Optimalists know that, work with it, and have produced some of the most exciting ecology of the 1980s by sticking to their rusty old guns.

The recent publication of a book on foraging behavior and ecology (Stephens et al. 2007) suggests to me that the study of foraging theory may have found new legs and that a revival of this important field may be in the works, although this revival will probably place more emphasis on the mechanisms of decision making and less emphasis on simple models.

The early developers of optimal foraging theory (MacArthur and Pianka 1966, along with Emlen 1966) envisioned that it could provide a tool to predict consumer diets, thereby providing a mechanism to better understand consumer–resource interactions and community dynamics. Unfortunately, this potential application of OFT to community ecology got lost in the rush to develop and test optimal foraging models, although a few studies (e.g., Gleeson and Wilson 1986; Fryxell and Lundberg 1994; Ma et al. 2003) showed how optimal foraging models could be used to explore predator and prey dynamics. A more recent and exciting development is the application of OFT to predict the potential feeding links in size-based food webs (Beckerman et al. 2006; Petchey et al. 2008; Thierry et al. 2011). By modeling prey handling time as an increasing function of the ratio of prey size to predator size, and by assuming that prey energy value is a positive function of prey size, Petchey and colleagues (2008) were able to correctly predict up to 65% of the predator–prey links in four real-world food webs, based on the criteria that predators prefer to eat the most energetically rewarding prey. This application of OFT to predict who eats whom in nature comes close to achieving the goals envisioned by MacArthur and Pianka (1966) when they first introduced OFT.

We will consider the application of optimal foraging theory to food webs in more detail in Chapter 10. In the meantime, we should keep in mind two important insights gained from OFT as we explore the interactions between predators and prey:

1. All else being equal, predators should prefer to eat the most profitable prey (those that provide the highest energy gain per handling time).

2. As the overall density of prey in the environment decreases, predator diets should broaden to include more prey types. Conversely, as the overall density of prey in the environment increases, predator diets should narrow (predators should become more specialized).

Consequences of selective predation for species coexistence

The fact that predators prefer to eat some prey and not others has important consequences for the structure of prey communities. For example, in a classic study of intertidal communities along the New England coast, Lubchenco (1978) showed that selective predation may increase prey diversity when predators feed preferentially on prey species that are competitively dominant, but that it decreases diversity when predation falls more heavily on inferior competitors (**Figure 6.6**). Likewise, Holt (1977) and Holt et al. (1994) showed that shared predation among *noncompeting* prey species may produce mutually negative interactions between those species, and that the consequences for biodiversity of this "apparent competition" depend (in part) on whether predator preferences change with changing prey densities. We will consider these and other consequences of selective predation and shared predation in more detail in Chapters 7, 10, and 14, after we have examined interspecific competition. However, before leaving our study of predator and prey, there is one more component of this relationship that we need to consider: the nonconsumptive effects of predators and

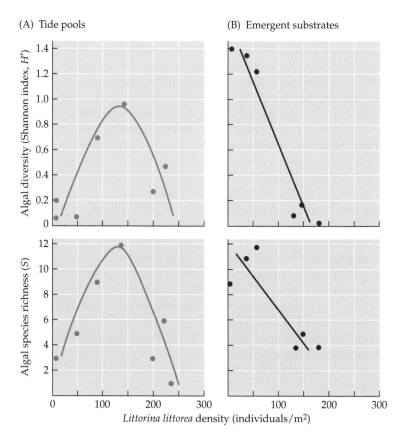

Figure 6.6 Selective predation affects the coexistence and diversity of competing prey. This classic study by Lubchenco (1978) illustrates how selective predation by the snail *Littorina littorea* may affect the diversity and species richness of algae growing along the rocky New England coast. Each point represents a single study site. (A) In tide pools, where the snails grazed preferentially on the competitively dominant algal species, both algal diversity (top) and algal species richness (bottom) peaked at intermediate snail densities. (B) On emergent substrates, where the snails grazed preferentially on the inferior algal competitors, increasing snail density had a negative effect on species richness and diversity. (After Lubchenco 1978.)

the adaptive responses of prey. Or, as some have more colorfully termed it, "the ecology of fear."

The Nonconsumptive Effects of Predators

Prior to 1980, most ecological theory viewed predator–prey interactions from the simple perspective of **consumptive effects** (i.e., predators kill and eat their prey). However, we now know that this view is incomplete. In recent years, ecologists have amassed a wealth of evidence showing that prey may respond to the threat of predation by changing their behaviors, morphologies, physiologies, and/or life histories. These nonlethal, or **nonconsumptive effects** of predators may act in concert with the direct consumption of prey to influence prey abundance and predator–prey dynamics (Werner and Peacor 2003). Further, as noted in Chapter 5, flexible antipredator traits (e.g., increased vigilance or refuge use in the presence of more predators) can make the predator's functional response predator-dependent; that is, the presence of more predators causes prey to be less "available," leading to a reduction in each predator's feeding rate. In this way, flexible antipredator traits may promote the stability of predator–prey interactions (Abrams 1982).

An excellent example of how inducible defenses may change the form of the predator's functional response is provided by Hammill et al. (2010), who studied predation on the protozoan *Paramecium aurelia* by the flatworm *Stenostomum virginianum*. In the presence of the predator, *Paramecium* reduced their average swimming speed and increased their body width (**Figure 6.7A, B**). These inducible defenses had a marked effect on the predator's attack rate and handling time, causing the predator's functional response to change from a type II (for undefended prey) to a type III (for defended prey). This change was brought about by low attack rates at low prey densities. Simulations using a Lotka–Volterra predator–prey model parameterized for the observed type II and type III functional responses showed that inducible defenses in the *Paramecium* population reduced the death rate at low prey densities and increased the stability of the predator–prey interaction (**Figure 6.7C, D**).

Figure 6.7 Induced antipredator defenses can stabilize predator–prey dynamics. Changes in the swimming speed (A) and body width (B) of *Paramecium* after exposure to chemical cues from a predator, the flatworm (*Stenostomum*). Means ±1 SE. Simulated prey and predator population densities from a Lotka–Volterra predator–prey model incorporating functional responses derived from the empirical data for undefended prey (C; type II functional response) and defended prey (D; type III functional response). Induced defenses against predators stabilized the predicted predator–prey dynamics. (After Hammill et al. 2010.)

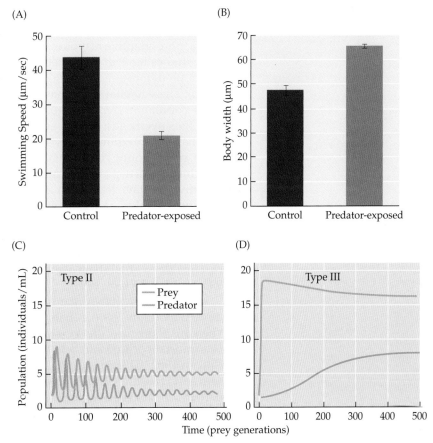

A variety of terms have been used to describe the consumptive and nonconsumptive effects of predators on predator–prey dynamics and the interactions between predators and prey at different trophic levels. These terms include "trait-mediated indirect effects" and "density-mediated indirect effects" (Abrams 1995; Abrams et al. 1996); "trait-mediated indirect interactions" and "density-mediated indirect interactions" (TMII and DMII; Peacor and Werner 1997); and "trait-mediated interactions (TMI; Bolker et al. 2003) and density-mediated interactions (DMI; Bolnick and Preisser 2005; Preisser et al. 2005). As Abrams (2007) notes, the introduction of these different terms has led to some confusion and ambiguity. Therefore, we will stick to the simpler terms "consumptive effects" and "nonconsumptive effects" to describe the different impacts that predators have on their prey. In the remainder of this chapter, we will discuss a variety of nonconsumptive effects, organizing them into four general categories: (1) habitat shifts, (2) life history evolution, (3) activity level, and (4) morphological changes. We will then ask how important these nonconsumptive effects are relative to the impact of predators consuming their prey. As we shall see, this is a critical question that is difficult to answer. In subsequent chapters on food webs and the indirect effects of interactions (see Chapters 10 and 11), we will consider the consequences of both consumptive and nonconsumptive effects for multispecies interactions.

Habitat shifts

The ostrich may (mythically) stick its head in the sand, but most prey show a more adaptive response when predators threaten: they seek refuge. Many studies have documented shifts in habitat use by prey in the presence of predators—more than 70 studies were cited in a review by Lima (1998b). These habitat shifts may be short term and have little effect on the prey's population dynamics (e.g., the mouse that hides in the grass as a hawk flies overhead). However, other types of predator-induced habitat shifts have far-reaching effects on prey dynamics and life histories. For example, small (young) individuals tend to be most vulnerable to predators. As a consequence, predators may restrict vulnerable size (age) classes to protective habitats, leading to a change in the prey's habitat use and diet (if resources differ among habitats) as it grows. These **ontogenetic niche shifts** (*sensu* Werner and Gilliam 1984) may reduce competition between size (age) classes within a species (Mittelbach and Osenberg 1993; Werner et al. 1983), but may increase interspecific competition among species and size (age) classes that share a refuge (Mittelbach and Chesson 1987; Persson 1993; Diehl and Eklöv 1995).

In choosing habitats, organisms often face a trade-off between foraging gain and mortality risk (i.e., habitats with more resources tend to be riskier). When this is the case, the "optimal" habitat choice depends on the relative costs (mortality risks) and benefits (energy gains) available in each habitat (Gilliam 1982; Gilliam and Fraser 1987; Ludwig and Rowe 1990; McNamara and Houston 1994). Moreover, if predation risk varies temporally or spatially, or changes over the life history of an organism, we would expect organisms to respond by changing their habitats. An excellent example is vertical migration by some zooplankton. In the open water of lakes and oceans, small crustaceans (zooplankton) that feed on algae (phytoplankton) show a pronounced daily migration cycle, spending the daylight hours in the dark, cold depths and migrating to warm surface waters at dusk, then returning to the depths at dawn. These diel (day–night) vertical migrations can cover tens of meters in freshwater lakes and hundreds of meters in the open ocean; such movements come at a substantial energetic cost. For years, scientists puzzled over the explanation for these migrations. We now know that, in most cases, vertical migrations are a response to temporal variation in predation risk. During the day, mortality rates from fish predation are high in the warm, well-lit surface waters, whereas mortality rates are substantially lower in the cold, dark depths. At night, predation pressure at the surface drops substantially because most fish are visual feeders. A strong vertical gradient in energy gain and developmental rate exists as well: food (phytoplankton, or algae) is often more abundant near the surface, and (more importantly) the warm surface water shortens zooplankton development times and increases birth rates, leading to higher population growth rates (Stich and Lampert 1981; Lampert 1987). Thus, by migrating vertically, zooplankton may balance the conflicting selection pressures of reducing predation risk and increasing energy gain and reproductive rate.

Pangle et al. (2007) showed that the degree of vertical migration of zooplankton in Lake Michigan and Lake Erie was correlated with the abundance of their invertebrate predator, *Bythotrephes longimanus*. If we assume that zooplankton are migrating in response to the presence of *Bythotrephes*, then we can estimate the impact of the nonconsumptive effect on population growth rate (due to a reduction in birth rate caused by migration into the cold hypolimnion) separately from the consumptive effect on population growth (due to direct mortality). In their analysis, Pangle et al. (2007) found that the nonlethal effects of the predator on zooplankton population growth were often of similar magnitude to the lethal effects.

Diel vertical migration in zooplankton is one particularly well-documented example of a very general response of prey to spatial or temporal variation in predation risk. There are literally hundreds of other examples that show how prey modify their habitat use in response to the threat of predation (reviewed by Lima 1998b). Werner (1986) has extended our thinking about the nonconsumptive effects of predators to include an even more dramatic type of habitat shift: metamorphosis (described in the following section). In the examples given above and those reviewed by Lima, the adaptive nature of habitat choice seems clear. However, it should be noted that very few studies have attempted to compare an organism's observed habitat choices with predictions based on a quantitative theory of optimal habitat selection (e.g., Gilliam and Fraser 1987).

Life history evolution

Life history theory predicts that organisms should undergo metamorphosis when the fitness that can be achieved in the larval habitat (e.g., by a tadpole in fresh water) drops below the fitness that could be achieved by shifting to the adult habitat (e.g., by an adult frog in terrestrial habitat) (Werner 1986). In the simple case of a stable population under no time constraints, Gilliam (1982) showed that the optimal size to switch from the larval to

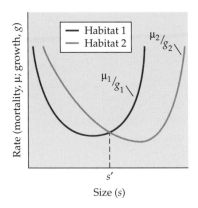

Rate (mortality, μ; growth, g)

— Habitat 1
— Habitat 2

μ_2/g_2

μ_1/g_1

s'

Size (s)

Figure 6.8   A graphic model illustrating the optimal body size at which to switch habitats as a function of individual growth rates (g) and mortality rates (μ) in each habitat. The optimal size to switch from using habitat 1 to using habitat 2 is the point at which the ratio μ/g in habitat 1 exceeds that ratio in habitat 2 (designated by s'). Predictions of this model may also be applied to the optimal size at which to metamorphose from one life stage to another (e.g., from a tadpole to a frog). (After Werner and Gilliam 1984.)

the adult habitat occurs when μ/g in the larval habitat rises above μ/g in the adult habitat, where μ is a species' size-specific mortality rate and g is its size-specific growth rate (**Figure 6.8**). Intuitively, minimizing μ/g at each size permits growth at the minimal mortality cost and maximizes the probability of reaching reproductive size.

Gilliam's rule of "minimize μ/g" strictly applies only under special conditions, yet as a heuristic tool, it has proved useful in addressing a variety of questions concerning optimal behavioral decisions for organisms under predation risk, including habitat selection, activity level, and life history evolution. Ludwig and Rowe (1990), Rowe and Ludwig (1991), and Abrams and Rowe (1996) have advanced the theory to show how the optimal size at metamorphosis (or size at maturity) is affected by mortality risk in seasonal environments. They show that increased mortality risk in the juvenile stage should favor earlier maturity, which, given a fixed growth rate, also means maturing at a smaller size (Abrams and Rowe 1996). Thus, we have the general prediction that increased predation risk in the larval or juvenile stage should lead to earlier metamorphosis and metamorphosis at a smaller body size.

Bobbi Peckarsky and her colleagues tested this prediction by studying the impacts of trout predation on mayfly (*Baetis*) populations in streams in the Rocky Mountains of the western United States (Peckarsky et al. 1993; Peckarsky et al. 2001; Peckarsky et al. 2002). Mayflies spend most of their lives as larvae, grazing algae from the stream bed and growing through a series of instars until they metamorphose into winged adults, reproduce, and die. Adult mayflies do not feed (in fact, they lack mouthparts) and live for only two days. Thus, all the energy needed to develop eggs and reproduce must be gained in the larval stage. Peckarsky and colleagues noted that mayflies from streams containing trout emerged earlier, and at much smaller adult sizes, than conspecifics found in streams without trout (Peckarsky et al. 2001). Thus, their observations fit the theoretical predictions outlined above. However, because fish prefer to eat larger prey (as we saw earlier in this chapter), an alternative hypothesis is that size-selective predation was responsible for the difference in the sizes of mayflies hatching from fish-inhabited and fishless streams. To test this hypothesis, the researchers dripped water from tanks containing live trout into fishless streams and compared the size of metamorphosing mayflies in these fish-cue treatment streams with the size of mayflies from fishless streams that received water from tanks without fish (the control). The addition of the fish-cue water (containing chemical signals released by trout) resulted in a 13%–20% reduction in adult mayfly size and an estimated 24%–35% loss of fecundity

TABLE 6.1 Effect of a chemical cue ("fish cue") emitted by predatory trout on size and fecundity in mayfly (*Baetis*) populations[a]

Baetis generation	Sex	Mean size (dry mass, mg)			Mean female fecundity		Reduction (%)
		Experimental	Control	Reduction (%)	Experimental	Control	
Summer 1999	M	0.632	0.802	21	—	—	—
	F	0.927	1.161	20	357.9	547.6	35
Winter 2000	M	0.896	1.104	14	—	—	—
	F	1.259	1.455	13	638.1	838.9	24

Source: Peckarsky et al. 2002, *Ecology* 83: 612–618.

[a]The experiment was carried out in the summer of 1999 and repeated in winter 2000. Researchers measured the size and fecundity of mayflies in two separate stream environments, neither of which contained predatory fish. In the experimental condition, "fish cue"—water from tanks containing predatory trout—was added to the stream water. The control stream received water from tanks that did not contain fish.

(**Table 6.1**; Peckarsky et al. 2002). Moreover, demographic analysis showed that the reduction in mayfly population growth caused by the nonlethal effects of trout on mayfly size at maturation exceeded the direct effects of consumption by the predator (**Table 6.2**; McPeek and Peckarsky 1998). Thus, in this remarkably complete series of studies, we see a clear example of how the lethal and sublethal effects of a predator combine to affect the population growth rate of its prey.

Activity levels

Prey also may reduce their exposure to predators by lowering their activity level (speed and extent of movement) and/or increasing the amount of time they spend in a refuge. A review of the literature shows that this response is ubiquitous across taxa: almost all species studied exhibited decreased movement, increased refuge use, or both in response to an increase in the

TABLE 6.2 Some estimated lethal and nonlethal effects of trout predation on mayfly larval and population growth

	Larval mortality rate	Larval final mass (mg)	Larval duration (days)	Population growth rate	Change in λ (%)
Natural mayfly population (no predation)	0.0378	0.2448	42	1.993	—
Growth (nonlethal) effect removed	0.0378	0.3665	42	4.264	114.0
Mortality (lethal) effect removed	0.0300	0.2448	42	2.765	38.8
Growth & mortality effects removed	0.0300	0.3665	42	5.916	197.0

Source: After McPeek and Peckarsky 1998, *Ecology* 79: 867–879.

Figure 6.9 Effects of predatory owl threat cues and a variety of other treatments on the activity levels of two gerbil species, *Gerbillus allenbyi* and *G. pyramidum*, in the field. Histograms show the ratio of gerbil activity levels in control plots relative to that in treatment plots. A ratio of 1.0 indicates no difference in gerbil activity in control and treatment plots; a ratio below 1.0 means that gerbil activity in the treatment plot was less than in the control. Note the strong inhibitory effects of owl flights and owl hunger calls on gerbil activity. *$P < 0.05$, **$P < 0.001$, ***$P < 0.0001$. (After Abramsky et al. 1996.)

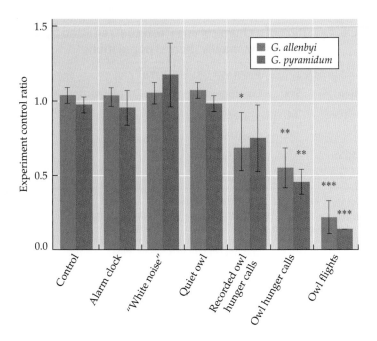

risk of predation (Lima 1998b; see **Figure 6.9** as an example). But this reduction in activity may come at the cost of lost feeding time, a lower rate of energy gain, and slowed growth. Thus, there is a growth rate–predation risk trade-off (Abrams 1990; Houston et al. 1993; Werner and Anholt 1993; McPeek 2004). Species differ in how they resolve this trade-off (e.g., Werner and McPeek 1994), and these interspecific differences are thought to be a major mechanism promoting species coexistence. We will explore the question of predator-mediated species coexistence in Chapters 8 and 11.

At the intraspecific level, it is difficult to quantify the impact of reduced prey activity levels on prey population dynamics, especially in comparison to the direct consumptive effects of predators on their prey (Preisser et al. 2005; Creel and Christianson 2008). Moreover, as McPeek (2004) has shown for damselfly larvae, reduced activity in the presence of predators may not directly affect prey feeding rates, but it may affect prey growth and/or survival through physiological processes. Finally, a predator-induced reduction in feeding rate may actually increase prey population growth if this reduction in feeding rate prevents overexploitation of the prey's resource (**Figure 6.10**; Abrams 1992; Peacor 2002). This effect would be seen as an example of "prudent predation" (see Chapter 5), although in this case, the predator is not acting consciously to increase prey resources.

Morphology

Predators have been shown to induce a variety of morphological defenses in their prey populations, as we saw earlier for *Paramecium aurelia* (Hammill et al. 2010). These inducible defenses include changes in toxicity, color, body shape, shell hardness, and the presence of spines (**Figure 6.11**), all of which may reduce a prey's chances of being eaten (see reviews in Harvell

(A)

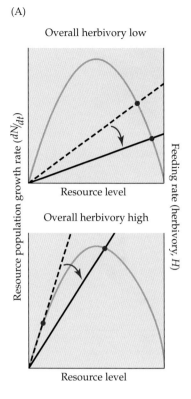

Overall herbivory low

Resource population growth rate (dN/dt)

Resource level

Feeding rate (herbivory, H)

Overall herbivory high

Resource level

(B)

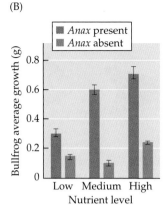

Bullfrog average growth (g)

■ *Anax* present
■ *Anax* absent

Low   Medium   High
Nutrient level

Figure 6.10 An illustration of the potential effect of a predator-induced reduction in consumer (prey) feeding rate on the equilibrium levels of the consumer's resource. (A) In this example (which assumes logistic growth for the resource), a reduction in the consumer's feeding rate (from dashed line to solid line in each panel) results in a decrease in equilibrium resource levels at a low feeding rate (top), but an increase in equilibrium resource levels at a high feeding rate (bottom). (B) Average growth rates (±SE) of consumers in an experiment in which the presence of nonlethal predators (caged dragonfly larvae, *Anax*) reduced the feeding rate of the consumers (large bullfrog tadpoles). In this experiment, the presence of *Anax* reduced tadpole feeding rates, resulting in higher standing stocks of algae (the tadpole's resource) and leading to higher growth rates of the tadpoles at medium and high nutrient levels. (After Peacor 2002.)

1990; Tollrian and Harvell 1999; Benard 2004). These phenotypically plastic responses can be so dramatic as to cause biologists to mistake the different body forms for different species. A case in point is the crucian carp (*Carassius carassius*). In lakes without piscivorous predators, crucian carp are found in dense populations of narrow-bodied individuals, but in lakes with piscivores (especially pike, *Esox lucius*), carp are few in number, and

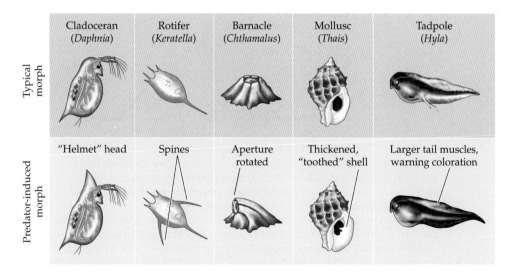

Figure 6.11 Examples of predator-induced defenses in some aquatic organisms.

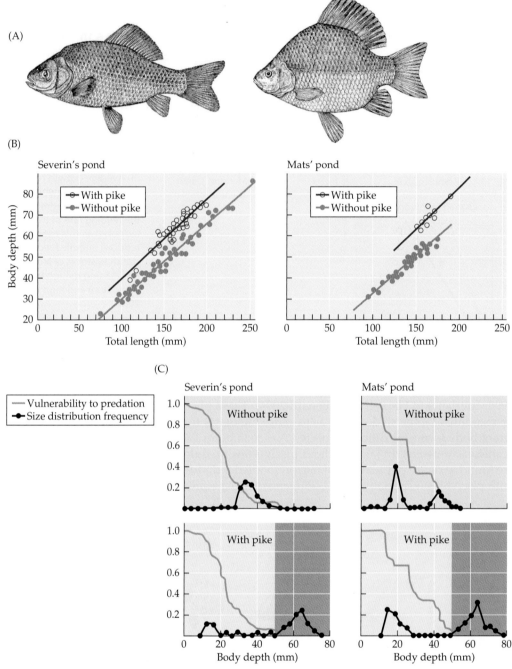

Figure 6.12 Differences in body form in the crucian carp (*Carassius carassius*) are an induced adaptation to reduce mortality by predation. (A) In the presence of predatory fish, crucian carp are very deep-bodied, whereas in the absence of predators, they are narrow-bodied. (B, C) In southern Sweden, Brönmark and Miner divided two small ponds in half with a fish-proof curtain. They introduced juvenile crucian carp and predatory northern pike (*Esox lucius*) into one half of each pond and juvenile carp only into the other half. (B) After 12 weeks, carp in the presence of pike were much deeper-bodied than carp in the absence of pike. (C) This increase in body depth reduced the carp's vulnerability to predation. Darker shaded areas denote carp that are too large to be vulnerable to predation. (After Brönmark and Miner 1992.)

individuals are deep-bodied (**Figure 6.12A**; Brönmark and Miner 1992). The two carp morphs were originally considered separate species until transplant experiments showed them to be the same (Ekström 1838; Brönmark and Miner 1992). Ecologists then hypothesized that it was variation in resource levels and growth rates that led to the development of the two morphs. However, a series of field and laboratory studies by Brönmark and colleagues showed that the differences in body form were in fact induced by the presence of predators.

In a field experiment, Brönmark and Miner (1992) introduced young crucian carp from the same population into two ponds. Each pond was divided in half by a plastic curtain, and one half of each pond also received 15–20 pike. After 12 weeks, the carp population had diverged in body shape between the predator and no-predator treatments: in the presence of the pike, carp had significantly deeper bodies (**Figure 6.12B**). This shift in body form allowed most of the carp to escape being eaten by their pike predators (**Figure 6.12C**). Importantly, laboratory experiments showed that the observed divergence in body shape could not be accounted for by differences in resource availability and growth rates. Experiments also showed that expression of the deep-bodied form resulted in about a 30% increase in energy expended when swimming compared with the shallow-bodied, more fusiform shape (Brönmark and Miner 1992; Pettersson and Brönmark 1997). Thus, by becoming deeper-bodied, the carp reduced their vulnerability to predation, but paid a significant energetic price. In general, we expect inducible defenses to evolve when predation risk varies temporally or spatially; when prey have the ability to detect predators via reliable cues; and when the defense against predators carries a fitness cost (Brönmark and Hansson 2005).

The Relative Importance of Consumptive and Nonconsumptive Effects

The above discussion illustrates the wealth of nonconsumptive effects that predators may have on their prey but, truth be told, we have only scratched the surface of this active and fascinating field. Despite all the excellent research in this area, however, perhaps the most significant questions remain unanswered: How important are nonconsumptive effects relative to consumptive effects in predator–prey interactions? and, How do we incorporate nonlethal effects into our existing theory of predator–prey interactions (Abrams 2010)? The challenge in addressing both of these questions comes from the fact that nonconsumptive and consumptive effects are often measured in very different units (e.g., growth versus mortality), as well as the fact that these effects may operate on different time scales (e.g., changes in prey behaviors occur much more rapidly than changes in prey densities).

One common experimental approach to assessing the importance of nonconsumptive effects relative to consumptive effects is to compare the impact of a functional predator with that of a nonlethal predator on the density or fitness (measured as fecundity or growth) of the prey population. Nonlethal predator treatments can be created using caged or disabled predators ("risk predators," such as a spider with nonfunctional mouthparts; Schmitz 1998) or by introducing a predator cue (e.g., water containing the chemical signal, or kairomone, of a predator). Many such experiments have been performed, and meta-analyses of the results show that the nonlethal effects of predators are often as great as or greater than the lethal effects (Bolnick and Preisser 2005; Preisser et al. 2005). These analyses give us an appreciation of the potential importance of nonconsumptive effects in predator–prey interactions. However, such simple comparisons sweep a number of complications under the rug. For example, the presence of a nonlethal predator in an experiment may result in prey behaviors that reduce feeding rates and growth, resulting in weakened and even starved prey. In nature, however, such weakened prey often fall victim to functional predators. Thus, consumptive and nonconsumptive effects of predators interact in nature, and sophisticated experimental designs are needed to accurately estimate their relative importance in predator–prey interactions (e.g., Werner and Peacor 2003). Okuyama and Bolker (2007), Abrams (2008), and Schmitz (2010) provide excellent discussions of these issues as well as guides to experimental design.

Looking Ahead

We have seen in this chapter how the dynamic responses of prey to their predators often lead to indirect effects. In a three-link food chain, for example, the presence of the predator may result in increased vigilance and a reduction in feeding rate in the prey, thus increasing the abundance of the prey's resource. We will discuss this type of indirect interaction (a cascading effect from the predator to the prey's resource) in much more detail in Chapter 11, where we will examine top-down and bottom-up control in food chains and food webs. Before we get there, however, we need to consider consumer–resource interactions from the point of view of competition (Chapters 7 and 8), as well as the beneficial interactions of mutualism and facilitation (Chapter 9).

Summary

1. Predator preference for a prey type can be defined as a difference between the proportion of that prey type in the predator's diet and the proportion of that prey type in the environment. Predator preference is influenced by (1) the probability that a prey item will be encountered, (2) the probability that an encountered prey item will be attacked, and (3) the probability that an attacked prey item will be captured and eaten.

2. Optimal foraging theory (OFT) is an approach to understanding predator diet choice by asking what set of foraging decisions will maximize a predator's energy gain. Optimal diet models predict a predator's diet choices.

3. The standard optimal diet model makes two general predictions and two more specific predictions about predator choice:
   - The two general predictions are (1) foragers should prefer the most profitable prey (i.e., prey that yield the most energy per unit handling time), and (2) as the overall density of profitable prey decreases, an efficient forager should broaden its diet to include more of the less profitable prey.
   - The two specific predictions of the standard optimal diet model are (1) prey types are either always eaten upon encounter or never eaten upon encounter (the zero-one rule), and (2) the inclusion of a prey type in the diet depends on its profitability and on the characteristics of prey types of higher profitability.

4. Tests of the optimal diet model support its predictions, but in most cases predators include some prey types in their diets that are not in the predicted optimal set.

5. Prey may respond to the threat of predation by changing their behaviors, morphologies, physiologies, or life histories. These nonconsumptive (or nonlethal) effects of predators act in concert with the direct consumption of prey to influence prey abundance and predator–prey dynamics.

6. In choosing habitats, organisms may face a trade-off between foraging gain and increased risk of predation. When this is the case, the "optimal" habitat choice depends on the relative costs (mortality risks) and benefits (energy gains) available in each habitat. Thus, for example, increased predation risk in the larval or juvenile stage is predicted to lead to the evolution of earlier metamorphosis and/or a smaller adult body size.

7. Many prey species reduce their activity level in the presence of predators, but these less-active prey may incur a reduction in growth (i.e., a growth rate–predation risk trade-off). Inducible morphological defenses may evolve when predation risk varies temporally or spatially; when prey have the ability to detect predators via reliable cues; and/or when the defense carries a fitness cost.

8. The relative importance of the nonconsumptive and consumptive effects can be addressed by performing experiments in which the effect on prey density or fitness of a nonlethal (e.g., caged or disabled) predator or a predator cue is compared with that of a functional predator. However, such experiments do not address the complicated interactions of these effects in nature.

7 Interspecific Competition

Simple Theory

[M]ultispecies stable equilibrial coexistence requires interspecific tradeoffs. If a species arose that was able to avoid such tradeoffs and be a superior competitor relative to all other species, it would eliminate its competitors.

David Tilman, 2007: 93

One hill cannot shelter two tigers.

Chinese proverb

Ecologists have long viewed interspecific competition as a major (perhaps *the* major) factor influencing community structure. In Chapter 1, for example, we saw how Gause's competitive exclusion principle provided a foundation for the development of later ideas about the role of interspecific competition and limiting similarity in determining the number of coexisting species. At about the same time that Gause (1934) was conducting his experiments on competitive exclusion, Alfred Lotka (1925) and Vito Volterra (1926) independently developed the first mathematical theory of interspecific competition, based on a simple extension of the logistic model of population growth. In this chapter, we will examine the working principles of interspecific competition. Much as we did in our discussion of predator–prey interactions, we will use simple mathematical theory and isocline analysis to develop this understanding. Although we will begin with the classic Lotka–Volterra model, we will devote most of our time to studying a more recent modeling approach that explicitly considers the interactions between species and their resources. In Chapter 8, we will examine the empirical evidence for interspecific competition and relate those findings to theoretical predictions from simple models. We will explore the effects of shared predators on interspecific competition in Chapters 10 and 11. Environmental variation in space and time can have important consequences for the outcome of interspecific competition, and we will explore some of those consequences in Chapters 12, 13, and 14.

Defining Interspecific Competition

It is difficult to come up with a rigorous and succinct definition of interspecific competition. Two definitions highlight the essentials:

> [Interspecific competition] is the interaction occurring between species when increased abundance of a first species causes the population growth of a second to decrease, and there is a reciprocal effect of the second on the first. (Grover 1997: 11)

> Interspecific competition between two species occurs when individuals of one species suffer a reduction in growth rate from a second species due to their shared use of limiting resources (exploitative competition) or active interference (interference competition). (Case 2000: 311)

Both definitions note that interspecific competition involves a reduction in the population growth rate of one species due to the presence of one (or more) other species. The definition by Case outlines the two general classes of mechanisms that may cause a competitive reduction in growth rate. **Exploitative competition** (also known as **resource competition**) occurs when a species consumes a shared resource that limits its and other species' population growth—thus making that resource less available to the other species. **Interference competition** (or **contest competition**) occurs when one species restricts another species' access to a limiting resource. Interference competition may involve overt aggression between individuals (e.g., territorial defense), or it may simply involve occupying a space to the exclusion of another individual. We will examine the definition of a limiting resource in more detail later in this chapter. First, however, we will consider the classic Lotka–Volterra model of interspecific competition.

The Lotka–Volterra Competition Model

Recall from our discussion of the logistic equation in Chapter 4 that we can write equations for the change in abundance of two species whose populations are growing according to a logistic model as

$$dN_1/dt = r_1 N_1 \left(\frac{K_1 - N_1}{K_1} \right) \qquad \text{Equation 7.1}$$

$$dN_2/dt = r_2 N_2 \left(\frac{K_2 - N_2}{K_2} \right) \qquad \text{Equation 7.2}$$

where (N_1, N_2) are the densities of species 1 and 2, respectively, (r_1, r_2) are their intrinsic growth rates, and (K_1, K_2) are their carrying capacities. Intraspecific competition is implicit in this formulation of population growth because the density of each species cannot exceed its carrying capacity, and the per capita rate of population growth (dN/Ndt) declines linearly with population density (see Figure 4.4C). To examine the competitive effect of species 2 on species 1 (and vice versa), Lotka (1925) and Volterra (1926) incorporated into the logistic equation a term for the density of the competing species and multiplied this density term by a constant (α_{ij}), the competition coefficient. Equations 7.3 and 7.4 represent the Lotka–Volterra competition model for two species:

$$dN_1/dt = r_1 N_1 \left(\frac{K_1 - N_1 - \alpha_{12} N_2}{K_1} \right) \qquad \text{Equation 7.3}$$

$$dN_2/dt = r_2 N_2 \left(\frac{K_2 - N_2 - \alpha_{21} N_1}{K_2} \right) \qquad \text{Equation 7.4}$$

The **competition coefficient** α_{ij} is the effect of an individual of species j on the per capita growth rate of species i relative to the effect of an individual of i on its own per capita growth rate (and vice versa for α_{ji}). Note that there is an implicit constant of 1 in front of the density term for the species whose growth rate is being modeled. The model can be extended to include as many species as desired by adding more terms of the form $\alpha_{ij}N_j$ in the numerator of species i's growth equation.

Note that the Lotka–Volterra model does not specify the mechanism of competition between the species; it simply states that the density of one species has a negative effect on the population growth rate of the second species. Negative effects in the L–V model could be the result of direct aggression between competitors, or they could be the result of indirect interactions through consumption of a shared resource (MacArthur 1970). The L–V model is a phenomenological description of competition. Later in this chapter, we will consider a more mechanistic model of exploitative competition that explicitly incorporates resource consumption. For now, however, we can use the L–V model and simple isocline analysis to develop an understanding of the potential outcomes of two-species competition. This approach is very similar to the one we used for predator–prey interactions in Chapter 5. In that chapter, we plotted the zero growth isoclines for predator and prey on a single graph and then examined the regions of population densities that yielded positive and negative growth rates to predict the outcome of the predator–prey interaction under different conditions (model assumptions). We can do much the same thing for interspecific competition by solving Equations 7.3 and 7.4 for points where the per capita growth rates of species 1 and species 2 are zero.

The combination of densities of species 1 and species 2 that results in a zero per capita growth rate for species 1 can be found by setting $dN_1/dt = 0$ in Equation 7.3; likewise for species 2 by setting $dN_2/dt = 0$ in Equation 7.4. Graphing this combination of densities on a plot whose axes are N_1 and N_2 yields a zero growth isocline for each species (**Figure 7.1**). At each point along the species' zero growth isocline, its population is at equilibrium and its density is unchanging. Taking species 1 as an example (Figure 7.1A), we know that its population growth rate will be zero when its density is at its carrying capacity (K_1). Species 1's growth rate also will be zero when the density of species 2 is equivalent to the carrying capacity of species 1 (i.e., when $N_2 = K_1/\alpha_{12}$). The zero growth isocline between these endpoints is linear because per capita density dependence is linear in the model. Densities below

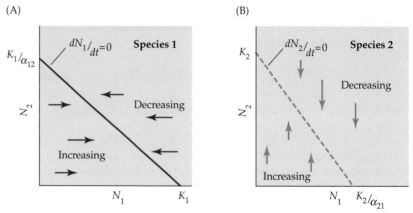

Figure 7.1   Zero growth isoclines for two-species Lotka–Volterra competition. (A) The combination of densities of species 1 (N_1; x axis) and species 2 (N_2; y axis) for which the population of species 1 shows zero growth ($dN_1/dt = 0$). For points to the left of the isocline, the population of species 1 is increasing (right-pointing arrows). Right of the isocline, the population of species 1 is decreasing (left-pointing arrows). (B) The zero growth isocline for species 2. For points below the isocline, species 2 is increasing, and for points above the isocline, species 2 is decreasing (upward and downward pointing arrows, respectively).

the isocline in Figure 7.1A result in positive growth of species 1, whereas densities above the isocline result in its negative growth. You can follow the same logic to construct the zero growth isocline for species 2 (Figure 7.1B).

Plotting the zero growth isoclines for species 1 and species 2 on the same graph yields four possible outcomes (**Figure 7.2**). When the isocline for species 1 is above the isocline for species 2, species 1 wins in competition and excludes species 2 (Figure 7.2A). We can understand this outcome by considering the population growth trajectories of the species in each of the three regions of the graph defined by the isoclines. In the lower region, each species exhibits positive growth; in the upper region, each species exhibits negative growth. In the middle region, between the two isoclines, the population growth of species 2 is negative and the population growth of species 1 is positive—the result being that the density of species 1 reaches a stable equilibrium at K_1 and the density of species 2 goes to zero. By the same logic, when the isocline for species 2 is above the isocline for species 1, species 1 is excluded and the density of species 2 equilibrates at K_2. In order for species to coexist, their isoclines must cross.

When the species' isoclines cross, their intersection point represents an equilibrium at which the densities of both species are positive and unchanging (Figure 7.2C, D). This equilibrium may be stable or unstable, depending on the orientation of the isoclines. When the isoclines are oriented as in Figure 7.2C, the two-species equilibrium is stable because the trajectories for both populations always point toward the equilibrium; if densities are perturbed from the equilibrium point, they will return to it. However, if the isoclines are oriented as in Figure 7.2D, the population trajectories point away from

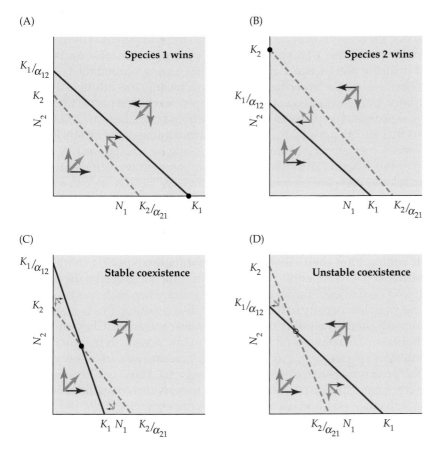

Figure 7.2 The four potential outcomes of two-species competition. Horizontal and vertical arrows show the trajectories of each species; diagonal arrows show the joint vectors of movement in the regions of state space defined by the two zero growth isoclines. (A) Species 1 outcompetes species 2. The equilibrium outcome when the species 1 isocline lies above the species 2 isocline is that the density of species 1 goes to its carrying capacity K_1 and the density of species 2 goes to zero. (B) Species 2 outcompetes species 1. When the isocline for species 2 lies above the isocline for species 1, the density of species 2 goes to its carrying capacity K_2 and that of species 1 goes to zero. (C) Stable coexistence. The isoclines cross and the population trajectories for each species point toward the equilibrium point. If densities are perturbed from the equilibrium point, they will return to it. (D) Unstable coexistence. The isoclines cross, but the population trajectories point away from the equilibrium point in two of the four quadrants. If densities are perturbed from the equilibrium point, one species or the other will win, depending on starting abundances. (After Gotelli 2008.)

the equilibrium in two of the four quadrants. Thus, population densities in these quadrants will go to fixation at either K_1 or K_2; that is, either species 1 or species 2 will win. The outcome of competition in this case depends on the starting densities of species 1 and species 2—a result that is sometimes called **founder control** because the initially more abundant species is likely to predominate.

Another way to look at the L–V competition model

The isocline analysis in Figure 7.2 illustrates the four potential outcomes of two-species competition: species 1 wins; species 2 wins; stable coexistence; and unstable coexistence. But honestly, I have always found the L–V competition isoclines to be confusing. They are certainly less intuitive then the predator–prey isoclines we studied in Chapter 5, which is part of the reason why we considered predator–prey interactions first. Chesson (2000b) suggests that we can get a more intuitive understanding of the factors leading to species coexistence or competitive exclusion by writing the L–V model (Equations 7.3 and 7.4) in terms of absolute competition coefficients rather than relative competition coefficients. Following Chesson, we can write the two-species L–V competition equations as

$$dN_i/dt = r_1 N_i (1 - \alpha_{ii} N_i - \alpha_{ij} N_j) \text{ when } i = 1,2, \, j \neq i \qquad \text{Equation 7.5}$$

where the quantities α_{ii} and α_{ij} are now, respectively, the **absolute intraspecific competition coefficient** and the **absolute interspecific competition coefficient** (i.e., competition coefficients that are not scaled by the species' carrying capacities). Species coexistence occurs when each species can increase from low density in the presence of its competitor; this is basis of the commonly used **invasibility criteria** for species coexistence (Chesson 2000b). In Equation 7.5, species i can increase from low density in the presence of competitor species j if $\alpha_{jj} > \alpha_{ij}$, and species j can increase from low density in the presence of competitor species i if $\alpha_{ii} > \alpha_{ji}$. Thus, as Chesson notes, the criteria for species coexistence in this two-species competitive interaction is that $\alpha_{11} > \alpha_{21}$ and $\alpha_{22} > \alpha_{12}$ or, in words, that *intraspecific competition is greater than interspecific competition*. That is, coexistence requires that each species have a greater negative effect on its own per capita growth rate than it does on the per capita growth rate of its competitor.

Modifications to the L–V competition model

The Lotka–Volterra model assumes that competitive effects are linear—i.e., that the per capita growth rate of each species declines linearly with increases in its own density or the densities of its competitors. However, it is unlikely that competitive effects in nature are truly linear. Curved competition isoclines may occur for a variety of reasons (Abrams 2008). Gilpin and Ayala (1973) first proposed a simple modification of the L–V model to accommodate nonlinear competitive interactions, where

$$dN_i/dt = r_i N_i \left[1 - \left(\frac{N_i}{K_i} \right)^{\theta_i} - \alpha_{ij} \frac{N_j}{K_i} \right] \qquad \text{Equation 7.6}$$

Although Equation 7.6 (often referred to as the "θ-logistic model") produces nonlinear competition isoclines when $\theta \neq 1$, the nonlinear competitive effects occur only in the intraspecific portion of the interaction. A more general approach to examining the causes and consequences of nonlinear competitive effects has been advanced (in a consumer–resource competition model) by Abrams et al. (2008), who showed that nonlinear competitive interactions have important consequences for predicting evolutionary responses to competition and the limiting similarities of coexisting species.

The Lotka–Volterra competition model may also be extended to include many competing species. The conditions for equilibrium in multispecies Lotka–Volterra type models are commonly written in matrix format (Levins 1968):

$$\mathbf{K} = \mathbf{AN^*}$$

where $\mathbf{K}$ is a column vector of the species' carrying capacities, $\mathbf{N^*}$ is a column vector of the species' equilibrium densities, and $\mathbf{A}$ is a square matrix of the interaction coefficients. For the case of three competing species, the matrices would look like this:

$$\mathbf{K} = \begin{pmatrix} K_1 \\ K_2 \\ K_3 \end{pmatrix} \qquad \mathbf{N^*} = \begin{pmatrix} N_1^* \\ N_2^* \\ N_3^* \end{pmatrix} \qquad \mathbf{A} = \begin{pmatrix} 1 & \alpha_{12} & \alpha_{13} \\ \alpha_{21} & 1 & \alpha_{23} \\ \alpha_{31} & \alpha_{32} & 1 \end{pmatrix}$$

The $\mathbf{A}$ matrix is commonly referred to as the **community matrix** because it specifies the intra- and interspecific per capita interaction strengths; intraspecific effects are on the diagonal and interspecific effects are on the off-diagonal (Levins 1968; Vandermeer 1970; MacArthur 1972). This matrix approach can be broadly applied to multispecies communities in which the interactions between species are governed by generalized Lotka–Volterra models that may include competition, predation, or positive effects (see Case 2000 for a clear presentation of this approach).

The community matrix provides an elegant way to analyze the equilibrium outcome of competition among many species (Levins 1968; Yodzis 1989) and it has been used extensively by theorists to explore the effects of interspecific competition on questions of community stability (e.g., May 1973a), community invasibility (e.g., Case 1990; Law and Morton 1996), and the equilibrium number of coexisting species (e.g., Vandermeer 1972). However, applications of the community matrix to natural systems have been rare (e.g., Seifert and Seifert 1976) for two important reasons.

First, this approach assumes that interactions between pairs of competing species are unaffected by the other species in the community (i.e., that the competition coefficients in the $\mathbf{A}$ matrix are independent of each other and thus species effects are additive). This is unlikely to be the case in nature (Neill 1974; Abrams 1983a). Nonadditive (nonlinear) effects are often referred to as **higher-order interactions** or **interaction modifications** (Neill 1974; Wootton 1994). The trait-mediated interactions that we examined in Chapter 6 are examples of higher-order interactions in predator–prey systems; we will discuss some of the factors causing higher-order interactions in competitive systems in Chapters 12 and 13 on ecological networks. Second, in any reasonably diverse community, it is a challenge to estimate all the competition coefficients (the α's in the $\mathbf{A}$ matrix). In a community of 10 species, for example, there are 45 unique pairwise species interactions. Thus, although generalized Lotka–Volterra models and the community matrix continue to be useful theoretical tools, they have severe limitations when applied to natural communities. More recent models that explicitly incorporate resources into the mechanics of competition have come to the forefront, in part because they provide a stronger connection to empirical work. These **consumer–resource models** are the subject of the next section.

Consumer–Resource Models of Competition

The Lotka–Volterra model shows that species coexistence and competitive effects depend critically on the relative strengths of intraspecific and interspecific competition. The L–V model does not explicitly include resources, however, and thus is of limited value in understanding exploitative competition, or in providing predictions that can be directly tested by manipulating resources. To better study exploitative competition, we need a model that incorporates how competing consumers affect their resources and how consumer population growth rates are in turn affected by resource densities.

MacArthur (1968, 1970, 1972) was the first to include resource dynamics in a model of competition for multiple resources (although Volterra had examined the case of competition for a single resource much earlier). Subsequent models were developed by May (1971), Phillips (1973), Schoener (1974, 1976), León and Tumpson (1975), and others. These pioneering efforts helped shift the focus away from the phenomenological Lotka–Volterra model and toward more mechanistic models of exploitative competition.

It was the graphical portrayal of consumer–resource competition models, however, that made the results of this theory accessible to most ecologists. León and Tumpson (1975) introduced a graphical analysis of resource competition models and applied it to different types of resources, yielding criteria for species coexistence. Tilman (1980, 1982) refined and greatly expanded on this approach and provided the first empirical tests of the model (Tilman 1976, 1977). The presentation that follows is based on a graphical analysis by Tilman (1980, 1982), which has been widely used and extended to other contexts (e.g., Grover 1997; Chase and Leibold 2003b). This simple graphical approach is less robust than the mathematical theory behind it. However, it is a useful heuristic tool, and students of community ecology should be familiar with it. For an alternative verbal explanation of the essence of exploitative competition, as well as a detailed mathematical analysis, readers should consult Abrams (1988).

What are resources?

Species compete when they share a resource whose availability in the environment limits a species' population growth. But what is a resource? Here are two definitions:

> A resource [is] any substance or factor which can lead to increased [population] growth rates as its availability in the environment is increased, and which is consumed by an organism. (Tilman 1982: 11)

> Resources are entities which contribute positively to population growth, and are consumed in the process. (Grover 1997: 1)

Note the two properties that define a resource: (1) a resource contributes positively to the growth rate of the consumer population, and (2) a resource is consumed (and is thus made unavailable to other individuals or species).

Resources may be further classified as biotic (living and self-reproducing) or abiotic (nonliving), a distinction that is important when thinking about how resources are supplied. Examples of abiotic resources include mineral nutrients (e.g., nitrogen, phosphorus), light, and space. Biotic resources include prey (for predators), plants (for herbivores), and seeds (for granivores). Although space is often considered a resource for sessile species such as barnacles and plants, procuring space may simply be a prerequisite for obtaining other resources (e.g., light, nutrients, planktonic prey); thus, thinking about space as a resource can be complicated. Finally, some things that are *not* resources include temperature (temperature may affect population growth rates, but it is not generally altered by the organism whose growth is affected) and oxygen (oxygen is consumed by heterotrophs, but it is generally not limiting to population growth, at least not in terrestrial ecosystems).

One consumer and one resource: The concept of R*

For most organisms, the per capita rate of population growth has an upper bound, which is approached in an asymptotic fashion as the abundances of all potentially limiting resources become high enough (**Figure 7.3**). This pattern of population growth can be described by a variety of mathematical functions [e.g., the Monod (1950) equation first developed for bacteria growing on organic substrates]. However, for heuristic purposes, we can skip the formal mathematics and simply graph the per capita birth rate of a consumer population as a positive, decelerating function of the availability of a resource *R* (**Figure 7.4A**). If we assume that the consumer population experiences some per capita density-independent

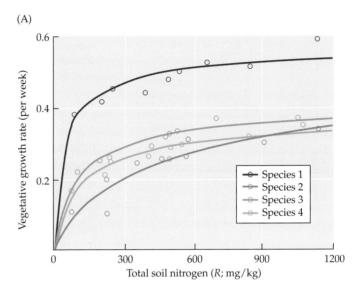

(A)

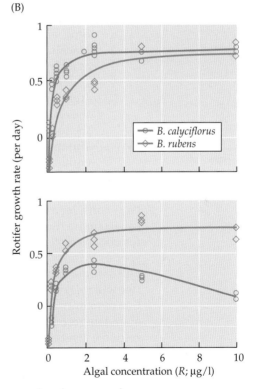

(B)

Figure 7.3 Examples of resource-dependent consumer population growth rates. Note that as resource availability rises in each case, the increase in growth rate reaches an asymptote. (A) Vegetative growth rates (measured as aboveground biomass) for four grass species as functions of soil nitrogen availability (*R*). (B) Population growth rates of two rotifer species (*Brachionus*) as functions of food (algal) concentration (*R*). *B. rubens* and *B. calyciflorus* were fed on the the algae *Chlamydomonas sphaeroides* (above) and *Monoraphidium minutum* (below). (A after Tilman and Wedin 1991a; B after Rothhaupt 1988.)

(A)

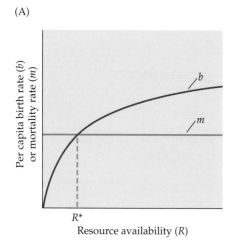

(B)

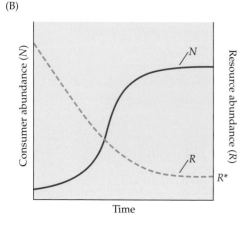

Figure 7.4 Graphic representation of a one consumer–one resource system. (A) The per capita birth rate (b) of the consumer is a positive, decelerating function of the availability of the resource (R), and the per capita mortality rate (m) is constant. (B) Change in consumer (N) and resource (R) abundances over time. The consumer population will increase and the abundance of the resource will decline, eventually reaching R^*.

mortality rate m (represented by the horizontal line in Figure 7.3A), then the consumer population will be at equilibrium ($dN/dt = 0$) where the lines in Figure 7.4A intersect. This intersection point specifies the resource availability (R^*) at which the consumer's per capita birth rate (b) equals its per capita mortality rate (m). Thus, R^* is the minimum equilibrium resource requirement needed to maintain a consumer population experiencing a given mortality rate (m). The R^* result is not dependent on the assumption of density-independent mortality, but we make that assumption here for simplicity in graphic presentation.

If a consumer is introduced to an environment with mortality rate m and $R > R^*$, the consumer population will increase over time, and the abundance of the resource will decline, eventually reaching R^*, the point at which both consumer and resource are at their equilibrium values (**Figure 7.4B**). For abiotic resources, R^* generally defines a stable equilibrium point (Grover 1997; but see Abrams 1989 for exceptions); for biotic resources, the stability of the equilibrium depends on the specific functions used to describe the consumer and resource growth responses (Hsu et al. 1977; Armstrong and McGehee 1980). For now, we will assume that a stable equilibrium between the consumer and its resource is possible.

Two consumers competing for one resource

Now consider a second consumer species that also requires resource R. We can examine the outcome of this interaction (i.e., two species competing for one resource) by graphing the per capita birth rates of both consumer species as functions of resource availability (**Figure 7.5A**). For simplicity, we will assume that both species have the same per capita mortality rate (m) (we will relax this assumption in a minute). **Figure 7.5B** shows the population dynamics of the two species, starting with low consumer densities and abundant resources. Initially, N_1's population grows faster, because at high resource levels its net rate of population growth (birth rate minus mortality rate) is higher. However, as resources are consumed and R declines, the growth rate of N_1's population slows, reaching zero when resource levels reach R_1^*. If N_1 was the only consumer species present, the system would equilibrate when $R = R_1^*$. The situation changes, however, when N_2 is present. As we can see in Figure 7.5A, N_2's birth rate is greater than its mortality rate at $R = R_1^*$. Thus, its population will grow and continue to deplete R until $R = R_2^*$ (Figure 7.5B). When $R = R_2^*$, N_2's birth rate matches its mortalilty rate (Figure 7.4A), and N_2's population growth ceases. Note, however, that $R_2^* < R_1^*$. Therefore, N_1 cannot maintain its population at a

Figure 7.5 Graphic representation of two consumers competing for one limiting resource. (A) The per capita birth rates (b_1, b_2) and per capita mortality rate ($m_{1,2}$) for consumer species N_1 and N_2 as a function of resource availability (R). Note that consumer species N_2 can maintain its population on a lower resource level than can consumer species N_1 ($R_2^* < R_1^*$). (B) Change in consumer and resource abundances over time. Note that consumer species N_1 is driven extinct by competition with consumer species N_2.

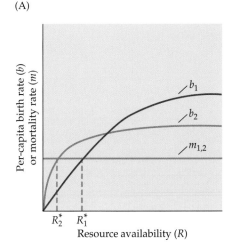

(A)

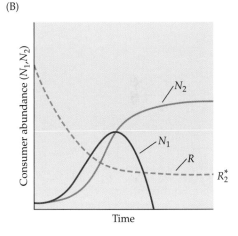

(B)

resource level equal to R_2^* and its population declines over time. Eventually, N_1 is eliminated by competition with N_2 (Figure 7.5B).

The analysis in Figure 7.5 suggests the following results: (1) two species cannot coexist on one limiting resource, and (2) the species that can maintain positive population growth at the lowest resource level (lowest R^*) will win in competition. Hsu et al. (1977, 1978) and Armstrong and McGehee (1980) showed that we need to modify these results slightly to the following: (1) two species cannot coexist on one limiting resource *at constant population densities*, and (2) the species with the lowest R^* will win. This **R^* rule** provides a powerful and intuitively satisfying way to think about the outcome of competition for a single limiting resource.

In the example above, we assumed that each species had the same fixed mortality rate m. However, we can easily incorporate different morality rates into our graphic analysis. We can also consider how differences between consumer species' reproductive rates may affect the outcome of competition. In Figure 7.5, the species' birth rate curves cross, such that one species has a relatively greater population growth rate at low resource levels and the other species has a relatively greater population growth rate at high resource levels. These two consumer types have been termed **gleaners** and **opportunists**, respectively (Frederickson and Stephanopoulos 1981; Grover 1990). Species 2 is a gleaner because it grows relatively well when the resource is scarce. Species 1 is an opportunist—it grows relatively poorly at low resource levels but grows well when resources are abundant. Thus, the reproduction curves in Figure 7.5A illustrate a trade-off that has important consequences for species coexistence.

Coexistence on a single, fluctuating resource

In a constant environment, we would expect species 2 in Figure 7.5 to win in competition because it has the lower R^*. However, Armstrong and McGehee (1976, 1980; see also Koch 1974; Hsu et al. 1977, 1978) showed that we need to consider another possibility: If resource levels fluctuate between high and low abundance, then the two species may coexist because species 1 has a growth advantage at high resource abundance and species 2 has a growth advantage at low resource abundance. If R is an abiotic resource, then exogenous temporal variation in the physical environment (i.e., seasons) may cause fluctuations in resource abundance and allow the species to coexist (Grover 1997). If R is a biotic resource, then the interaction between consumers and resources (think predators and their prey) may generate endogenous temporal variation in resource abundance that may also allow species coexistence (Armstrong and McGehee 1980). Recall from Chapter 5 that predator and prey populations have a natural tendency to cycle, especially when predators are efficient at exploiting their prey. Under many conditions, these fluctuations in predator and prey densities may be sustained indefinitely as a stable limit cycle (May 1974; Nisbet and Gurney 1982).

Armstrong and McGehee (1980) showed that when predator and prey densities fluctuate in a stable limit cycle, then two (or more) predator (consumer) species may coexist indefinitely on a single resource. The conditions allowing for this coexistence are (1) one of the species' resource-dependent growth functions (or equivalently, their functional responses) must be nonlinear (as in Figure 7.5); and (2) there must be a trade-off between performance at high and low resource levels (i.e., the species with the higher growth rate at low resource densities experiences a lower growth rate at high resource densities—the gleaner–opportunist trade-off). Abrams (2004b) further examined the effects of temporal resource fluctuations on species coexistence and showed that species coexistence is most easily promoted when the superior competitor causes fluctuations in a biotic resource, but the inferior competitor is not affected as much by those fluctuations. Huisman and Weissing (1999) provide examples of species coexistence in models of competition for multiple resources due to nonequilibrium population dynamics, and Grover (1997) summarizes empirical examples of microorganisms that appear to coexist via temporal fluctuations in the density of a single resource.

Competition for Multiple Resources

We can extend the graphic analysis of exploitative competition outlined in the previous section to the case of two species competing for two limiting resources, thereby gaining insight into the nature of competition for multiple resources. The graphic treatment that follows is based on Tilman's (1982) analysis; rigorous mathematical treatments of the results can be found in Hsu et al. (1981) and Butler and Wolkowicz (1987). Tilman considered many different types of resources in his analysis, but most studies of exploitative competition have focused on either essential or substitutable resources, and we will limit our consideration to these two resource types. **Essential resources** are required for growth and one essential resource cannot substitute for another. For example, plants require about 20 different mineral resources (e.g., nitrogen, phosphorus, *and* potassium) in order to grow and reproduce. On the other hand, if one resource can take the place of another, each is considered a **substitutable resource**. For example, any one of several different prey species (e.g., rabbits, squirrels, *or* mice) may provide a complete diet for a carnivore. A more extensive discussion of resource types may be found in León and Tumpson (1975), Tilman (1982), and Grover (1997). Elser et al. (2007) and Harpole et al. (2011) provide recent discussions and examples of nutrient co-limitation and other synergistic interactions between multiple limiting resources.

Competition for two essential resources

Let's assume that the environment contains two essential resources, R_1 and R_2, whose abundances we can graph on the x and y axes, respectively (**Figure 7.6**). Given a consumer with some density-independent mortality rate, we can plot the levels of R_1 and R_2 at which the consumer's per capita birth rate matches its per capita mortality rate and its population is at equilibrium (i.e., $dN_1/dt = 0$). These points define the consumer's **zero net growth isocline** (or in common shorthand, its **ZNGI**; Tilman 1982). For essential resources, the ZNGI is represented by straight lines that run parallel to both axes of the graph, forming an L-shaped curve. At all points along the ZNGI, the consumer population is at equilibrium. At resource levels below the ZNGI, the consumer population goes extinct (birth rate < mortality rate). At resource levels above the ZNGI, the consumer population increases (birth rate > mortality rate).

Note that for essential resources, there is a level of each resource at which any further increase in abundance will not increase the consumer population's growth rate (i.e., growth rate is limited by the availability of the other essential resource). In Figure 7.6, the consumer population is limited by the availability of R_2 at points along the ZNGI to the right of the corner of the L-shaped curve; at points along the ZNGI above the corner, the consumer population is limited by the availability of R_1. As mentioned above, the consumer's ZNGI represents the resource levels at which its population is unchanging. More information is needed, however, to determine whether the system as a whole is at equilibrium (i.e., both consumer and resource populations are unchanging). As Tilman (1982: 61) noted,

> [F]our pieces of information are needed to predict the equilibrium outcomes of resource competition. These are the reproductive or growth response of each species to the resource, the mortality rate experienced by each species, the supply rate of each resource, and the consumption rate of each resource by each species. An equilibrium occurs when the resource dependent reproduction of each species exactly balances its mortality and when resource supply exactly balances total resource consumption for each resource.

The ZNGI defines the conditions (i.e., abundances of R_1 and R_2) under which the consumer's reproductive rate balances it mortality rate. Thus, we have half of the information we need to determine the outcome of competition. Next we need to consider the dynamics of the two resources, R_1 and R_2.

Resources will be at equilibrium when the rate of consumption by the consumer matches the rate of resource supply. We can represent consumption at any point on the ZNGI by plotting vectors pointing toward the resource axes; each of these **consumption vectors** (C in Figure 7.6) has a slope equal to the relative amounts of R_1 and R_2 consumed. Think of these consumption vectors as indicating the rate and direction in which the consumer is drawing down the two resources. Counterbalancing consumption is the rate at which resources are supplied, which can be represented as a vector pointing toward the **resource supply point** (S in Figure 7.6)—the expected standing

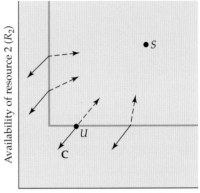

Figure 7.6   The zero net growth isocline (ZNGI) for a consumer species feeding on two essential resources (R_1, R_2). S is the resource supply point, representing the availability of resources R_1 and R_2 in the environment if there were no consumers. The rates at which R_1 and R_2 are consumed are indicated by consumption vectors (**C**; solid arrows pointing from the ZNGI toward the resource axes); four of many possible consumption vectors are shown here. The slope of each of the consumption vectors is the ratio of R_2:R_1 consumed, and this ratio is determined by the traits of the consumer. The rate of resource supply is indicated by the supply vectors (dashed lines), which point from the ZNGI toward the resource supply point, S. Resource consumption matches resource supply when the consumption and supply vectors are directly opposed (at point U). At this point, consumer and resource populations are in equilibrium. Note that at the equilibrium point U in this example, the consumer population is limited by the availability of resource 2.

stock of resources 1 and 2 in the absence of consumption. At one and only one point (U in Figure 7.6) along the consumer's ZNGI is the consumption vector in exact opposition to the supply vector. At this point, the system is at equilibrium. Resource consumption matches resource supply, and (because the equilibrium point is on the consumer ZNGI) the consumer's birth rate matches its death rate. For abiotic resources, the equilibrium point is stable (León and Tumpson 1975; Tilman 1980). For biotic resources, stability is not assured; the outcome depends on the resource growth functions and on any interactions between the resources (Grover 1997). For our goal of understanding the potential outcomes of competition for two resources, we will assume that a stable equilibrium between consumer and resources is possible at point U.

A second consumer species (N_2) may be added to the system by graphing its ZNGI along with that of consumer species N_1. There are four possible orientations of the consumers' ZNGIs and their consumption vectors, and these orientations dictate the outcome of two-species competition (Tilman 1982; Grover 1997). The descriptions of these outcomes given below follow closely from Tilman (1982).

SPECIES 1 WINS   In **Figure 7.7A**, the ZNGI for species 1 is always "inside" the ZNGI for species 2 (i.e., it is closer to the origin of the graph). Therefore, species 1 requires less of both resources than species 2 in order to maintain its population at equilibrium (birth rate = mortality rate). The positions of the ZNGIs in Figure 7.7A define three regions of potential resource availabilities. Consider what would happen to species 1 and species 2 if they were introduced into an environment where the resource supply points fell within each of these regions. Environments with resource supply points in region 1 would have insufficient resources to support either consumer species (birth rate < mortality rate). Thus, both species would go extinct. In region 2, the supply of resources would be sufficient to maintain the population of species 1, but not that of species 2. Thus, only species 1 would survive in an environment whose resource supply point fell within region 2. Finally, in region 3, there would be enough resources for each species to survive by itself. However, if both species were introduced into an environment with a resource supply point in region 3, the population growth of species 1 would reduce resources to a point on species 1's ZNGI, which is below the ZNGI of species 2. Therefore, species 1 would outcompete and exclude species 2.

SPECIES 2 WINS   In **Figure 7.7B**, the ZNGI for species 2 is always "inside" the ZNGI for species 1. As in Figure 6.6A, the winner of the competition will be the species that can tolerate the lowest level of resource supply. Neither species can survive in environments with resource supply points in region 1, only species 2 can survive in region 2, and species 2 will outcompete species 1 in region 3.

SPECIES 1 AND 2 COEXIST AT A STABLE EQUILIBRIUM POINT   In **Figure 7.7C**, the ZNGIs for the two species cross. The point at which the two ZNGIs cross is a two-species equilibrium point because at this point, each species' birth rate matches its mortality rate. Whether this equilibrium point is stable or

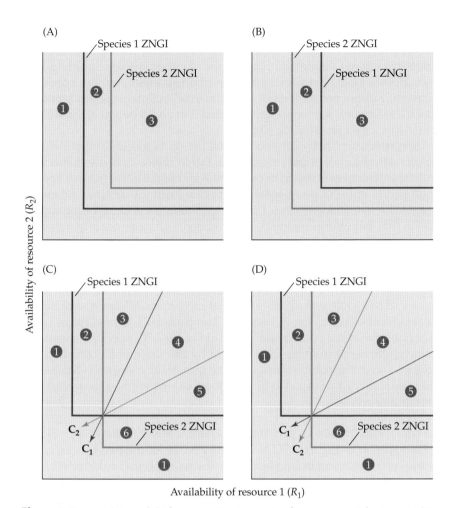

Figure 7.7 Graphic model of two species competing for two essential resources (R_1, R_2), showing four outcomes of resource competition. (A) Species 1's ZNGI lies inside species 2's ZNGI; species 1 outcompetes species 2. (B) Species 2's ZNGI lies inside species 1's ZNGI; species 2 outcompetes species 1. (C, D) The ZNGIs for species 1 and 2 cross. The intersection of the ZNGIs defines an equilibrium point at which the densities of species 1 and species 2 are unchanging (i.e., births = deaths for both species). **C_1** and **C_2** are the consumption vectors for the two species. (C) The equilibrium point is locally stable because each species consumes more of the resource that most limits its own growth. For environments with resource supply points in region 4, species 1 and species 2 coexist. (D) Consumption vectors for the two species are reversed from those in (C); the equilibrium point is unstable because each species consumes more of the resource that most limits the growth of the other species. For environments with resource supply points in region 4, the outcome of competition depends on initial conditions, and either species 1 or species 2 may win. (After Tilman 1982.)

not depends in part on the orientation of the consumption vectors for each species. (Note that the outcomes of competition in the cases represented by Figures 7.7A and 7.7B are unaffected by the orientation of the species' consumption vectors, and therefore we did not include consumption vectors in those graphs.) In Figure 7.7C, the species' ZNGIs and consumption vectors define six regions into which resource supply points may fall. Environments with resource supply points that fall in region 1 have insufficient resources to maintain populations of either species. Supply points that fall in region 2 are sufficient to maintain species 1, but not species 2. Likewise, supply points that fall in region 6 can maintain species 2, but not species 1. Regions 3–5 have supply points that could potentially maintain either species when growing alone.

In competition, however, species 1 will exclude species 2 from environments with resource supply points that fall in region 3, because species 1 will draw resources down to its own ZNGI and to the left of the two-species equilibrium point (a point at which species 2 cannot survive). Likewise, species 2 will exclude species 1 from environments with resource supply points that fall within region 5. Environments characterized by resource supply points that fall in region 4 may support the stable coexistence of both species 1 and species 2 because the joint consumption of resources by both species will draw resource availabilities down to the equilibrium point, where the ZNGIs intersect. *This two-species equilibrium point is locally stable because at this point, each species consumes proportionally more of the resource that most limits its own growth* (León and Tumpson 1975; Tilman 1980). To see this, note that at the equilibrium point in Figure 7.7C, species 1 consumes relatively more of resource 2 (as shown by the slope of its consumption vector), and at the equilibrium point the growth of species 1 is limited by the availability of resource 2. Similar logic can be applied to species 2 at the equilibrium point.

SPECIES 1 AND 2 COEXIST, BUT THE EQUILIBRIUM POINT IS UNSTABLE **Figure 7.7D** is similar to Figure 7.7C in that the ZNGIs for the two species cross. However, the consumption vectors for the two species are reversed from those in Figure 7.7C. This reversal causes the two-species equilibrium point to be locally unstable. At equilibrium, species 1 is limited by resource 2 and species 2 is limited by resource 1. However, the slopes of the consumption vectors show that species 1 consumes relatively more of resource 1 and species 2 consumes relatively more of resource 2. Thus, at the equilibrium point, each species consumes relatively more of the resource that most limits the growth of the *other* species, not the resource that most limits its own growth. This situation causes the equilibrium to be unstable; any deviation away from the equilibrium point will be magnified over time until either species 1 or species 2 outcompetes the other. Thus, environments with resource supply points in region 4 will lead to dominance by either species 1 or species 2 (and the extinction of the other species); the winner depends on the starting conditions. Environments with resource supply points in regions 1, 2, 3, 5, or 6 will lead to competitive outcomes identical to those in Figure 7.7C.

Competition for two substitutable resources

If resources are perfectly substitutable, such that each resource supplies the full complement of substances required for a consumer's growth and one resource can substitute completely for another, then the consumer ZNGIs can be represented by straight lines in resource state space (**Figure 7.8**). Most

Figure 7.8 Graphic model of two species competing for two perfectly substitutable resources (R_1, R_2). In this example, the ZNGIs for species 1 and 2 cross, and the intersection of the ZNGIs defines an equilibrium point at which the densities of species 1 and species 2 are unchanging (i.e., births = deaths for both species). C_1 and C_2 are the consumption vectors for the two species. The equilibrium point is potentially stable because each species consumes more of the resource that most limits its own growth. In environments with resource supply points in region 4, species 1 and species 2 coexist.

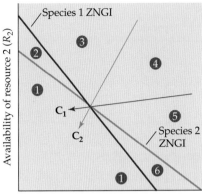

heterotrophs (e.g., predators) consume resources that are best viewed as substitutable. León and Tumpson (1975) suggested that a fundamental difference between competition among plants and competition among animals is the difference between essential and substitutable resources. Thus, by considering competition for essential and for substitutable resources, we have covered the two major types of exploitative competition found in natural communities. A number of factors may cause resources to be imperfectly substitutable, resulting in nonlinear ZNGIs. However, as long as the ZNGIs decrease monotonically with a decline in resource abundance, the general conclusions below will apply (Grover 1997).

If the ZNGIs of two consumer species do not intersect, then the species whose ZNGI is closest to the origin will win in competition and exclude the other species (i.e., it will have a lower R^* for both resources). Thus, just as in the case of competition for essential resources, coexistence among two species competing for substitutable resources requires that the species' ZNGIs cross (i.e., there must be a trade-off in the species' abilities to grow on the two resources). The criteria for stable coexistence on two substitutable resources are also similar to those for essential resources, although there are some important distinctions. *For substitutable resources, stable coexistence can occur if each species consumes proportionally more of the resource that most limits it own growth and if equilibrium resource densities are stable and positive* (MacArthur 1972; León and Tumpson 1975; Grover 1997). In the case of biotic resources that are themselves competing, the factors determining stable positive resource populations can be complex (see discussions in Abrams 1988; Grover 1997; Abrams and Nakajima 2007).

Spatial heterogeneity and the coexistence of multiple consumers

The above examples show how two species may coexist on two resources. Coexistence requires that the species' ZNGIs cross, or in other words, that the species exhibit a trade-off in their abilities to compete for the two resources. In the example shown in Figure 7.7C, species 1 is a better competitor (has a lower R^*) for resource 1, but it is a poorer competitor for resource 2. In general, stable coexistence requires that each species consume more of the resource that most limits its growth. Tilman (1980, 1982) showed how this argument can be extended to explain the coexistence of multiple species on two resources if the species follow the same trade-off curve such that a species that is better at competing for resource 1 is inferior at competing for resource 2. In this case, there will be regions of the two-resource state space in which different pairs of species may coexist (**Figure 7.9A**).

(A)

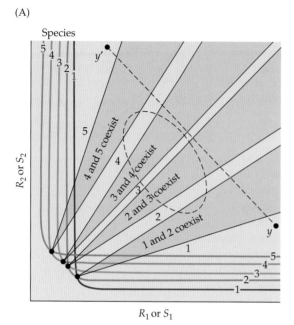

Species

(B)

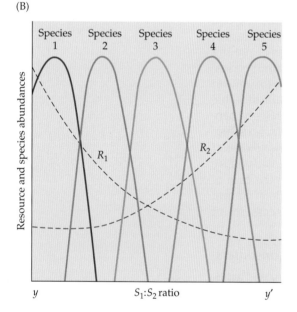

Figure 7.9 Spatial heterogeneity in resources may promote the coexistence of multiple competing species. (A) ZNGIs for five consumer species (1–5) utilizing two resources (R_1, R_2). In this example, there is an interspecific trade-off such that species 1–5 are each progressively better competitors for resource 1 and progressively poorer competitors for resource 2 (i.e., species 1 is the best competitor for R_2 but the worst competitor for R_1). The four dots indicate two-species equilibrium points (where two species' ZNGIs cross). The line from y to y' represents a gradient in resource supply points (S_1, S_2), and the ellipse represents a habitat that has spatial heterogeneity in resource supply points. (B) Exploitative competition causes the five competing species to become separated in space along the resource supply gradient (y to y'). (After Tilman 2007.)

Moreover, if the environment varies spatially in the supply rates of the two resources, this spatial heterogeneity may allow the coexistence of multiple consumer species in the community. For example, spatial variation in resource supply could be modeled as a gradient in the supply rate of resource 1 and resource 2 (the dashed line from resource supply point y to resource supply point y' in Figure 7.9A). In this case, competitors will be separated in space along the resource supply gradient running from y to y' (**Figure 7.9B**). If spatial heterogeneity in resource supply is instead modeled as a mixture of resource patches varying in their supply rates of R_1 and R_2 (dashed ellipse in Figure 7.9A), then species would coexist as one- or two-species pairs intermingled in space (Tilman 1982, 2007).

Apparent Competition

We have seen that competition occurs when two or more species share a limiting resource. Species that share a predator also may have negative indirect effects on each other, an interaction that has been called **apparent competition**. The simple mechanism behind this negative interaction is a numerical response by a predator to its prey: an increase in the abundance of one prey species may lead to an increase in predator abundance, which in turn may have a negative effect on the abundance of other prey species that the predator consumes. The predator's numerical response may be behavioral and occur quickly (e.g., predators may aggregate in areas of high prey density), or it may result from slower demographic processes (e.g., predator numbers increase due to higher birth rates or lower death rates in response to increased prey density).

MacArthur (1970: 21) noted the potential for shared predation to lead to negative indirect effects on prey species; however, it was Holt (1977) who developed the theory and explored its consequences for species coexistence. Holt referred to the negative effect of shared predation as "apparent competition" because the negative interaction between species sharing a predator resembles the negative interaction between species sharing a limiting resource (exploitative competition). Apparent competition has the potential to be a strong interaction in most food webs. In a survey of 23 marine intertidal communities, Menge (1995) found that apparent competition occurred in about 25% of the food webs studied (which made it the second most common indirect effect).

We can understand the action of apparent competition by using a version of the simple predator–prey models developed in Chapter 5. Following Holt et al. (1994), we can define the population growth rates for two prey species i ($i = 1, 2$) as

$$dN_i/dt = r_iN_i - a_iN_iP,$$

Equation 7.7

where N_i is the abundance of prey species i, r_i is the intrinsic growth rate for prey species i, P is the abundance of the predator, and a_i is the predator's attack rate on prey species i (number of prey eaten per predator, per prey). You should recognize Equation 7.7 as a multispecies version of the familiar Lotka–Volterra model of predator–prey dynamics (see Equation 5.5).

Like the L–V model, Equation 7.7 assumes no self-limitation in the prey population and assumes that the predator has a linear functional response. We can define the population growth rate for the predator as

$$dP/dt = \left(\sum_{i=1}^{2} f_ia_iN_iP - qP\right) + I - EP$$

Equation 7.8

where f_i is the number of predators born as a result of the consumption of each prey item and q is the predator's density-independent death rate. Equation 7.8 differs from the standard Lotka–Volterra formulation for the predator population (e.g., Equation 5.7) in that it includes terms for the rates of immigration (I) and emigration (E) of predators. Holt et al. (1994) included an immigration rate for the predator to help stabilize the predator's population dynamics. For our present purposes, we don't need to be concerned with these predator migration terms in order to understand the impact of apparent competition on species interactions.

In the simple model described above, the prey species that dominates in apparent competition is the species that withstands (and maintains) the highest density of the shared predator relative to other prey species. We can see this by noting that the per capita rate of change in the prey population is

$$dN_i/N_i\, dt = r_i - a_iP$$

Equation 7.9

Thus, prey species i will have a positive per capita growth rate when $dN_i/N_i dt > 0$, which is true when $r_i > a_iP^*$ (where P^* is the equilibrium density of predators in the system). The prey species that can maintain a positive population growth rate at the highest equilibrium predator density, defined as

$$P^* = r_i/a_i$$

Equation 7.10

will exclude all other prey species from the system. Holt et al. (1994) labeled this statement the **P^* rule**. The P^* rule is analogous to the R^* rule for interspecific competition, which states that the species able to maintain its population at the lowest resource level (R^*) will dominate all other species when competing for a single resource. Likewise, when two (or more) species share a common predator, the winning species will be the one that can maintain its population at the highest density of predators (P^*). Equation 7.10 emphasizes the two attributes that allow prey species to persist at high predator densities: (1) a high intrinsic growth rate for the prey (r_i) and (2) a low vulnerability to attack by the predator (a_i). Predator preference and prey vulnerability are two major factors influencing attack rates. Abrams (1987, 2004a) and others discuss a variety of traits of predators and prey that may affect attack rates.

The impact of apparent competition between prey will be weakened when (1) the predator is not food-limited but is limited instead by other resources (breeding sites, for example); (2) prey species have spatial or temporal refuges from predation; or (3) the predator exhibits **prey switching** behavior (i.e., feeding preferentially on the more abundant prey; Murdoch 1969). If predators concentrate their feeding on the most abundant prey species, then the impact of apparent competition between prey species will be reduced. Such switching behavior may even lead to a short-term indirect mutualism between prey species.

There is a well-developed theory for apparent competition that includes a number of refinements beyond the simple L–V type model presented above (e.g., Holt 1984; Abrams 1987; Holt et al. 1994; Leibold 1996; Noonburg and Byers 2005; Křivan and Eisner 2006). For many years, theory outpaced empirical examples of apparent competition, and the classic study by Schmitt (1987) on apparent competition in subtidal reef communities was often cited as one of the few field experiments showing its effect on species distributions and abundance. However, the impact of apparent competition in natural communities is now well documented (see reviews by Holt and Lawton 1994; Chaneton and Bonsall 2000), as is its role in applied ecology (van Veen et al. 2006). It has been applied to predator–prey and host–parasitoid interactions involving insect pests (e.g., Müller and Godfray 1997; Murdoch et al. 2003; Östman and Ives 2003), to the biocontrol of pests through predator introductions (e.g., Willis and Memmott 2005; Carvalheiro et al. 2008), and to studies of the impact of invasive species on native species (e.g., Norbury 2001; Rand and Louda 2004; Orrock et al. 2008; Oliver et al. 2009). In both the basic and applied literature, numerous studies have noted that species interactions involving shared enemies often show asymmetrical effects—that is, one species is more strongly impacted by a shared predator than is the other (Chaneton and Bonsall 2000). This asymmetrical impact, predicted by the theory of apparent competition (Holt and Lawton 1994), may lead to the exclusion of species from areas where they might otherwise exist (e.g., Meiner 2007; Oliver et al. 2009). Finally, species may compete for resources as well as sharing a predator, leading in the simplest case to a diamond-shaped food web that reflects a mixture of exploitative competition and apparent competition (see Figure 10.13C). We will explore some of the consequences of this "diamond interaction web" in Chapter 10, when we discuss food webs, keystone predation, and ecosystem regulation.

Conclusion

This has been quite a long chapter, yet we have only scratched the surface of a vast literature describing the conceptual foundation of interspecific competition. (Interested readers are referred to Grover 1997; Case 2000; and Chesson 2000 for further introduction.) But despite our limited exposure to competition theory, we have learned a great deal about the nature and outcomes of interspecific competition. We have seen that both the phenomenological Lotka–Volterra model and the more mechanistic consumer–resource model specify general conditions for two-species competition that lead either to competitive exclusion (one species wins and the other is excluded) or to the coexistence of species (with either a stable or an unstable equilibrium point). In both models, we see that *stable coexistence is predicted to be possible when intraspecific competition is stronger than interspecific competition*. The consumer–resource model specifies this condition for stable coexistence mechanistically, based on the species' relative consumption of their limiting resources. This consumer–resource model is sometimes referred to as a "resource-ratio" model of competition because the outcome of competition depends in part on the ratios of resources consumed and supplied (see Figure 7.9B). In the L–V model, when each species has a stronger effect on the growth of its competitor than it does on its own growth, the result is an unstable equilibrium and competitive displacement, with the winner determined by the initial conditions (founder control). The same outcome is predicted by the consumer–resource competition model when each species consumes more of the resource that limits the other species at the equilibrium point. Thus, *an unstable equilibrium and founder control can occur when interspecific competition is stronger than intraspecific competition*.

Conditions for the competitive displacement of one species by another are most easily visualized with the consumer–resource model. In the case of a single limiting resource, the species with the lowest $R*$ wins and excludes the other species. In the case of two species competing for two resources, the winner will be the species whose ZNGI falls inside that of the other species (i.e., the species that can maintain a population on a lower level of both resources). Conditions for competitive exclusion in the L–V model are less intuitive, being expressed as the ratios of the species' competition coefficients and carrying capacities.

The rule that coexistence of n competing species requires at least n limiting resources (or n limiting factors) may be relaxed if resources are allowed to vary spatially or temporally (Grover 1997). It is important to note that species coexistence via temporal or spatial resource heterogeneity requires trade-offs in the abilities of different species to use resources. In the case of temporal variation in resource availability (driven by either exogenous or endogenous factors; e.g., limit cycles), species coexistence requires a trade-off such that species that grow relatively well when resources are rare ("gleaners") grow relatively poorly when resources are abundant (and "opportunist" species do the reverse). This trade-off can be expressed as a crossing in the species' nonlinear resource-dependent growth functions (e.g., Figure 7.5A). In Chapter 14, we will examine how temporal variation in the environment may promote species coexistence by another mechanism, the "storage effect" (Chesson 2000a).

In the case of spatial resource heterogeneity, coexistence of multiple species requires a trade-off in a species' ability to grow on different resources (e.g., Figure 7.9A). As Tilman (2007: 93) notes, "Except for the diversity explanation based on neutrality (Hubbell 2001), all other explanations for the high diversity of life on Earth require tradeoffs." The concept of trade-offs is fundamental to ecology, and we will consider ecological trade-offs in a variety of contexts in later chapters. Although interspecific trade-offs are required for stable, equilibrial species coexistence, we will see in later chapters that equalizing mechanisms can make species coexistence easier to achieve (Chesson 2000a). We will explore equalizing and stabilizing mechanisms of species coexistence in more detail in Chapter 14, and we will consider one particular model of species coexistence via equalizing mechanisms (Hubbell's neutral theory, alluded to in Tilman's quote above) in detail in Chapter 13. For now, however, we will conclude our study of the fundamentals of interspecific competition in Chapter 8, where we will examine the evidence for the importance of interspecific competition in nature, as well as the results of experiments testing the predictions of simple competition theory.

Summary

1. Interspecific competition is an interaction between species in which an increased abundance of one species causes the population growth of another species to decrease. Exploitative competition (resource competition) occurs when a species consumes a shared resource that limits its own and other species' population growth, thus making that resource less available to other species. Interference competition (contest competition) is when a species restricts another species' access to a limiting resource.

2. The Lotka–Volterra competition model shows that two-species competition has four potential outcomes: species 1 wins, species 2 wins, stable coexistence, or unstable coexistence. The criterion for stable species coexistence in the Lotka-Volterra model is that each species have a greater negative effect on its own per capita growth rate than it does on the per capita growth rate of its competitor. The Lotka–Volterra competition model is a "phenomenological model" in that it does not specify the mechanism by which one species has a negative effect on another.

3. The Lotka–Volterra competition model may be extended to include more than two competing species by using a community matrix approach. Although theorists have made extensive use of the community matrix approach to study multispecies communities, this approach is difficult to apply to natural communities and generally fails to take higher-order interactions into account.

4. Consumer–resource competition models mechanistically link the outcome of interspecific competition to the consumption of resources, specifying how competing consumers affect their resources and how consumer population growth rates are affected by resource abundances.

5. Resources are entities which contribute positively to the growth rate of the consumer population and which are consumed. Essential resources are those required for an individual's growth and survival; one essential resource cannot substitute for another. If one resource can take the place of another, each one is considered a substitutable resource.

6. R^* is the minimum amount of a resource R that consumer species N requires to maintain a stable population at a given mortality rate. At R^*, the consumer's per capita birth rate matches its per capita mortality rate and its population is unchanging.

7. The R^* rule in consumer-resource competition models states that (1) two species cannot coexist on one limiting resource (at constant densities), and (2) the species that can maintain positive population growth at the lowest resource abundance (lowest R^*) will win in competition.

8. If resource abundance fluctuates over time, two species may coexist on a single limiting resource if species 1 has a growth advantage at high resource abundance ("opportunist" species) and species 2 has a growth advantage at low resource abundance ("gleaner" species). Such trade-offs between species types can have important consequences for species coexistence.

9. The abundances of two resources at which a consumer population will be at equilibrium can be plotted as that consumer's zero net growth isocline (ZNGI). At resource levels below the ZNGI, the consumer goes extinct. At resource levels above the ZNGI, the consumer population increases. If a ZNGI for a second consumer species is added to such a plot, we see that the consumer that can survive at the lower level of resources wins in competition.

10. Two consumer species can coexist on two resources at a stable equilibrium if the species' ZNGI's intersect and at the intersection point each species consumes proportionally more of the resource that most limits its own growth. The equilibrium is unstable, however, if each species consumes relatively more of the resource that most limits the growth of the *other* species.

11. Spatial heterogeneity in the supply of resources within a community can allow more than two species to coexist on two limiting resources.

12. Apparent competition occurs when two (or more) species share a predator and an increase in the abundance of one prey species leads to an increase in shared predator abundance, which may have an indirect negative effect on the abundance of other prey species. The prey species that can maintain its own population at the highest density of the shared predator is favored under apparent competition.

8 Competition in Nature

Empirical Patterns and Tests of Theory

The literature on competition is as vast and diverse as beetles in the biosphere.
Paul Keddy, 1989: ix

It is easy, and relatively uninformative, to assert that resource competition is important in nature, sometimes and somewhere. It is far harder, though essential, to delineate when, and where it is important. … This must ultimately be determined by incisive experiments and insightful observations.
James Grover, 1997: 282

The perceived importance of interspecific competition has waxed and waned in community ecology over the past 50 years. In the 1960s, competition was the dominant paradigm explaining species richness and diversity within communities. The strength of interspecific competition was thought to vary predictably with the degree of species' overlap in resource use and, by extension, the number of species that could coexist within an ecological guild was thought to be determined by limits to their similarity in resource use. This paradigm shifted in the 1970s, when the evidence linking patterns of species diversity to competitive interactions was severely challenged by work with "null" or "neutral" models showing that many of the observed patterns of species distributions or niche relationships appeared to have little to do with the action of interspecific competition (Simberloff and Boecklen 1981; see also Chapter 1). The null-model approach of the 1970s was found to be too restrictive in excluding the role of interspecific competition in many cases; nevertheless, the study of competition went through a near-death experience after the null model debate of the 1970s and early 1980s. Slowly, however, the study of interspecific competition has been revitalized due to (1) the development of new theory incorporating the mechanisms of resource competition; (2) new and more rigorous analyses of observed patterns; and (3) the accumulation of experimental evidence demonstrating the action of interspecific competition in nature.

In this chapter, we examine the empirical evidence for interspecific competition. First, we will look at the predictions of consumer–resource competition theory described in Chapter 7 and ask how well those predictions are supported by experimental studies. As we shall see, the results of laboratory experiments strongly confirm the predictions of resource competition theory, but the evidence from field experiments is more limited and their results are mixed. This difference in the outcome of laboratory and field experiments provides a good opportunity to consider what it means to "test a theory" in ecology and why it is important to take our predictions into the field. We will then examine the different approaches used to study competition in nature, including observational and experimental methods, and look at some examples of each. Finally, we will review the evidence for the importance of interspecific competition in structuring communities. We will look at the weight of the evidence assessed by comprehensive reviews and meta-analyses and will consider the role of competition in determining species ranges, as well as the impact of predation and other sources of mortality on the outcome of competitive interactions.

Testing the Predictions of Competition Theory

One of the strengths of resource competition theory is that it provides a mechanistic link between consumers and resources, allowing us to test the theory's predictions using measures of resource use and resource supply. In contrast, more phenomenological competition models (e.g., the Lotka–Volterra model) require estimates of the relative strengths of the interactions within and between species (the competition coefficients in the L–V model) in order to predict the long-term competitive outcomes. In the simple case of multiple species competing for a single limiting resource, resource competition theory predicts that the species with the lowest equilibrium resource requirement (R^*) will exclude all other competing species. The first step in testing this prediction is to measure each species' R^*, either (1) by growing the species in monoculture on the resource and measuring the resource concentration when the consumer and resource are at steady state, or (2) calculating R^* from the species' growth kinetics (Wilson et al. 2007). When the species are then grown in mixture and allowed to compete, theory predicts that the species with the lowest R^* will win.

A number of laboratory experiments have tested the R^* prediction using two species in competition for a single limiting resource (**Figure 8.1** shows one example). Wilson et al. (2007) summarized the results from 43 laboratory competition studies with bacteria, phytoplankton, and zooplankton and concluded that 41 of the studies were consistent with the R^* prediction, while the two remaining studies were inconclusive. Miller et al. (2005) arrived at a somewhat different conclusion regarding support for the R^* prediction after examining a smaller number of experimental studies, concluding that the results were mixed and that the theory had been insufficiently tested. However, I believe that the Wilson et al. review convincingly demonstrates that "at least in competition experiments with bacteria, phytoplankton, and zooplankton, R^* is almost always a good guide to competitive outcome"

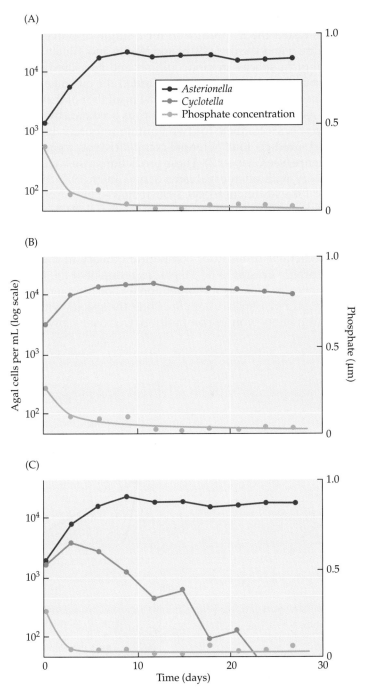

Figure 8.1 A test of the R^* prediction for two species of freshwater phytoplankton (algae) potentially competing for a limiting nutrient. Under the conditions tested in the experiment, both algal species were limited by a single nutrient—phosphorus—and the *Asterionella* species had a lower R^* for phosphorus than did the *Cyclotella* species. (A, B) Each species grew successfully in monoculture. (C) When the two species were grown in competition, *Asterionella* eliminated *Cyclotella* . (After Tilman 1977.)

(Wilson et al. 2007: 704). However, only one set of experiments (that I am aware of) has tested the R^* prediction for a single resource outside the laboratory. Tilman and Wedin (1991b) and Wedin and Tilman (1993) grew five perennial grass species in monocultures and measured their R^*s for the limiting soil resource (nitrogen), then tested the outcome of competition among species in pairwise experiments. In all four of the competition pairings, the species with the lowest R^* won, as predicted by the theory.

When two species compete for two limiting resources, resource competition theory predicts four potential competitive outcomes, including stable coexistence (see Chapter 7). These predictions also have been tested in the laboratory with microorganisms (e.g., Tilman 1977, 1981; Kilham 1986; Sommer 1986; Rothhaupt 1988, 1996; reviewed in Wilson et al. 2007). The results of these lab experiments support the theory (Wilson et al. 2007; see **Figure 8.2** for an example), although the tests are again taxonomically limited to phytoplankton and zooplankton. Dybzinski and Tilman (2007) tested the predictions of resource competition theory in the field using six perennial prairie grasses in an 11-year experiment at the Cedar Creek Reserve in Minnesota looking at competition for soil nitrogen and light. In six of the eight pairwise species combinations, the species with the lowest resource requirements won in competition. Coexistence was predicted

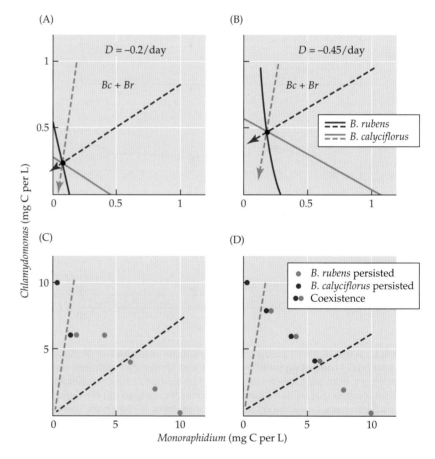

Figure 8.2 Tests of the predictions of resource competition theory for two rotifer species (*Brachionus rubens* and *B. calyciflorus*) competing for two algal resources (*Monoraphidium* and *Chlamydomonas*) in the laboratory. As shown in Chapter 7 (see Figure 7.2B), when growing in monoculture *B. rubens* has the highest growth rate at all densities of *Monoraphidium*, while *B. calyciflorus* has the highest growth rate at all densities of *Chlamydomonas*. (A, B) Predicted outcomes (ZNGIs and consumption vectors) at two different dilution (mortality) rates (*D*). The two rotifers are predicted to coexist at resource concentrations defined by the sector between the dashed lines (*Bc + Br*). (C, D) Actual competitive outcomes. Theory correctly predicted the outcome of interspecific competition in 11 of the 12 cases. (After Rothhaupt 1988.)

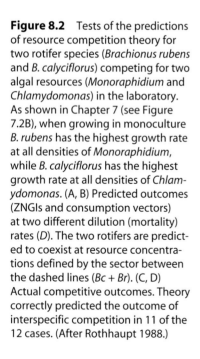

and observed in two pairwise combinations of species. However, it was unclear whether species coexistence in these cases was due to a trade-off between competition for nitrogen and competition for light (the limiting resources). Other, more indirect tests of resource competition theory have looked for changes in species composition and in resource concentrations across ecosystems that differ naturally in the supply of limiting nutrients. Tilman (1982) found some support for the theory in the distribution of plant species across nutrient gradients in freshwater and terrestrial ecosystems, but Leibold (1997; see also Chase and Leibold 2003a) found no support for the theory's prediction that nitrogen and phosphorus availability should covary negatively across freshwater ecosystems. Thus, predictions for two species coexisting on two limiting resources are nicely supported by laboratory studies with microorganisms, but there is only limited evidence from field experiments.

It is not surprising that nearly all the direct tests of resource competition theory have come from the laboratory and not the field. In the laboratory, the assumptions of the theory can be met far more easily, and the outcome of competition can be observed over short time scales using microorganisms. It is satisfying to see the predictions of theory generally supported under these controlled conditions. However, the paucity of field tests leaves open the question of whether the theory is useful for predicting the outcome of competitive interactions in the field or the distribution and abundance of organisms in nature (Miller et al. 2005). It is interesting to note that Dybzinski and Tilman (2007: 317), the authors of one of few field tests of resource competition theory, concluded that "while competition for multiple resources might maintain some diversity in natural (plant) communities, other mechanisms are almost certainly operating to maintain the remainder." We will consider other diversity-maintaining mechanisms in subsequent chapters when we expand on the study of competitive systems to include such factors as environmental variation, variable nutrient requirements and stoichiometric effects, and species interactions in food webs (e.g., Chesson 2000a; Sterner and Elser 2002; Diehl 2007; Cherif and Loreau 2009; Hall 2009). However, this is a good point in our study of the empirical evidence for interspecific competition to reflect briefly on what it means to "test" a theory and to consider what the results of different types of "tests" tell us.

Some general comments about testing theory

A theory in ecology provides an explanation for an observed phenomenon of nature. As an example, let's consider one of the first theories of competition: the competitive exclusion principle. Repeated observations of species interacting in simple, homogeneous environments showed that species tended to exclude one another, such that there was (nearly always) a single winner. These observations led to the hypothesis that two (or more) species cannot coexist on one limiting resource due to interspecific competition (Gause's competitive exclusion principle; see Chapter 1). The competitive exclusion principle is an example of a verbal hypothesis or a verbal theory. Stating a theory verbally is often the first step in providing an explanation for the phenomenon in nature. Testing that theory can involve a number of subsequent steps; I suggest that these steps fall into three broad categories.

1. IS THE THEORY LOGICALLY SOUND?   This step usually involves translating the verbal theory into a set of mathematical equations. Writing out the theory in formal mathematics achieves two things: first, it makes clear the assumptions that underlie the verbal model; and second, it allows us to express the theory in a way that illuminates its properties and allows us to state its predictions accurately. For example, expressing the idea of the competitive exclusion principle in the mathematical framework of resource competition theory allows us to predict that two (or more) species cannot coexist on one limiting resource *and* that the species with the lowest R^* for the limiting resource will win in competition.

Expressing a theory mathematically may also lead to additional insights. For example, Armstrong and McGehee (1980) and others showed that nonlinear functional responses may lead to internally driven consumer–resource cycles (limit cycles; see Chapter 5) and that these cycles may allow the coexistence of more than one consumer species on a single, but fluctuating, resource. Thus, expressing the theory mathematically and exploring its dynamics showed that the competitive exclusion principle needed to be modified to state that two species cannot coexist *at constant population densities* on one limiting resource.

2. DO OBSERVATIONS MATCH THE THEORY'S PREDICTIONS WHEN THE ASSUMPTIONS OF THE THEORY ARE MET?   This step of theory testing usually involves experiments conducted in the controlled environment of a laboratory or greenhouse, where the properties of the organisms and the environment can be matched to the theory's assumptions. Examining the predictions of resource competition theory and the R^* rule in laboratory experiments with rapidly growing microorganisms is a perfect example of this step of theory testing. The fact that these organisms behave in laboratory experiments as predicted by resource competition theory demonstrates that the theory captures important aspects of competitive interactions. A more jaded interpretation might be that such experiments simply demonstrate that in an extremely artificial environment, we can find some organisms that match our mathematical assumptions well enough to serve as living analogs of the mathematical equations. But even if we were to take this extreme view, there is no denying that our theory successfully predicts the responses of real organisms.

3. IS THE THEORY A POWERFUL EXPLANATION OF NATURE?   We often hear that ecological theories need to be tested in the field. Such field tests may involve experimental manipulations (with appropriate controls) or comparative/observational studies (Diamond 1986). Why do ecologists place such great emphasis on testing theories in the field? I think the main reason is to determine whether our theory validly reflects a powerful force in nature. For example, if we test the R^* rule repeatedly in the field and find that our predictions are confirmed, we can conclude that resource competition theory is a good explanation of natural phenomena (e.g., species distributions). On the other hand, if our theory fails to predict the outcome of interspecific competition or the distribution of species in relation to resources, we will "reject" the theory. But what are we rejecting?

We are not rejecting the logic behind the theory, if in fact the mathematical foundation is sound. And we are not rejecting the theory as not applying to the real world, because our laboratory tests showed that real organisms *do* behave as predicted if the assumptions of the theory are met. Rather, *we are rejecting the theory as a powerful explanation for a phenomenon in nature*— for example, as a predictor of the distribution and abundance of species within a community.

Why might an ecological theory fail to make accurate predictions in the field, even if it is logically sound and has been shown to work in the lab? One reason might be that the assumptions underlying the theory are not met in nature. But field conditions will rarely match all the assumptions of theory. A more productive way to think about the failure of a theory in the field is to view such a failure as revealing that there are important processes operating in nature that are absent from the theory, and that the action of these unaccounted-for processes may overwhelm the responses predicted by our theory. Or, simply stated, our theory is incomplete, and we need to consider other factors. In the case of competition theory, these other factors may include spatial effects, disturbance, variable environments, interactions with higher trophic levels (e.g., predation), and the indirect effects of other species.

Observational Approaches to Studying Competition in Nature

Natural variation in species distributions over space and time can provide valuable opportunities to study the outcome of interspecific competition over broad spatial and temporal scales (Diamond 1986). Such studies are important because the outcomes of interspecific competition can take generations to play out, and some of its consequences are expected to occur only over evolutionary time. Competition may also be intermittent in time and space (Wiens 1977). As MacArthur (1972: 21) notes, "The better competitor may exclude the other species even though in a habitat where both normally coexist an observer might only witness severe competition 1 year in 20. This is the main reason most evidence for competition is from biogeographers. The ecologist watching the populations may well not see them competing severely although the biogeographer has strong evidence that competition must sometimes occur."

Changes in a species' diet or habitat use in the absence or the presence of a competitor are referred to as **niche shifts**. Of course, niche shifts may occur for other reasons besides interspecific competition (such as changing predation pressures; recall the discussion of ontogenetic niche shifts in Chapter 6). Thus, inferring competition from observed niche shifts is more powerful when additional information ties a niche shift to competition. Researchers have compared the traits of species occurring in sympatry (together) and in allopatry (apart) to look for competition-driven niche shifts. For example, in an early study of niche shifts, Diamond (1970) showed that bird species on New Guinea and other smaller islands in the South Pacific expanded their distributions up and down mountainsides when

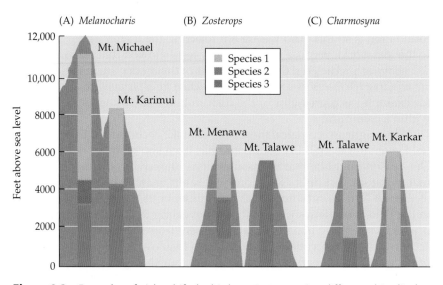

Figure 8.3 Examples of niche shifts by bird species occupying different altitudinal ranges in the mountains of New Guinea and other smaller Pacific islands. (A) Altitudinal ranges of three similar, congeneric berrypeckers (*Melanocharis*) on two New Guinea mountains. On Mt. Michael, three species of *Melanocharis* are present and have mutually exclusive ranges, as illustrated by the different shades on the histogram. On the smaller and more isolated Mt. Karimui, the middle species is absent and the other two species have expanded their ranges to fill the gap. (B) Altitudinal ranges of two congeneric white-eyes (*Zosterops*) on two Pacific islands. On Mt. Menawa in New Guinea, both species are present and have mutually exclusive ranges. On Mt. Talawe on the island of New Britain, only one species is present, and it has expanded its range to take over the range of the missing high-altitude congener. (C) Altitudinal ranges of two species of lorikeets (*Charmosyna*) on two different islands. On Mt. Talawe on New Britain, both species are present and have mutually exclusive ranges. On Mt. Karkar on the island of Karkar, only one species is present, and it has taken over the range of its missing low-altitude congener. (From Diamond 1970.)

their (presumed) competitors were absent (**Figure 8.3**). At a smaller spatial scale, Grace and Wetzel (1981, 1998) found that two species of cattails (*Typha*) were separated along a depth gradient in freshwater ponds. Transplant experiments further showed how the realized niches (post-competition depth distributions) of these two cattail species differed from their fundamental niches (pre-competition depth distributions). In this example, the species occupying deeper water undergoes a strong niche shift in the presence of the shallow-water species (**Figure 8.4**).

In some cases, a species' functional morphology (e.g., its beak shape or body size) differs in the presence versus the absence of a competitor (Brown and Wilson 1956; Grant 1972). This phenomenon, known as **character displacement**, has long interested both ecologists and evolutionary biologists because it demonstrates the potential for interspecific competition to drive species divergence in morphology and ultimately to result in **adaptive radiation** (the adaptive diversification of evolutionary lineages through time; Schluter 2000). In short, ecologically similar species are hypothesized

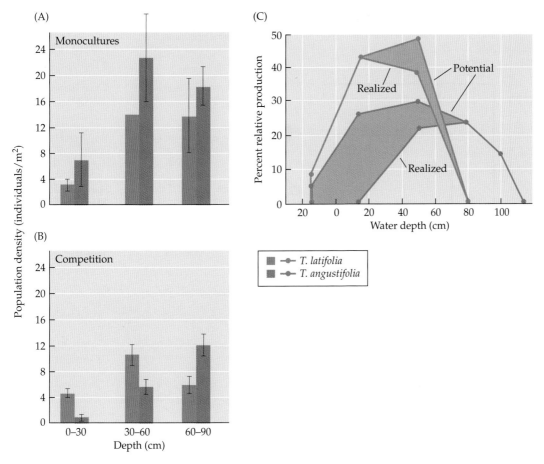

Figure 8.4   Niche shift driven by interspecific competition in two species of cattails (*Typha*). (A, B) Average water depth distributions of *Typha latifolia* and *T. angustifolia* grown in five experimental ponds at the Kellogg Biological Station in southwestern Michigan. (A) In monoculture, the two species show similar distributions across a depth gradient. (B) In competition, the species exhibit habitat segregation. *T. latifolia* is more abundant at shallow depths and *T. angustifolia* is more abundant in deeper water. Histograms show means ±1 SE. (C) Potential and realized distributions of the two *Typha* species as revealed by transplant experiments. Shaded areas represent the loss of production caused by the presence of the other species. (A, B after Grace and Wetzel 1998; C after Grace and Wetzel 1981.)

to compete strongly, creating selection pressure for divergence in resource use to reduce the impact of competition. Such selection pressure, extended over evolutionary time, may lead to a diversification of forms and adaptive radiation, a subject that will be discussed in detail in Chapter 15. Interspecific competition may also lead to character convergence (Abrams 1996); few empirical examples of convergence exist, however, perhaps because there has been less effort to look for them (see Sidorovich et al. 1999).

Classic examples of character displacement include differences in beak morphology among Darwin's finches (*Geospizinae*) on the islands of the

(A)

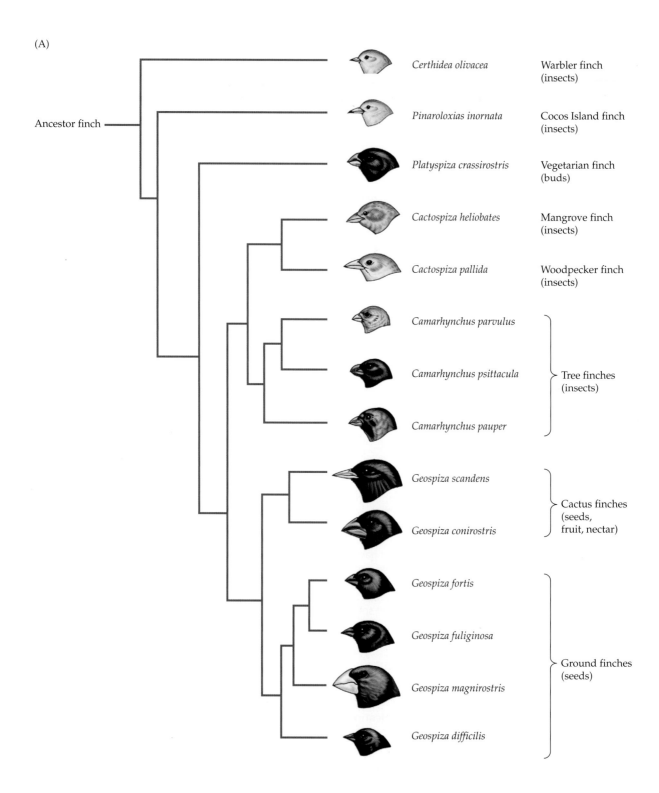

(B)

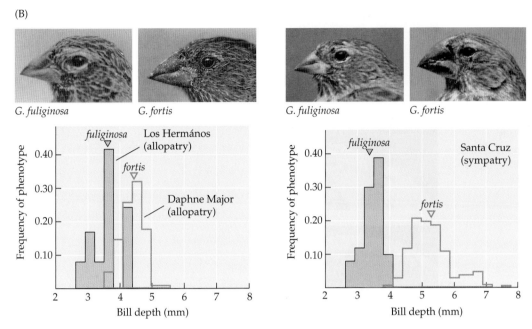

Figure 8.5 Adaptive radiation and character displacement in Darwin's finches. (A) The diversity of beak sizes and shapes found in 14 species of Darwin's finches on the Galápagos archipelago, where they have undergone adaptive radiation from a single ancestral species. Beak sizes and shapes reflect diverse feeding habits. (B) Character displacement in *Geospiza fortis* and *G. fuliginosa*. Beak depth is strongly correlated with the sizes of seeds the birds are able to eat; mean beak depth is indicated by solid triangles. Note that on Santa Cruz, where the two species are sympatric (co-occur), beak depth distributions are displaced relative to their distributions on Daphne Major and Los Hermános, where each species is found alone (allopatry). Such enhanced difference in a character trait (here, beak depth) between sympatric species is one response to competition for a resource. (A after Schluter 2000 and Grant and Grant 2008; B from Grant 1986.)

Galápagos archipelago (**Figure 8.5**; Lack 1947; Schluter et al. 1985; Grant 1986) and differences in body size between mud snails (*Hydrobia*) along the coast of Denmark (Fenchel 1975). Interspecific competition also appears to have driven adaptive radiation in *Anolis* lizards on islands in the Greater Antilles. In this case, independent adaptive radiations are known to have occurred on Cuba, Jamaica, Puerto Rico, and Hispaniola, with interspecific competition driving the diversification of lizards into a number of distinct niches based on habitat type (**Figure 8.6**; Losos 2009, 2010).

Community-wide character displacement is predicted to produce differences in trophic morphologies among groups of species exploiting the same set of resources ("ecological guilds," *sensu* Root 1967) and may lead to the **overdispersion** of traits associated with resource acquisition (Dayan and Simberloff 2005). Example of such overdispersion in niche space include

Crown giant
Large body, large toe pads

Cuba: *Anolis equestris*
Hispaniola: *A. ricordii*
Jamaica: *A. garmani*
Puerto Rico: *A. cuvieri*

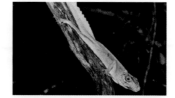

Trunk–crown
Medium body, large toe pads

Cuba: *Anolis allisoni*
Hispaniola: *A. chlorocyanus*
Jamaica: *A. grahami*
Puerto Rico: *A. evermanni*

Twig
Short body, slender legs and tail

Cuba: *Anolis angusticeps*
Hispaniola: *A. insolitus*
Jamaica: *A. valencienni*
Puerto Rico: *A. occultus*

Trunk
Vertically flattened body, long forelimbs

Cuba: *Anolis loysianus*
Hispaniola: *A. distichus*
Jamaica: none found
Puerto Rico: none found

Trunk–ground
Stocky body, long hindlimbs

Cuba: *Anolis sagrei*
Hispaniola: *A. cybotes*
Jamaica: *A. lineatopus*
Puerto Rico: *A. gundlachi*

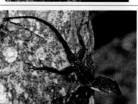

Grass–bush
Slender body, very long tail

Cuba: *Anolis alutaceus*
Hispaniola: *A. olssoni*
Jamaica: none found
Puerto Rico: *A. pulchellus*

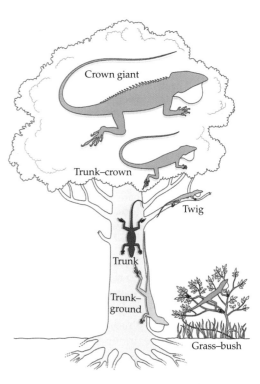

Figure 8.6 Adaptive radiation among *Anolis* lizards in the islands of the Greater Antilles. Interspecific competition has led to a variety of niche specialists (inset drawing; lizard body outlines are drawn to approximate scale relative to one another). Lizards adapted to corresponding niches on the different islands look substantially similar, although radiation has been independent on each island and thus the species listed are not closely related. Photos illustrate one species (named in color type) exemplifying each niche. (From Losos 2009, 2010; photographs courtesy of Jonathan Losos.)

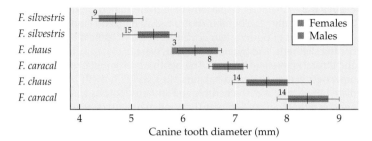

Figure 8.7 Community-wide character displacement as illustrated by canine tooth diameters in a feeding guild of three small cat species (Felidae) in Israel: the wildcat (*Felis sylvestris*), jungle cat (*F. chaus*), and caracal (*F. caracal*). Each of the three sympatric species is sexually dimorphic in size; thus, each sex can be viewed as a different "morphospecies," yielding six distinct morphospecies that potentially compete with one another. Cats are specialized predators that kill their prey using their canine teeth to separate the neck vertebrae; cats with larger canine teeth thus may be more efficient at killing larger prey. The graph shows the mean canine tooth diameters (vertical lines) for males and females of each species, as well as sample sizes (numbers), ranges (color blocks), and standard deviations (horizontal lines). Tooth diameters of the six morphospecies are very evenly spaced on a log-scaled line (i.e., the size ratios between adjacent morphospecies in a size ranking are nearly equal), and the pattern of spacing is far more regular than expected by chance. These results support the hypothesis that competition has led to character displacement in tooth size in this guild of predators. (After Dayan et al. 1990.)

the body sizes of competing desert rodents (see Figure 1.3; Brown 1975; Bowers and Brown 1982) and the tooth sizes of co-occurring felid carnivores (**Figure 8.7**; Dayan et al. 1990). In both of these cases, there is a functional link between the measured niche parameter (body size or tooth size) and a species' resource use, implicating interspecific competition as the factor driving the niche differences among co-occurring species. Furthermore, the observed patterns are significantly different from expectations based on null models of no competition (Bowers and Brown 1982; Dayan et al. 1990)—that is, the traits are "overdispersed" in niche space in comparison to random patterns.

More recent work on character displacement has combined observational studies of morphologies with experiments and genetic analyses. One of the best examples is the work of Dolph Schluter and colleagues on character displacement in a small fish, the three-spined stickleback (*Gasterosteus aculeatus*). Some lakes in coastal British Columbia contain two morphs of the three-spined stickleback, whereas others contain only a single morph. The benthic morph is relatively large bodied and feeds on invertebrates found in the nearshore vegetation of the littoral zone; the limnetic morph is more slender and feeds primarily on zooplankton in the open water (**Figure 8.8A**). The limnetic morph also has more closely spaced gill rakers than the benthic morph (**Figure 8.8B**), which allows it to capture small zooplankton more effectively. Sticklebacks in the single-morph lakes are intermediate in phenotype between the two morphs (Figure 8.8B) and are less efficient at using either benthic or limnetic resources than are the

(A)

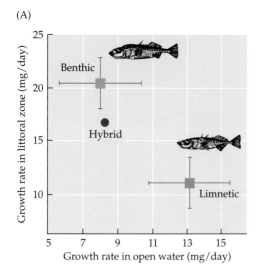

(B)

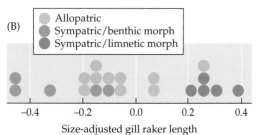

Figure 8.8 Character displacement in three-spined sticklebacks. (A) Drawings illustrate the benthic (bottom-dwelling) and limnetic (open-water) morphs of *Gasterosteus aculeatus*. In a 3-week field experiment, benthic, limnetic, and hybrid individuals were enclosed in open-water and littoral (shoreline) habitats. The benthic and limnetic morphs each grew best in their preferred habitats; the hybrid's growth was intermediate between that of the benthic and limnetic morphs and was not as good in either habitat. Symbols show mean growth rates ±1 SE. (B) The distribution of average gill raker lengths in sympatric and allopatric populations. Gill raker length is a functional trait that influences prey capture success; longer gill rakers better capture limnetic prey, while shorter gill rakers are better for capturing benthic prey. The mean morphologies of stickleback forms have diverged much more in sympatry than in allopatry. (After Schluter 1996.)

specialized morphs (Figure 8.8A; Schluter and McPhail 1992; Schluter 1996). Genetic analysis indicates that benthic and limnetic morphs evolved independently in each lake following colonization by the ancestral marine stickleback near the end of the Pleistocene (about 10,000 years ago; Taylor and McPhail 2000). Further, the two morphs are reproductively isolated by strong assortative mating and sexual selection against hybrid males (Nagel and Schluter 1998; Boughman 2001; Boughman et al. 2005). Here, then, is a case where divergence in resource use, driven by interspecific competition, has led to the evolution of two morphologically distinct phenotypes that are reproductively isolated—an example of recent speciation and adaptive radiation.

Studies of character displacement provide some of the strongest observational evidence for the evolutionary and ecological importance of interspecific competition in nature (Schluter 2000; Dayan and Simberloff 2005; Losos 2009). As noted by Dayan and Simberloff (2005: 888), "Several dozen well analyzed cases now support the notion that close competitors coevolve to increase size differences between them, or that ecological guild members tend to be overdispersed in sizes of their trophic apparatus." This summary of the evidence for character displacement carries extra weight, in part because Simberloff was one of the staunchest critics of the early evidence for community-wide character displacement (e.g., Simberloff and Boecklen 1981), and his critiques were instrumental in making the analysis of observational evidence more rigorous. However, all observational studies are limited in their ability to determine the causal agents behind a pattern. To get at causality, we need to do experiments.

Experimental Approaches to Studying Competition in Nature

In the field, the most direct approach to studying competition experimentally is to remove a potential competitor and look for a response in the remaining species. Connell (1961) pioneered this approach in his classic study of the factors determining the vertical distribution of two barnacle species on the rocky intertidal shore of Scotland. He removed individuals of *Balanus balanoides*, the most abundant species in the lower intertidal, and found that *Chthamalus stellatus*, a barnacle species occupying the upper intertidal, was able to extend its vertical distribution downward. Importantly, Connell also made observations on the mechanism of this competitive interaction and found that *Balanus* was able to exclude *Chthamalus* by growing faster and overtopping the larval *Chthamalus* that settled in the lower intertidal. In turn, *Balanus* was excluded from the upper intertidal by its sensitivity to desiccation.

Factors affecting experimental design

Connell's experiments, along with studies by Paine (1966) on the effects of predation on species diversity in the intertidal zone, set the stage for a wealth of experimental studies on competition and other species interactions. Most of these experiments employed density manipulations (e.g., species removals or additions). Although this approach to studying competition and other species interactions seems simple, in reality there is much to consider in choosing and designing the proper experimental study (e.g., see Bender et al. 1984; Scheiner and Gurevitch 1993; Resetarits and Bernardo 1998). It is comparatively easy to alter the density of sessile organisms (e.g., barnacles, plants); as John Harper (1977: 515) so famously remarked, "plants stand still and wait to be counted." Manipulating the density (or even the presence/absence) of mobile organisms, however, usually involves constructing some type of enclosure or exclosure, which can lead to experimental artifacts if the enclosure/exclosure affects factors such as resource dynamics, the physical environment, the action of predators, and the behavior of competitors. Thus, researchers must be cognizant of enclosure effects and apply appropriate controls. In addition, enclosure and exclosure structures tend to have limited life spans, thus reducing the time that an experiment can run.

Time scale in general poses significant challenges in experimental studies of competition. Ideally, we would like to study the long-term, multigenerational effects of competition, for such long-term results most readily connect to theory (see Chapter 7). However, long-term experiments pose two major obstacles. First, time and resources (e.g., money) are always limited, which means most experimental studies that measure population dynamics are conducted on small, relatively short-lived organisms. Competition experiments using larger, longer-lived species must generally measure short-term, individual-level responses to density manipulations (e.g., individual growth rates, survival, or feeding rates). Researchers are then forced to infer the long-term consequences of these short-term responses for species abundances. Second, the long-term effects of the removal of one species on the dynamics of a second (potentially competing) species include both the direct effects of removing the competitor and any indirect effects mediated through changes in the densities of other, interacting species (e.g., resources, competitors, predators, mutualists). Bender et al. (1984) published an important paper on the use of short-term (**pulse**) and long-term (**press**) experiments to examine interspecific interactions. We will discuss the issue of time scale in detail when we look at the indirect effects of interactions in food webs and other ecological networks in Chapters 10 and 11.

Designing competition experiments

Experimental designs to study competition fall into three general categories, based on how densities are manipulated: substitutive, additive, and response surface designs (Underwood 1986; Silvertown 1987; Goldberg and Scheiner 1993). **Substitutive designs** (or replacement series) hold total density constant and vary the relative abundances of each of the species (**Figure 8.9A**). This design was developed and widely used by agronomists to study plant yields in monoculture and in mixture (Harper 1977). (We discussed substitutive experiments previously in another context when we looked at experimental studies of the effects of species diversity on ecosystem functioning in Chapter 3.) The substitutive design, however, should be avoided in studying competition because it tests only for the relative intensities of intraspecific and interspecific competition (since intraspecific and interspecific densities are varying simultaneously; see Connolly 1986; Underwood 1986; Goldberg and Scheiner 1993).

The simplest way to test for the occurrence of interspecific competition and measure the magnitude of the interspecific effect is an **additive design** (**Figure 8.9B**). With an additive design, the density of one species is held constant (controlling for any intraspecific density effects) and the density of the hypothesized competitor species is varied. Note that the species removal experiments discussed earlier are a variation of an additive design

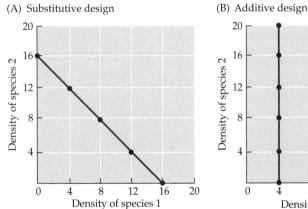
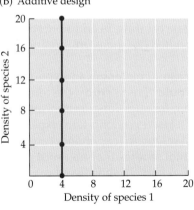
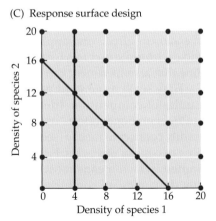

(A) Substitutive design

(B) Additive design

(C) Response surface design

Figure 8.9 Three potential designs for competition experiments, in which the densities of two species (or two genotypes, or two groups of species) are manipulated. Each point represents a single experimental treatment. (A) A substitutive design holds total density constant while varying the relative abundances of both species. (B) Additive designs hold the density of one species constant and vary the density of a hypothesized competitor species. (C) Response surface designs vary the density of each species separately so that the magnitudes of inter- and intraspecific effects can be compared. Solid lines in (C) identify the entire experimental designs for (A) and (B). (After Goldberg and Scheiner 1993.)

TABLE 8.1 Minimal (2-species) response surface design[a]

	Treatment				
	1	**2**	**3**	**4**	**5**
Population of species A	N	$2N$	N	0	0
Population of species B	0	0	N	$2N$	N
Intraspecific effects: Treatment 1 vs. treatment 2; treatment 4 vs. treatment 5					
Interspecific effects: Treatment 3 vs. treatment 1; treatment 3 vs. treatment 5					

Source: After Underwood 1986, *Community Ecology Pattern and Process*, pp. 240–268.

[a]A response surface (or addition series) design is used to compare the magnitudes of intraspecific vs. interspecific effects. All treatments should be replicated.

in which the density of one species is held constant among replicates (or as close to constant as possible in nature) and the density of a second species is reduced to zero. Despite clear presentations by Underwood (1986), Goldberg and Scheiner (1993), Gibson et al. (1999), and others on the relative merits of additive versus substitutive designs for studying competition, I am surprised at how often students continue to put forth a substitutive design as their first approach to studying interspecific competition. I hope the above discussion will serve as a reminder to check the literature on this topic before venturing out to set up experimental plots or enclosures.

Although an additive design allows us to test for interspecific competition, ideally, we would like to compare the magnitudes of inter- and intraspecific effects. To do this, we need to conduct an experiment in which we vary the density of each species separately using a **response surface design**, or addition series (**Figure 8.9C**) (Harper 1977; Silvertown 1987; Goldberg and Scheiner 1993; Inouye 2001). The challenge with this design is the large number of experimental treatments needed; however, the number of treatments can be pared down from that shown in Figure 8.9C (Inouye 2001). **Table 8.1** presents the minimal design needed to compare inter- and intraspecific competitive effects.

Ideally, we would like to vary the densities used in competition experiments around the average densities of organisms observed in nature. However, field densities can differ greatly from site to site or year to year; thus, choosing the appropriate "natural" densities for an experiment can be tricky. A variation of the additive design, the **target–neighbor design**, addresses this issue by incorporating a wide range of competitor densities into the design (Goldberg and Werner 1983; Goldberg 1987). In a target–neighbor experiment, a focal individual (or a small, fixed number of focal individuals) of the "target" or "focal" species is exposed to varying numbers of a "neighbor" or "associate" (presumed competitor) species. The response of the focal individual to competition is generally measured in terms of growth, survival, or reproductive output, and this response is plotted as a function of competitor density (e.g., **Figure 8.10**). Regressing the focal species response

Figure 8.10 A target–neighbor experiment to study the interspecific competitive effects of juvenile bluegill sunfish (*Lepomis macrochirus*) on juvenile pumpkinseed sunfish (*L. gibbosus*) in a Michigan lake. Two pumpkinseeds were introduced into each experimental enclosure; the number of bluegill introduced per enclosure ranged from 0 to 10. Mean growth of individuals of each species in each enclosure over the course of the 50-day experiment was used as a measure of competitive effect. Lines are significant least squares regressions with negative slopes. (After Mittelbach 1988.)

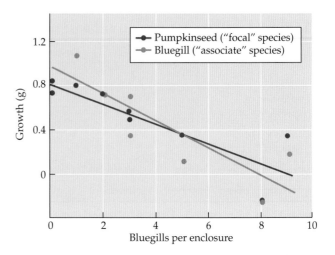

against competitor density (or competitor biomass) measures the per capita (or per biomass) effect of interspecific competition (i.e., the per capita or per biomass effect is the slope of the regression line). The shape of the relationship also provides information on whether per capita (or per biomass) competitive effects are constant (i.e., the relationship is linear) or whether they vary with density (or biomass); i.e., a nonlinear relationship. Recall from Chapter 7 that many theoretical treatments of interspecific competition (such as the L–V model) assume constant per capita effects; the target–neighbor design provides a good test of this assumption. The target–neighbor design can be easily altered to look at the combined competitive effects of a group of species (i.e., the neighboring community) on the performance of a focal species, an approach that is commonly used in studies of plant competition (e.g., Callaway et al. 2002; discussed in Chapter 9).

Finally, note that the additive design, response surface design, and target–neighbor design are all variations on the same theme: hold the density of one species constant and vary the densities of its (presumed) competitor. The designs differ in their ability to compare intra- and interspecific effects and to visualize the shape of the competitive relationships (linear or nonlinear per capita effects), as well as in their approaches to statistical analysis (ANOVA versus regression). Each design has advantages and disadvantages; Goldberg and Scheiner (1993) and Inouye (2001) provide excellent discussions of the relative merits of each approach for measuring competitive effects and for comparing the results to different mathematical models of competition.

Regardless of their design, experimental studies of exploitative competition are stronger when they consider the responses of the resources as well as the competitors, because an examination of resource dynamics and resource use provides insight into the mechanisms of competition. For example, according to resource competition theory, the predicted winner in competition for a single limiting resource is the species with the lowest R^* (see Chapter 7). Thus, experimental tests of this theory have focused specifically on the resource requirements for the competing species. Similarly,

(A)

(B)

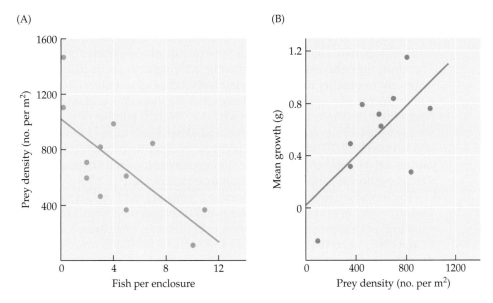

Figure 8.11 Effects on and responses to prey resources in an experiment to study competition between juvenile bluegill and pumpkinseed sunfish (see Figure 8.10). (A) If exploitative competition is occurring, one would expect to find a negative relationship between the density of fish in an enclosure and the density of prey remaining at the end of the experiment, as is the result here. (B) A positive relationship between the abundance of prey in an enclosure and fish growth rates is also expected when competition is occurring. In both cases, the relationship (shown here by the best-fit linear regression line) has a slope significantly different from zero. (After Mittelbach 1988.)

whether a species' response to a change in the density of another species is due to competition or some other factor (e.g., shared predation) can be tested by looking at the resources used by each species. Demonstrating resource competition requires (at a minimum) that there be overlap in resource use between the competing species and that an increase in competitor density result in a decrease in resource availability. **Figure 8.11** shows an example of this approach as applied to a target–neighbor competition experiment with two species of sunfish.

Experimental Evidence for the Importance of Competition in Nature

How important is interspecific competition in nature? A number of studies have reviewed the literature on field competition experiments to address this question. The first two reviews (Connell 1983; Schoener 1983), published in the midst of the great null model debate, showed that interspecific competition occurred in a high percentage (70%–90%) of the studies examined. Goldberg and Barton (1992) followed up these reviews with a synthesis of field competition experiments with plants, concluding that competition occurred in the vast majority of experimental studies designed to test for its effect in plants. Goldberg and Barton noted, however, that documenting interspecific competition in an experiment does not necessarily imply that this competition has important consequences for the structure of ecological communities. Therefore, they also asked how competition influences species distributions across environments and species relative abundances and diversities within environments. The number of studies examining these effects of interspecific competition on communities was much lower than the number of studies that simply tested for interspecific competition. Despite this limitation, Goldberg and Barton (1992: 771) found evidence that "competition always had significant effects on distribution patterns (five experiments), on relative abundances (two experiments) and on diversity (four experiments)."

The literature reviews by Connell, by Schoener, and by Goldberg and Barton consist of narrative summaries and counts of the number of studies showing significant competitive effects in different categories. Gurevitch et al. (1992) were the first to apply the tools of formal meta-analysis to the evidence for interspecific competition in the field, using the biomass responses to competition as a measure of effect size (d) for each experiment. In their study, d was calculated as the difference between the means of the experimental and control groups divided by their pooled standard deviation. When all experiments ($n = 217$) were considered together, Gurevitch and colleagues found that competition led to a large and significant reduction in species biomass, with a grand mean effect size $d_+ = 0.80$ and a 95% confidence interval of 0.77–0.83. In other words, the experimental groups experiencing competition had a mean biomass that was eight-tenths of a standard deviation less than that of the control groups. Thus, the meta-analysis provided strong additional support for the conclusion that interspecific competition is common in nature. Gurevitch et al. (2000) followed up their review of competition with a meta-analysis of the interaction between competition and predation in field experiments, and Denno et al. (1995) reviewed the effects of competition in phytophagous insects (they concluded that the evidence for competition in this insect group was clear).

Surprisingly, no systematic reviews of competition experiments have been published since 2000. Perhaps most ecologists feel that that the evidence for interspecific competition in field experiments is undeniable, and thus "the case is closed." In one sense, this is true—the vast majority of field experiments that have tested for competitive effects have found them. However, as Goldberg and Barton (1992) and others have noted, perhaps the more interesting questions for community ecology are how often (or to what extent) does competition influence the distribution of species across the landscape or determine the abundance of species within communities. At the time of their review, Goldberg and Barton found only five studies that examined the impact of interspecific competition on the distribution of plant species (all five studies showed a competitive effect). No comprehensive reviews of this question have appeared since Goldberg and Barton's in 1992. However, there is much current interest in the effects of competition and other biotic interactions on species distributions and range limits, driven in part by the desire to understand species responses and movements in relation to climate change (e.g., Geber 2011).

Interspecific competition, climate, and species distributions

The idea that interspecific competition may affect the distribution of species dates back to Darwin (1859), and one of the first experimental studies of interspecific competition—Connell's experiment on barnacles (1961)—demonstrated that interspecific competition affected the vertical distribution of species in the rocky intertidal zone. Connell found that the lower distribution of *Chthamalus* was set by competition with *Balanus*, whereas the upper distribution of *Balanus* was set by the abiotic environment (desiccation). At broader spatial scales, species ranges often correlate with environmental conditions, and climate has long been considered the major abiotic factor driving species distributions (Gaston 2003). However, climate and biotic interactions may combine to determine species ranges if the effects of competitors or predators vary across a species' range (e.g., Gross and Price 2000). For example, MacArthur (1972) hypothesized that the high-latitude limits of species ranges are often determined by physiological tolerances for cold climates, whereas the low-latitude limits may be set by competitive interactions. In a sense, this is Connell's result played out on a geographic scale: abiotic factors tend to set range limits in harsh environments, whereas in more benign environments, competitive interactions are more important. MacArthur noted that from about the central United States southward to the equator, not only do winters become milder, but summer climates become more equable as well (since tropical climates experience lesser extremes of both cold and hot temperatures than temperate climates do). Thus, MacArthur (1972: 133) asked, "Why does any animal or plant have a southern edge of its distribution between central United States and the equator?" He reasoned, based on the distributions of bird species, that interspecific competition was the most likely answer. Indirect evidence supporting MacArthur's hypothesis is found in the shapes of species distributions at the edges of their northern and southern ranges: high-latitude range edges are geometrically more regular than low-latitude range edges, which would be expected if abiotic factors correlated with latitude played a stronger role in determining high-latitude range limits than low-latitude range limits (Kaufman 1998). Interestingly, there was no difference between northern and southern range edges in the tropics, as would be predicted. So far, evidence for evaluating MacArthur's hypothesis is very limited (reviewed in Gaston 2003). Nevertheless, the goal of disentangling the factors that determine species distribution, including the role of interspecific competition, remains central to ecology (Cunningham et al. 2009; Geber 2011).

Case et al. (2005) explored a number of mathematical models of competitive interactions and concluded that "interspecific competition has the capacity to produce abrupt range limits in either homogeneous or heterogeneous spaces" (Case et al. 2005: 29). Such abrupt range limits may arise in heterogeneous environments when competing species respond inversely to environmental gradients (e.g., Roughgarden 1979), or when species respond similarly to the abiotic environment, but at different rates (MacLean and Holt 1979; Case et al. 2005). If interspecific competition is greater than intraspecific competition (i.e., $\alpha > 1$ in Lotka–Volterra type models), then distinct range limits between species may arise even in a homogeneous environment, depending on initial conditions and the movement rates of species (Levin 1974; Yodzis 1978; Case et al. 2005). In patchy environments characterized by metapopulation dynamics (described in Chapter 12), interspecific competition may also produce range limits through a variety of processes (Holt and Keitt 2000; Holt et al. 2005). Price and Kirkpatrick (2009) showed that species range limits should remain stable under a much greater range of conditions when competition is involved than when abiotic factors alone set the range limit.

The theoretical studies cited above show how competitive interactions may mold species range limits. These results, however, are based largely on models of pairwise competitive interactions; how well their results apply to more complex, multispecies communities remains to be seen. A number of empirical studies suggest that interspecific competition plays an important role in determining species distributions and range limits (e.g., Jaeger 1970; Heller and Gates 1971; Loehle 1998; Bullock et al. 2000; Gross and Price 2000; Sax et al. 2002; Cunningham et al. 2009). For example, Price et al. (2011) examined the factors setting northerly range limits in more than 500 species of birds occupying the Bhutan and Kashmir regions of the Himalaya. By comparing the characteristics of species restricted to more southerly latitudes with those of species that extend across the entire geographic gradient, Price et al. (2011) found that climate- and competition-mediated resource distributions were the most important factors setting species' northerly range limits. More studies of this type are needed to determine how biotic interactions (e.g., competition) act in combination with abiotic factors (e.g., climate) and constraints on adaptive evolution (e.g., niche conservatism; Wiens et al. 2010) to limit species ranges. Determining what sets range limits is fundamental to understanding broad-scale patterns of biodiversity and to predicting how species are likely to respond to global environmental change (Parmesan et al. 2005; Geber 2011).

The interaction between competition and mortality

A common notion among ecologists is that factors that increase consumer mortality (e.g., predation, disturbance, environmental harshness), and which thereby reduce consumer densities, may lower the intensity of interspecific competition and thereby promote the coexistence of species (see discussion in Chase et al. 2002). The intermediate disturbance hypothesis (Connell 1971; discussed in detail in Chapter 14) is a prominent example of this perspective. However, the impact of mortality on the outcome of interspecific competition depends critically on the type of mortality imposed and on the criteria used to measure competition (Chesson and Huntly 1997; Abrams 2001b; Chase et al 2002). If competitive intensity is measured simply by the short-term responses of consumers to resources (e.g., by an increase in survival or growth rate), then any increase in consumer mortality should increase the resources available to individual consumers and decrease competitive intensity. Gurevitch et al. (2000) measured such short-term competitive responses in their meta-analysis of experiments looking at the interaction between competition and predation. However, we are more often interested in how mortality affects the long-term coexistence of competing species.

In this case, the impact of mortality is more complex because increasing mortality not only increases resources per individual, but also decreases an individual's chance of survival—that is, mortality's positive effect on lowered competition for resources is countered by its negative effect on population growth.

Chesson and Huntly (1997), Abrams (2001b), and Chase et al. (2002) provide excellent summaries of the theoretical impacts of predation (and other sources of mortality) on the outcome of competition. They show that mortality may enhance the coexistence of competing species in the following ways:

1. Mortality falls selectively on the competitive dominant. Two classic studies of predator effects in the marine intertidal (Paine 1966; Lubchenco 1978; see Figure 6.6) nicely illustrate this result and show how predation on a dominant competitor can increase species richness in the community.

2. Mortality falls selectively on the more abundant species (i.e., mortality is frequency-dependent; Roughgarden and Feldman 1975; Vance 1978). Frequency-dependent predation (prey switching; Murdoch 1969) is a potentially powerful coexistence mechanism because it shifts mortality onto the more abundant species, allowing species to increase their populations when rare. However, the evidence for frequency-dependent predation in nature is sparse (Kuang and Chesson 2010).

3. Mortality opens up habitat patches, and there is a trade-off between species' ability to colonize patches and their ability to compete within patches. This important interaction will be covered in Chapter 12, where we discuss the dynamics of metapopulations.

4. Mortality affects species differentially because of niche differences, leading to nonlinear dynamics (Chesson and Huntly 1997). We will discuss nonlinear responses to variable environments and their impacts on species coexistence in Chapter 14.

The above cases show that predation can be an important factor modifying the outcome of competition and that selective mortality may act to promote the coexistence of competitors. Mortality that is nonselective and density-independent, however, is unlikely to promote the coexistence of competing species (e.g., Cramer and May 1971; Abrams 1977; Yodzis 1977; Holt 1985; Chase et al. 2002). Thus, the commonly held idea that the impact of disturbance or environmental harshness will make competition weak and unimportant is incorrect; "no credence can be given to the idea that disturbance promotes stable coexistence by simply reducing population densities to levels where competition is weak" (Chesson 2000a: 354). However, Abrams (2001) showed that density-independent mortality may promote the coexistence of species competing for multiple resources if that mortality reduces apparent competition among species sharing a predator.

Although predation is often viewed as a factor modifying the outcome of competition, Chesson and Kuang (2008; see also early work by Holt 1984) suggested that competition and predation should be placed on more equal footing with regard to promoting or hindering species coexistence. Competition-based coexistence "works" by intensifying intraspecific competition relative to interspecific competition (as shown in Chapter 7). Predation may similarly promote species coexistence when it intensifies intraspecific apparent competition relative to interspecific apparent competition. Chesson and Kuang (2008) arrived at this conclusion by examining the fitness of consumer species that occupy the middle trophic level in a model three-trophic-level system based on MacArthur's (1970, 1972) consumer–resource model. But we are getting ahead of ourselves. Before we consider the implications of species interacting across multiple trophic levels, we need to think more broadly about the nature of ecological networks and how webs of interacting species may be constructed and analyzed. We will "put the pieces together" in Chapters 11 and 12, where we will see how interactions viewed from the framework of predation, competition, and mutualism can be combined into ecological networks, the most familiar of these being the food web. Before we get there, however, we have one more type of pairwise interaction to consider: the beneficial effects that arise from mutualism and facilitation, the subjects of Chapter 9.

Summary

1. Laboratory studies with microorganisms provide strong experimental support for the predictions of resource competition theory and the R^* rule. However, only a few studies have tested the theory in the nature.

2. Testing an ecological theory involves multiple steps, including: (1) translating a verbally stated theory into a set of mathematical equations to examine its logical soundness; (2) testing the theory with experiments conducted in a controlled environment; and (3) testing the theory in the field. Failure of a theory in the field may reveal important processes operating in nature that are not accounted for by the theory.

3. Observational studies comparing the diets or habitat use of species occurring in sympatry and in allopatry can provide evidence of niche shifts resulting from interspecific competition.

4. Character displacement—the difference in a species' functional morphology in the presence versus the absence of a competitor—demonstrates how interspecific competition may drive species divergence in morphology and may ultimately result in adaptive radiation. Community-wide character displacement can lead to the overdispersion of traits associated with resource acquisition by producing differences in trophic morphologies within ecological guilds.

5. Experimental designs to study competition fall into three general categories based on how densities are manipulated:
   - In substitutive experiments, total density is held constant and the relative abundances of species are varied. This design measures only the relative intensities of intraspecific and interspecific competition.

- An additive design can test for the presence of interspecific competition and the magnitude of its effect. In an additive experiment, the density of one species is held constant and the density of a hypothesized competitor species is varied. Species removal experiments are a variation on the additive design in which the density of one species is varied to zero.
  - A response surface experiment, in which the density of each species is varied separately, allows comparison of the magnitudes of inter- and intraspecific effects.
6. Experimental studies of exploitative competition, regardless of design, are stronger when they consider the responses of resources as well as competitors. Examination of resource dynamics provides insight into the mechanisms of competition.

7. Meta-analyses document that a high percentage of field experiments designed to test for competitive effects have found them.

8. Although competitive effects are common in nature, we know relatively little about how often (or to what extent) competition influences the distribution of species across the landscape or the abundance of species within communities. Abiotic factors are hypothesized to set species' range limits in harsh environments, whereas in more benign environments, competitive interactions may be more important.

9. A common notion among ecologists is that factors that increase consumer mortality (thereby reducing consumer densities) lower the intensity of interspecific competition and thus promote species co-existence. However, the impact of mortality on the outcome of interspecific competition depends critically on the type of mortality imposed and on the criteria used to measure competition.

9 Beneficial Interactions in Communities

Mutualism and Facilitation

From the algae that help power reef-building corals, to the diverse array of pollinators that mediate sexual reproduction in many plant species, to the myriad nutritional symbionts that fix nitrogen and aid digestion, and even down to the mitochondria found in nearly all eukaryotes, mutualisms are ubiquitous, often ecologically dominant, and profoundly influential at all levels of biological organization.

Edward Allen Herre et al., 1999: 49

[T]he consequences of positive interactions for species diversity are poorly understood.

Kevin Gross, 2008: 929

What role do beneficial interactions—notably facilitation and mutualism—play in community ecology? Are they as important in determining species diversity or regulating the abundances of individuals or species as are the negative interactions of competition and predation? The answer is, we don't know. Mutualism has long been the "bastard child" of community ecology. In contrast to competition and predation, there is little general theory on how mutualisms may affect patterns of species distribution and abundance. Why is this so? It is not because mutualisms are uncommon; as the above quote from Herre et al. (1999) notes, mutualisms are everywhere, and most organisms are involved, either directly or indirectly, in a variety of positive interactions. Nor is it because mutualisms are uninteresting; some of the most fascinating examples in natural history involve mutualistic interactions, especially those found among organisms in the species-rich tropics. And it is not because mutualisms were only recently recognized; Aristotle described mutualisms in the fourth century BCE.

Some have argued (e.g., Bruno et al. 2003) that the failure to include beneficial interactions in modern ecological theory is the result of the dominant role that interspecific competition (and to a lesser extent predation) played in the conceptual development of community ecology, from the formulation of Gause's competitive exclusion principle and Lotka's and Volterra's competition models to Hutchinson's and MacArthur's work on the niche and limiting similarity. However, while interspecific competition and predation have indeed provided much of the foundation for the development of community ecology (as described in Chapter 1), I disagree with the idea that ecology's early focus on negative species interactions somehow biased the thinking of current researchers, causing them to ignore the importance of mutualism and facilitation. Rather, there are fundamental reasons why ecologists have been slow to incorporate positive species interactions into general ecological theory. Before we describe these reasons or discuss the role of mutualism and facilitation in communities, however, we need to consider some definitions.

Mutualism and Facilitation: Definitions

Most ecologists would agree that **mutualism** can be defined as a reciprocally positive interaction between species (Bronstein 2009), but the definition of facilitation is less clear. Callaway (2007) suggests that **facilitation** can include any pairwise interaction in which at least one species benefits. Under this definition, mutualism would be a subset of facilitation. However, there are important distinctions between mutualism and facilitation (Bronstein 2009). For example, facilitation generally requires close proximity between the interacting species, whereas mutualisms may link species over long distances (e.g., seed dispersal and pollination mutualisms). In addition, most mutualisms involve multiple trophic levels, whereas facilitative interactions generally occur within a trophic level. Thus, Bronstein (2009: 1160) suggests that facilitation be defined as "an interaction in which the presence of one species alters the environment in a way that enhances growth, survival, or reproduction of a second, *neighbouring* species" (my italics). Examples of facilitation include nurse plants that aid the establishment of seedlings of other species by reducing thermal or evaporative stress (Went 1942; Bertness and Callaway 1994), nitrogen-fixing plants that benefit non-nitrogen-fixing neighbors (Vandermeer 1989; Loreau and Hector 2001; Temperton et al. 2007), and salt marsh plants that provide shade and reduce evapotranspiration, thus reducing soil salinity and allowing less salt-tolerant species to survive (**Figure 9.1**; Bertness and Hacker 1994; Hacker and Bertness 1999).

The most pervasive examples of facilitation involve species whose very presence so modifies the environment that whole communities of organisms can develop there. Some species provide structural habitat for other species; these are often termed **foundation species** (e.g., corals, kelp, mangroves, and trees; Dayton 1975). Species that modify the environment so as to make it more or less suitable for other species are often termed **ecosystem engineers** (Jones et al. 1994; Wright and Jones 2006). The construction of ponds by beavers is the quintessential example of ecosystem engineering. In this

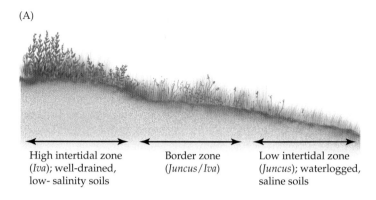

Figure 9.1 Facilitation in two salt marsh plants. (A) The three major vegetation zones, along with their physical environments, in a typical New England salt marsh. Notice the separation in distribution of *Iva* and *Juncus* species and their zone of overlap. (B–D) Results of transplant experiments in which adult *Iva* and *Juncus* plants were transplanted into each of the three vegetation zones with and without the other species. *Iva* performed better in the low intertidal and border zones when transplanted with *Juncus*; *Juncus* presumably provides shade and reduces evapotranspiration, thus lowering soil salinity and allowing the less salt-tolerant *Iva* to survive. At higher marsh elevations, where soil salinity is low and environmental conditions are less harsh, the interaction between *Iva* and *Juncus* shifts from facilitation to competition. (After Bertness and Hacker 1994.)

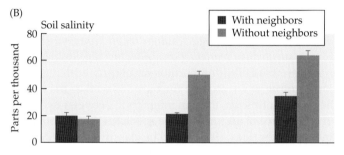

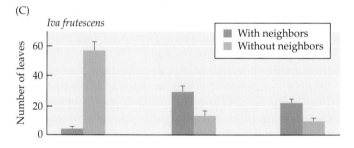

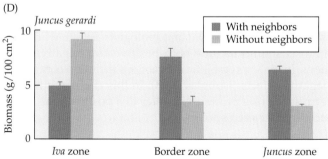

case, it is clear how beavers facilitate the coexistence of some species but not others. Because all organisms affect and are affected by their physical environment, ecologists have debated the use and relevance of the terms "foundation species" and "ecosystem engineers," questioning when and how to apply these concepts and how to distinguish organisms that do and do not fall into these categories (e.g., Jones et al. 1997; Power 1997; Reichman and Seabloom 2002). There is no doubt, however, that facilitation by means of habitat modification plays an important role in most communities.

Some ecologists have suggested that the definition of facilitation should be broadened to include positive effects between two species that act through a third species—that is, indirect effects. "After all, although rarely recognized as such, a trophic cascade is simply an indirect facilitation" (Bruno et al. 2003: 124). However, while indirect effects may often lead to positive interactions between nonadjacent species, I believe we gain little by relabeling coupled consumer–resource interactions that produce positive but indirect effects as facilitation. I would prefer to limit use of the term "facilitation" to describing a particular type of pairwise interaction, just as we use the terms "competition" and "predation." Thus, I will adopt Bronstein's definition of facilitation throughout this chapter: *facilitation is an interaction in which the presence of one species alters the environment in a way that enhances the growth, survival, or reproduction of a second, neighbouring species*. I recognize that even this definition permits ambiguity with regard to indirect effects. For example, if we include the biotic as well as the abiotic environment in the above definition, then an interaction between two plant species in which the presence of an unpalatable species reduces herbivory on a nearby palatable species (Hay 1986) could be classified as facilitation under Bronstein's definition, even though the interaction is an indirect effect involving another trophic level (the herbivore). We will discuss indirect effects (positive and negative) of species interactions in detail in Chapter 10. For now, it seems we will simply have to accept some "wiggle room" in our definition of facilitation.

Symbiosis refers to an intimate, persistent interaction between species. Mutualisms often involve some form of symbiosis, such as endosymbiosis, where one species lives inside another (e.g., *Rhizobium* bacteria living within the root nodules of leguminous plants) or ectosymbiosis, where one species lives on a body surface of another species (e.g., chemosynthetic bacteria living on the surfaces of shrimp and crabs in deep-sea hydrothermal vent communities). We will discuss examples of mutualistic endo- and ectosymbiotic relationships in the next section. The definition of symbiosis, like that of facilitation, is somewhat controversial and biologists debate whether a definition of symbiosis should include both positive and negative (e.g., parasitic) interactions. In reality, symbiotic interactions include a continuum of positive and negative effects (Douglas 2008). In this chapter, however, we will use the term symbiosis primarily to refer to a positive, persistent, and intimate association between species.

A Brief Look at the Evolution of Mutualism and Facilitation

How mutualisms evolve and how they are maintained has been an active area of study for decades, dating back to Darwin (1859). In contrast, the study of facilitation, and particularly the evolution of facilitative interactions, began only recently (Bronstein 2009). The evolution of mutualisms appears to depend on the relative costs and benefits of the interaction to the participating species. The current thought is that mutualistic interactions are best viewed as reciprocal exploitation that nonetheless provides net benefits to each partner (e.g., Leigh and Rowell 1995; Doebeli and Knowlton 1998; Sachs et al. 2004). Mutualisms can therefore be thought of as biological markets, where members of each species exchange resources or services (Noë and Hammerstein 1994; Schwartz and Hoeksema 1998; McGill 2005; Akcay and Roughgarden 2007). However, such markets are vulnerable to exploitation when individuals have conflicting interests.

What factors help align mutual interests and limit cheating? Partner fidelity and partner choice are two important mechanisms that can constrain exploitation. Partner fidelity is enhanced by the vertical transmission of symbionts from parent to offspring (Fine 1975; Axelrod and Hamilton 1981). However, there are many examples of the movement of mutualistic symbionts (e.g., pollinators, marine algae, mycorrhizal fungi) between unrelated host individuals (horizontal transmission). For some cases of horizontal transmission, there is evidence that hosts preferentially select among strains of symbionts, benefiting those strains that also benefit the host. Thus, partner choice may play an important role in reducing the impact of "cheating" in mutualistic interactions (Sachs et al. 2004).

Simms et al. (2006) provide a particularly nice example of partner choice in the yellow bush lupine (*Lupinus arboreus*). Lupines and other leguminous plants maintain a symbiotic association with bacteria of the genus *Rhizobium* that may benefit both species. Rhizobia occupy nodules in the plant's root system (**Figure 9.2A**). Within these nodules, the rhizobia capture atmospheric nitrogen, which they may exchange for photosynthates (carbon compounds) produced

(A)

(B)

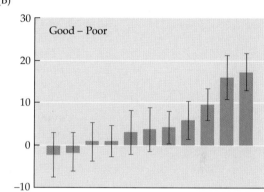

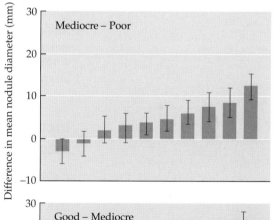

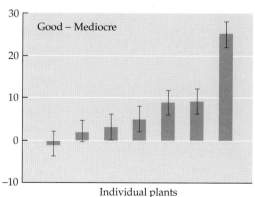

Figure 9.2 Partner choice in a legume–*Rhizobium* mutualism. (A) Legumes produce root nodules that house nitrogen-fixing *Rhizobium* bacteria. (B) The yellow bush lupine (*Lupinus arboreus*) exhibits partner choice by providing larger root nodules to more beneficial strains of rhizobia. Each histogram represents a single plant that was inoculated with two different strains of *Rhizobium*. The average difference in quality between the inoculated *Rhizobium* strains was: good to poor (top panel), mediocre to poor (middle panel), and good to mediocre (bottom panel). The size of the histogram bar indicates the difference between the average sizes of nodules housing the best and the worst strain in a single plant. In general, plants allocated resources to producing larger nodules for the more beneficial *Rhizobium* strain. (Photo by David McIntyre; B from Simms et al. 2006.)

by the plant. However, this interaction is not always cooperative, and there is clearly the potential for cheating on the part of the rhizobia. A number of authors have hypothesized that rhizobial cooperation could be promoted by plant traits that would differentially allocate resources to those nodules harboring the most cooperative strains of rhizobia (e.g., Denison 2000; Simms and Taylor 2002; West et al. 2002). Simms et al. (2006) showed that yellow bush lupine plants did exactly that. When inoculated with different strains of *Rhizobium*, the plants allocated more resources (provided larger nodules) to those strains that provided the greatest benefits to the plants (**Figure 9.2B**). These results suggest that post-infection partner choice may be an important mechanism constraining exploitation in this mutualistic interaction. Other examples of partner choice in mutualisms can be found in Sachs et al. (2004).

The study of the evolution of facilitation lags well behind the study of evolutionary relationships in mutualisms. Bronstein (2009: 1166) explains, "There has been minimal consideration to date of how facilitative interactions arise either de novo or in transition from other forms of interaction. Nor, to my knowledge, have the conditions that favour their evolutionary persistence or that lead them to break down been considered." Like the evolution of mutualism, the evolution of facilitation appears to depend on the relative costs and benefits to the participating species. Also like mutualism, facilitation ranges from relatively specific to highly generalized interactions (Callaway 1998; Bronstein 2009). Progress in understanding the evolution of facilitative relationships, and in determining whether these relationships have evolved via changes in just one of the partners (e.g., the facilitated species have evolved traits that promote physical proximity to the facilitator species) or in both partners (coevolution of traits), will require experimental work on the costs and benefits to the associated species, ideally within a known phylogenetic context. Such studies are currently in their infancy (Bronstein 2009).

Incorporating Beneficial Interactions into Community Theory

As mentioned at the start of this chapter, beneficial interactions have received short shrift in ecology textbooks. Most of the focus on pairwise interactions and, in particular, on the presentation of theory centers on interspecific competition and predator–prey interactions. Bruno et al. (2003) suggest that this focus reflects historical bias because models of competition and predation were developed first. However, over 70 years ago, Gause and Witt (1935) explored adding positive interactions to the ecological theory of their time: the Lotka–Volterra model of two-species interactions. According to Bruno et al. (2003: 121), "As Gause and Witt demonstrated in 1935, changing the Lotka–Volterra competition model into a mutualism model by switching the sign of the interaction coefficients predicts that mutualisms can result in a stable equilibrium, where the densities of both species can be greater when they co-occur." However, this statement glosses over a major problem with simple pairwise models of mutualism or facilitation (May 1981).

The equations below show how the Lotka–Volterra competition model can be rewritten as a model for positive species interactions simply by changing the sign of the interspecific effect from negative to positive:

Lotka–Volterra competition

$$dN_1/dt = rN_1\left(\frac{K_1 - N_1 - \alpha_{12} N_2}{K_1}\right)$$

$$dN_2/dt = rN_2\left(\frac{K_2 - N_2 - \alpha_{21} N_1}{K_2}\right)$$

Lotka–Volterra mutualism

$$dN_1/dt = rN_1\left(\frac{K_1 - N_1 + \alpha_{12} N_2}{K_1}\right)$$

$$dN_2/dt = rN_2\left(\frac{K_2 - N_2 + \alpha_{21} N_1}{K_2}\right)$$

In this Lotka–Volterra type model of mutualism, species coexistence is stable only if the product of the mutalism coefficients is less than unity (i.e., α_{12} $\alpha_{21} < 1$) or is less than the product of the intraspecific competition coefficients (May 1975; Holland et al. 2006). To see this, consider the situation in which the two species populations are at some density and one individual is added to each species' population (N_1 and N_2). If $\alpha_{12} = \alpha_{21} = 1$, then the *intraspecific* negative density-dependent effect of adding one individual is exactly countered by the *interspecific* positive density-dependent effect of adding one individual. Therefore, both species populations continue to grow to infinity (Gause and Witt 1935; **Figure 9.3**). Thus, May (1981: 95) concluded that modeling positive interactions using the Lotka–Volterra equations is unrealistic:

> [S]imple, quadratically nonlinear Lotka–Volterra models that capture some of the essential dynamical features of prey-predator (namely, a propensity to oscillation) or competition (namely, a propensity to exclusion), are inadequate for even a first discussion of mutualism, as they tend to lead to silly solutions in which both populations undergo unbounded exponential growth, in an orgy of mutual benefaction. Minimally realistic models for two mutualists must allow for saturation in the magnitude of at least one of the reciprocal benefits.

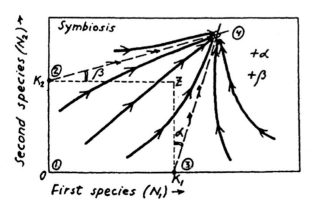

Figure 9.3 Gause and Witt (1935) modified the Lotka–Volterra competition model to consider conditions in which two species benefit each other ("symbiosis") by changing the signs of the L–V interaction coefficients (α and β in this figure, which correspond to α_{12} and α_{21} in the text) from negative (competition) to positive (mutualism). The results are illustrated in this phase-plane diagram taken from Gause and Witt's 1935 paper. As the authors stated, when the interaction coefficients are positive, both species together attain larger biomasses than separately, and a "stable knot" (equilibrium point) occurs when $\alpha\beta < 1$: "As coefficients of mutual aid α and β increase the 'knot' continuously moves up and rightwards and finally with $\alpha\beta = 1$ passes into infinity" (Gause and Witt 1935: 603–604). Thus, Gause and Witt showed that considering mutualisms within the simple framework of two-species Lotka–Volterra interactions is unrealistic.

Thus, a simple exploration of the Lotka–Volterra model demonstrates that for stable species coexistence to occur, the positive effects of mutualism must be balanced by negative density dependence in some other area. This negative density dependence could come from strong intraspecific competition. Alternatively, negative density dependence could occur via feedbacks from other life stages (e.g., nurse plants benefit seedlings, but not adults), or from fluctuating environments (sometimes good, sometimes bad for participants in a mutualism), or from other members of the food web.

The above analysis provides some insight into why beneficial interactions may have been "ignored" in the development of early ecological theory (Bruno et al. 2003). More recently, models have shown how mutualisms could evolve from parasitic interactions (Roughgarden 1975; de Mazancourt et al. 2005) and even from competition (de Mazancourt and Schwartz 2010). In the latter case, pairs of coexisting competitors could evolve toward a mutualistic "trading partnership" if the two competing species require the same resources, and if each species consumes in excess the resource that least limits its growth, then trades this excess resource to a partner whose growth is more limited by that resource. This mechanism works when the cost of excess resource acquisition and resource transfer between species is negligible (i.e., when there are no costs to cooperate, there is no benefit to cheating; de Mazancourt and Schwartz 2010).

Mutualisms may be context-dependent

In many cases, understanding the evolution and maintenance of positive interactions in communities requires that we consider the broader web of interactions in which mutualisms are embedded. The legume–*Rhizobium* mutualism is a good example: the existence of this positive interaction between leguminous plants and bacteria appears to depend on the net benefits accrued to legumes, making them stronger competitors against other, non-nitrogen-fixing plant species. Likewise, the positive effect of mycorrhizal fungi on plant growth is context-dependent (Johnson 2010).

The mutualism between plants and mycorrhizal fungi is probably as old as the evolution of land plants themselves, as structures resembling arbuscular mycorrhizae have been found on the fossil remains of plant roots from hundreds of millions of years ago. In general, plants deliver carbon compounds to their associated fungi and in return receive mineral nutrients—most commonly phosphorus, and in some circumstances nitrogen. In nutrient-poor ecosystems, plants may obtain up to 80% of their requirements for nitrogen and up to 90% of their phosphorus from mycorrhizal fungi (van der Heijden et al. 2008). However, this positive interaction between plants and fungi may shift to an antagonistic interaction under high-nutrient conditions, in which there is often no benefit to the association and plant growth may even be slightly reduced due to the carbon demand of the fungus (Johnson et al. 1997; Smith et al. 2009).

Johnson (2010) presents an excellent review of the ecology and evolution of mutualistic interactions between plants and arbuscular mycorrhizae, casting this relationship in the context of nutrient stoichiometry (Sterner and Elser 2002) and trading partnerships (e.g., Schwartz and Hoeksema 1998; Hoeksema and Schwartz 2003; Grman et al., in press). Johnson (2010:

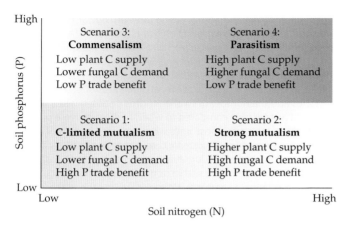

Figure 9.4   A trade balance model showing how the interaction between plants and mycorrhizal fungi is expected to shift from mutualism to commensalism to parasitism as supplies of nitrogen (N) and phosphorus (P) in the soil vary. Four scenarios are predicted based on the relative abundance of N and P. Scenario 1: A carbon-limited mutualism is predicted when levels of both N and P are low. Even though plant growth is limited by low nutrient levels and little carbon (C) is available for trading, a limited trade of C for P is still favorable to both partners. Scenario 2: A strong mutualism is predicted when N is abundant but P is rare. The plant has abundant carbon to trade, making the C-for-P trade extremely beneficial to both partners. Scenario 3: Commensalism is predicted when P is abundant and plants have little to gain from the C-for-P trade. In this case, mycorrhizal fungi gain by receiving carbon from plants, but fungal demand for C is low because fungal growth is limited by low N. Scenario 4: Parasitism is predicted when both N and P are abundant and plants gain no benefit from a C-for-P trade. Fungal demand for carbon is high, however, because N is abundant; thus the fungus is predicted to parasitize the plant. (After Johnson 2010.)

2) suggests that the dynamics of this mutualistic relationship can best be understood by "defining plants and fungi as chemical entities which must obey the laws of definite proportion and the conservation of mass and energy" (i.e., the stoichiometric perspective), and by assuming that "the symbiotic dynamics of mycorrhizal trading partnerships is defined by two factors: resource requirements and the ability to acquire resources." **Figure 9.4** presents a conceptual trade balance model for the sharing of carbon, nitrogen, and phosphorus between plant and fungal trading partners. The model shows how the relative abundances of nitrogen and phosphorus in the soil may shift the plant–fungus interaction from mutualism to commensalism to parasitism. Following these scenarios, global climate change and the concomitant enrichment of above- and belowground resources (e.g., nitrogen deposition and fertilization; CO_2 enrichment) is expected to affect the nature of the beneficial relationships between plants and mycorrhizae (Johnson 2010). The model of partner trade between plants and mycorrhizal fungi provides an excellent example of how the ecological concepts of stoichiometry, optimal foraging theory, and shared costs and benefits may be applied to understanding the factors controlling the nature of beneficial interactions under different environmental conditions (Johnson 2010; Grman et al., in press).

Figure 9.5 In an ant–acacia mutualism, acacia trees provide ants with housing and food, and in return the trees receive protection from herbivores. (A) Ants may live and raise their broods within the tree's specialized swollen thorns (domatia), and often are nourished by the tree's extrafloral nectaries and Beltian bodies. (B) Studies done in East Africa showed that excluding large herbivores caused trees of the whistling-thorn acacia (*Acacia drepanolobium*) to reduce the level of rewards offered to ants. Both the number of nectaries and the number of swollen thorns declined when herbivores were excluded (gold histograms) compared to when herbivores were present (orange histograms). (C) In the same study, most trees were occupied by ants of either of the species *Crematogaster mimosae* or *C. sjostedti*. *C. mimosae* aggressively defends host acacias from herbivores and relies on domatia to raise its broods; *C. sjostedti* is a less aggressive defender and does not live within domatia. Under natural conditions, *C. mimosae* is the most abundant ant symbiont. However, when herbivores were excluded, the better mutualist (*mimosae*) decreased in abundance and the less-beneficial ant species (*sjostedti*) increased. (A © A. Wild, Visuals Unlimited and M. Moffett, Minden Pictures; B, C after Palmer 2008.)

Ant–plant mutualisms provide another example of the role played by the broader environment (this time the biotic environment) in driving the evolution and maintenance of a mutualism. Mutualisms among a variety of species of ants and acacia trees throughout the tropics are a textbook example of a beneficial interaction (Janzen 1966; Heil and McKey 2003). The trees provide the ants with shelter (e.g., hollow thorns) and food (extrafloral nectaries and Beltian bodies; **Figure 9.5A**). The ants, in turn, protect the trees from herbivores and may also reduce competition with other trees (by clipping plants near the acacia). This well-studied example shows clear benefits that each species in the mutualism receives from the other. However, the positive nature of this interaction depends critically on the presence of other, antagonistic species (predators, competitors) in the community. Palmer et al. (2008) showed that the experimental exclusion of large mammalian herbivores from an African savanna shifted the balance away from a strong mutualism and toward a suite of less-beneficial behaviors by the interacting plant and ant species (**Figure 9.5B–D**). Ten years of herbivore exclusion lead to a reduction in the nectar and housing provided by the trees to the ants, increased antagonistic behavior by a mutualistic ant species, and shifted competitive dominance from a nectar-dependent mutualistic ant species to a less-beneficial ant species. Trees occupied by the less-beneficial ant species grew more slowly and experienced higher mortality than trees occupied by the more mutualistic ant species. Thus, a change in the herbivore community resulted in a rapid breakdown of the ant–plant mutualism. Again, we can see from this example that realistic models of mutualistic interactions often depend on the inclusion of other species and the recognition of a balance between positive and negative effects.

Interactions may change from positive to negative

As we have seen above, both the physical and the biotic environment can affect the nature of the interaction between species and may change the overall sign of the interaction from negative to positive. In addition, the signs and strengths of interspecific interactions can change across life stages or in the presence of exotic species, as illustrated by the following examples.

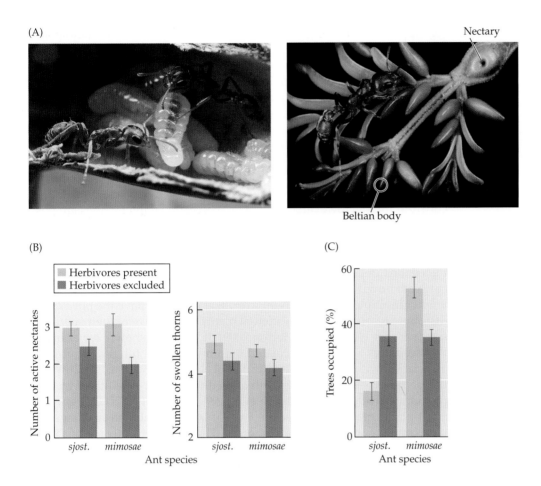

(A)

Nectary

Beltian body

(B)

Herbivores present
Herbivores excluded

Number of active nectaries

sjost.   mimosae

Number of swollen thorns

sjost.   mimosae

Ant species

(C)

Trees occupied (%)

sjost.   mimosae

Ant species

NURSE PLANTS   Plant recruitment in harsh environments is often facilitated by the presence of **nurse plants**: species whose canopies provide a protective microclimate promoting the germination and growth of other species. Particularly dramatic examples of the "nurse plant syndrome" can be seen in deserts, where trees such as paloverde (*Cercidium* spp.), mesquite (*Prosopis* spp.), and ironwood (*Olneya tesota*) shade the substrate, reduce water loss by evaporation, and/or increase soil fertility, thereby increasing seed germination and seedling survival rates for other plant species (see Shreve 1951; Hutto et al. 1986; Tewksbury and Lloyd 2001). However, the net benefit of association with a nurse plant may be positive or negative depending on the environmental context as well as the life stage of the "nursed" plant.

Studies have shown that nurse plants provide greater benefits to individuals in more water-stressed (xeric) environments (e.g., Franco and Nobel 1989). In addition, nurse plants have been shown to have the greatest impact on plant community structure and species richness when water stress is high (Tewksbury and Lloyd 2001). Other studies, however, have shown that temporal variation in the extent of drought stress did not change the beneficial impacts of nurse plants (e.g., Tielbörger and Kadmon 2000). Finally, the sign of the interaction may change over an individual's lifetime

(Schupp 1995; Callaway and Walker 1997; Miriti 2006). Individual plants may benefit from their interaction with nurse plants when they are young and small, but face increased competition with neighbors of the nurse plant species when older and larger; in other words, "the best place for a seedling to grow may not be the best place for a sapling to reach maturity" (Nuñez et al. 2009: 1067). Such ontogenetic shifts in the type and strength of species interactions are particularly common in size-structured or stage-structured species (Yang and Rudolf 2010).

MYCORRHIZAL ASSOCIATIONS   Earlier in this chapter, we discussed the fact that most plant species are engaged in symbiotic relationships with mycorrhizal fungi. However, not all plant–mycorrhizal interactions are positive, and the sign of the interaction may become neutral or negative under high-nutrient conditions (Johnson et al. 1997; Smith et al. 2009). Thus, the plant–mycorrhizal interaction provides another important illustration of the context dependence of facilitative interactions. Moreover, many mycorrhizal fungi are not host-specific, and these fungi may link the root systems of multiple plants. Thus, mycorrhizal networks have the potential to distribute resources among different plants irrespective of their species or size (van der Heijden and Horton 2009). It has been shown, for example, that seedlings are often inoculated with mycorrhizal fungi through their association with adult plants (Read et al. 1976; Dickie et al. 2002). When this occurs, the survival or growth of seedlings may be facilitated by their proximity to adults (e.g., Horton et al. 1999; Dickie et al. 2002). However, adult plants may also reduce the level of light available to seedlings and may compete with them for resources. Thus, the net outcome of the interaction between seedlings, adults, and their shared mycorrhizal fungi depends on the balance of positive and negative effects, and this balance may change spatially and with ontogeny (Dickie et al. 2005; van der Heijden and Horton 2009; Booth and Hoeksema 2010). In general, positive seedling–adult interactions shift to negative effects as seedlings mature, grow and compete with their neighbors.

POLLINATOR FACILITATION   It has been estimated that about 80% of all plant species are pollinated by animals. Plants clearly benefit from this mutualism. Less clear, however, is whether plants might facilitate each others' reproductive success by increasing pollinator visitation. For example, if one plant species attracts pollinators to the neighborhood of a second plant species, one or both of the species may benefit through increased seed set and higher reproductive rates (Schemske 1981; Rathcke 1988). Of course, neighboring plant species may also compete for pollinators or other resources, so the balance of their interaction may be positive or negative.

Population models show that pollinator facilitation is likely to occur only if a pollinator's visitation rate accelerates (at least initially) as the total number of flowers within a patch increases (i.e., the pollinator has a sigmoidal or type III functional response; see Chapter 5). Facilitation is unlikely when pollinators display a saturating (type II) functional response (Feldman et al. 2004). Thus, while facilitation between plant species is possible via pollinator sharing, the conditions for this to occur seem rather

Figure 9.6 Results of a meta-analysis showed that the identity of a neighboring plant species (exotic versus native) affects the pollination and reproductive success of a focal plant species. The graphs show the mean effect size (Hedges' *d*) and 95% confidence intervals for the effect of exotic or native neighboring plant species on pollinator visitation frequency (A) and reproductive success (B) of focal plant species. Exotic neighbors had a significant negative effect on both pollinator visitation and reproductive success (95% confidence intervals do not include zero mean effect size), whereas native neighbors had no significant overall effect (95% confidence intervals include zero mean effect size). Sample sizes are shown in parentheses. (From Morales and Traveset 2009.)

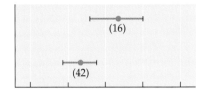

(A) Pollinator visitation frequency

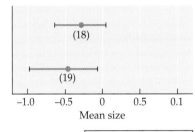

(B) Reproductive success

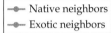

restrictive. Empirical studies looking at the effects of heterospecific floral density on pollinator visitation of a focal species have found positive effects (e.g., Laverty 1992; Ghazoul 2006; Hegland et al. 2009), negative effects (e.g., Thomson 1978; Campbell and Motten 1985), and no effects (e.g., Campbell and Motten 1985; Caruso 2002).

In a recent meta-analysis, Morales and Traveset (2009) examined the evidence for pollinator facilitation and competition for pollinators between native plant species and between native and exotic plant species in a total of 40 studies involving 57 focal species. They found that most shared pollinator interactions between native species had no effect on rates of pollinator visitation or plant reproductive success. However, exotic species had significant negative effects on pollinator visits to native species and on native species' reproductive success (**Figure 9.6**). Morales and Traveset (2009) suggest that exotic species are superior and more attractive competitors for pollinators than native co-flowering species on a per capita basis.

Combining positive and negative effects

Although the consequences of positive interactions for species diversity and community structure are still poorly understood, there have been some recent steps toward developing a general theory of species interactions that includes a combination of positive and negative interactions. In a notable example, Gross (2008) modified the standard consumer–resource competition model developed to study interspecific competition (see Chapter 7; Tilman 1982; Grover 1997) to include positive interspecific effects. In Gross's model, per capita mortality or the maintenance requirements of a resource competitor may be reduced by the presence of another competing species. Such positive interactions incorporate the general features of facilitation, in which one species ameliorates a stressful environment for a second species. In a two-species model, Gross (2008) found that both species could coexist on a single resource if facilitation by the superior resource competitor reduced the mortality rate of the inferior resource competitor enough to allow the "inferior" species to invade a community consisting solely of the superior competitor. Positive and negative effects might also be incorporated into the growth rate terms. The important point is that species may have both positive and negative effects on each other, and that considering these effects separately provides additional mechanisms for coexistence that are absent in models of the Lotka–Volterra type (in which the average effect of an interacting species is summarized in a single term, the interaction coefficient, α).

Gross's theoretical analysis shows how positive interactions could allow species coexistence and promote diversity even when they only partially counteract exploitative competition. Two empirical studies illustrate additional pathways by which positive interactions may mediate the effects of interspecific competition and thereby increase the probability that competing species will coexist. Bever (1999, 2002) showed that two competing plant species (*Plantago lanceolata* and *Panicum sphaerocarpon*) may coexist if the mutualistic mycorrhizal fungal species that confers the greatest growth benefit to these plants associates preferentially with *Panicum*—the otherwise inferior competitor. This result has interesting parallels to how shared predation may promote species coexistence if the predator selectively preys on the superior resource competitor (see Figure 6.6).

In the second study, Schmitt and Holbrook (2003) found an interesting wrinkle in the well-known mutualism between sea anemones and anemonefish, showing how this interaction could promote coexistence between competing fishes. Anemonefish of the genera *Amphiprion* and *Premnas* find shelter in the stinging tentacles of sea anemones (**Figure 9.7**); in return, the anemonefish drive off specialized anemone predators. Schmitt and Holbrook found that sea anemones grew faster and reached larger sizes when anemonefish were present. This increase in anemone size allowed a second fish species, the damselfish (*Dascyllus trimaculatus*), to occupy an anemone along with its superior competitor, the more aggressive anemonefish. Coexistence occurred in large anemones, but not in small anemones, because social interactions between anemonefish limit them to two individuals per anemone, regardless of anemone size. Thus, the intensity of intraspecific

Figure 9.7 Sea anemones grow faster and reach larger sizes when anemonefish—their mutualistic partners—are present. (Photograph © Jodi Jacobson/Shutterstock.)

competition between anemonefish remains constant regardless of anemone size; however, the intensity of interspecific competition between damselfish and anemonefish decreases with anemone size. Recall from Chapter 7 that species coexistence is promoted when the ratio of per capita intraspecific to interspecific competitive effects increases. Schmitt and Holbrook suggested that this mechanism may be prevalent in a variety of systems and thus adds another pathway by which mutualism may enhance diversity.

The stress gradient hypothesis for plant facilitation

If we assume that species interactions often represent a balance between positive and negative effects, we can ask what factors might tip the balance so that we see a preponderance of either net positive or net negative effects within a community. In plants, facilitation appears to be more important, or to occur more frequently, in harsh environments than in mild environments. **Primary succession**—the colonization of newly exposed substrates such as may be made available by receding glaciers or shoreline, or by volcanic activity—has long been viewed by plant ecologists as a facilitative process (Clements 1916; Connell and Slatyer 1977). During primary succession, early colonizing plant species may ameliorate the harsh physical conditions of the new habitat by reducing water stress or wind intensity or by promoting the accumulation of organic material, thereby allowing less hardy plant species to colonize (Chapin et al. 1994; Walker and del Moral 2003). Facilitation has also been frequently observed in harsh environments that are not undergoing succession, such as arid, alpine, and salt marsh environments. This general observation has led to a variety of conceptual models of facilitation in plant communities that can be collectively termed the **stress gradient hypothesis** (**SGH**) (Bertness and Callaway 1994; Callaway and Walker 1997; Brooker and Callaghan 1998; Callaway 2007; Butterfield 2009).

Simply put, the stress gradient hypothesis postulates that the relative importance of competitive versus facilitative interactions among plant species changes along a gradient of environmental harshness (**Figure 9.8**). Competitive interactions should dominate in benign environments, whereas facilitative interactions should become more important in stressful environments. One of the best empirical examples supporting the predictions of the SGH is the work of Callaway et al. (2002), in which a team of researchers from around the world compared the responses of alpine plant species to the removal of neighboring plant species at low elevations (subalpine vegetation) and at high elevations (alpine vegetation) at eleven different

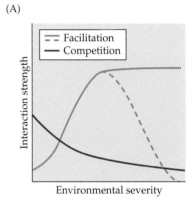

(A)

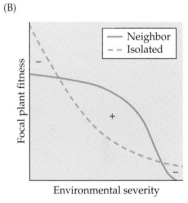

(B)

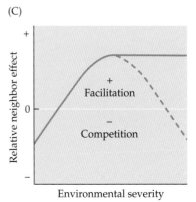

(C)

Figure 9.8 Predictions of the stress gradient hypothesis. (A) According to the SGH, the strength of interspecific competition decreases and the strength of facilitation increases with the severity of the physical environment. In very harsh environments, however, the strength of facilitation may decline (dashed line). (B) As a result of these relationships, the fitness of the focal plant species is expected to decline less in the presence of neighboring plants as environmental severity increases (C) The relative neighbor effect is expected to switch from negative (indicating competition) to positive (indicating facilitation) with increased environmental severity. (After Butterfield 2009.)

Figure 9.9   Empirical support for the stress gradient hypothesis. Removing neighboring plant species has a detrimental effect on alpine plant species located in harsh alpine habitats, but a positive effect on the same species located in milder, subalpine habitats. (A) Worldwide distribution of the experimental sites. At each experimental site, all neighboring plants within 10 cm of a focal plant were removed and two responses (survival and flowering) of the focal individuals were measured. RNE is the relative neighbor effect, where positive responses indicate facilitation (blue histograms) and negative responses indicate competition (red histograms). (B) The sign of the RNE changed from facilitation (positive RNE) to competition (negative RNE) as the harshness of the environment decreased (i.e., as June temperatures became warmer). (C) Neighbor removal had a negative effect on focal plant survival and reproduction (i.e., flower/fruit production) in high-elevation plots, but a positive effect at lower elevations. (After Callaway et al. 2002.)

mountainous sites across the globe (**Figure 9.9**). High-elevation sites were harsher environments than low-elevation sites. In addition, the shift from competition to facilitation was inversely correlated with one measure of environmental harshness, maximum summer temperature (Figure 9.9B). The pattern of response was very consistent at all eleven experimental sites: removing neighbors at low elevations resulted in an increase in target plant survival and in the proportion of target plants that flowered (evidence for interspecific competition), whereas removing neighbors at high elevations generally led to a decrease in target plant survival and flowering (evidence for facilitation; Figure 9.9C).

Although Callaway et al. (2002) and others have provided experimental evidence in support of the stress gradient hypothesis, a number of studies have found contradictory results (e.g., Tielbörger and Kadmon 2000; Pennings et al. 2003; Maestre and Cortina 2004). Moreover, a meta-analysis of field and common garden studies (Maestre et al. 2005) concluded that facilitation does not appear to increase steadily in importance with abiotic stress. Maestre and Cortina (2004) and Butterfield (2009), working in arid systems, have proposed that the facilitative effects of neighbors are likely to be a peaked function of environmental harshness when examined along the full gradient of environmental stress (see Figure 9.8). Butterfield (2009: 1194) notes that

> [R]egardless of the mechanism by which plants ameliorate environmental severity, there has to be some set of environmental conditions in which that amelioration is optimal, with reduced ameliorative effects in more or less severe environments. Evidence suggests that plants in many of the most extreme alpine environments approach or reach just beyond this optimal buffering ability (Callaway et al. 2002), whereas perennial plants in the Sonoran Desert lie beyond this optimal (Butterfield et al. 2010).

It remains to be seen whether the stress gradient hypothesis (including its proposed modifications; Maestre et al. 2009) describes a general feature of plant communities or not. Moreover, as theoretical models have shown, determining whether the net effect of a species interaction is positive or

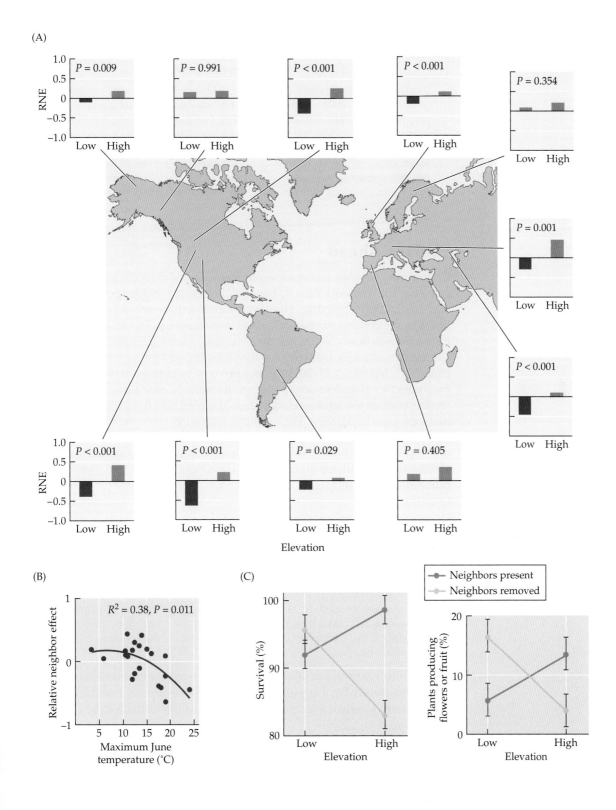

negative tells only part of the story (Gross 2008). What may be most important to species coexistence and species diversity is the joint action of positive and negative effects. As Gross (2008: 935) notes,

> Positive interactions can also drive coexistence when they only partially counteract exploitative competition. This latter effect is more subtle, but may be more common in nature. The fact that these more subtle positive interactions go undetected if interactions are classified only by their net effect emphasizes the need to understand all the components of an interaction between species (Callaway and Walker 1997). Even in communities that appear to be predominantly structured by resource competition, positive interactions may provide a key to explaining how many different species coexist.

Looking Ahead

Community ecology has made significant strides in recent years in understanding the consequences of beneficial interactions for species diversity and community structure. However, most of these advances have taken place in the context of studying local communities. The role of beneficial interactions in determining macroecological patterns, regional diversity, and species dynamics in metacommunities is essentially unknown (see Gouhier et al. 2011; Menge et al. 2011 for examples of how recruitment facilitation may affect local and regional species diversity in marine systems). Many studies suggest intriguing relationships between beneficial interactions and macroecological patterns. For example, ecologists since the time of Darwin have noted what seems to be a preponderance of mutualistic interactions in the tropics. May (1981: 98), based on his mathematical examination of the stability of mutualisms in different environments, speculated that obligate mutualisms are more likely to be favored in relatively stable environments (e.g., the tropics):

> These dynamical considerations [stability analyses] may go some way towards explaining why obligate mutualisms of this sort are less prominent in temperate and boreal ecosystems than in tropical ones. A good survey of empirical evidence bearing on this point is given in Farnworth and Golley (1974: 29–31), where it is observed that not only are there no obligate ant–plant mutualisms north of 24 degrees, no nectarivorous or frugivorous bats north of 33 degrees, no orchid bees north of 24 degrees in America, but also within the tropics mutualistic interactions are more prevalent in the warm, wet evergreen forests than in the cooler and more seasonal habitats.

Whether mutualisms are more prevalent in the tropics and whether beneficial interactions play a significant role in the development of the latitudinal diversity gradient are fascinating questions that remain to be explored (Schemske et al. 2009).

As we have seen in this and preceding chapters, interactions between species across trophic levels can result in positive indirect effects (e.g., predators may indirectly benefit plants by reducing the abundance or feeding rate of herbivores). In the following chapters, we will explore the consequences of indirect effects (positive and negative) in food chains and food webs, addressing questions of species diversity, stability, cascading effects, and the regulation of trophic level biomass. In addition to studying indirect effects in traditional food webs (which describe who eats whom within a community), ecologists are making significant progress in incorporating beneficial species interactions (mutualisms) directly into ecological networks (e.g., plant–pollinator networks, plant–seed disperser networks). In the next chapter, we will examine the properties of mutualistic networks and compare the properties of these networks of beneficial species interactions to ecological networks based upon feeding relationships (food webs).

Summary

1. Mutualism may be defined as a reciprocally positive interaction between species. Facilitation may be defined as an interaction in which the presence of one species alters the environment in a way that enhances the growth, survival, or reproduction of a second, neighboring species.

2. The evolution of mutualism and facilitation depends on the relative costs and benefits of these interactions to the participating species. Mutualistic interactions may best be viewed as reciprocal exploitation between species that nonetheless provides net benefits to each partner.

3. The consequences of beneficial interactions for the diversity and functioning of communities are still poorly understood. This failure to incorporate positive interactions into community theory is not simply the result of a historical bias in favor of studying competition and predation, but also stems from the complexity of these interactions.

4. The Lotka–Volterra competition model can be written as a model for positive species interactions simply by changing the sign of the interspecific effect from negative to positive; however, this model is not realistic because it can result in both species' populations growing without bound.

5. Recent models have shown the mutualisms may evolve from parasitic or competitive interactions. Pairs of coexisting competitors may evolve toward a mutualistic "trading partnership" if the two competing species require the same resources, and if each species consumes in excess the resource that least limits its growth, then trades this excess resource to a partner whose growth is more limited by that resource.

6. A change in the environment may affect the costs and benefits of an interspecific interaction. Realistic modeling of mutualistic interactions often depends on the inclusion of other species and on the recognition of a balance between positive and negative effects.

7. The type and strength of an interaction between species may be context-dependent. The extent to which it is beneficial versus antagonistic may change with the harshness of the abiotic environment, the presence of other species (competitors or predators), and the ontogeny or life stage of the interacting individuals.

8. Although the consequences of positive interactions for species diversity and community structure are still poorly understood, there have been some recent steps toward developing a general theory of species interactions that includes a combination of positive and negative interactions. One model shows that two species could coexist on a single resource if facilitation by the superior competitor reduced the mortality rate of the inferior competitor enough to allow it to invade a community consisting solely of the superior competitor.

9. The stress gradient hypothesis postulates that the relative importance of competitive and facilitative interactions between plant species changes along a gradient of environmental harshness: competitive interactions dominate in benign environments, whereas facilitative interactions become more important in stressful environments.

Putting the Pieces Together

Food Webs and
Ecological Networks

10 Species Interactions in Ecological Networks

It is interesting to contemplate a tangled bank, clothed with many plants of many kinds, with birds singing on the bushes, with various insects flitting about, and with worms crawling through the damp earth, and to reflect that these elaborately constructed forms, so different from each other, and dependent upon each other in so complex a manner, have all been produced by laws acting around us. There is grandeur in this view of life ... from so simple a beginning endless forms most beautiful and most wonderful have been, and are being evolved.

Charles Darwin, 1859: 489–490

Ever since Darwin contemplated the interdependence of the denizens of his "entangled bank," ecologists have sought to understand how this seemingly bewildering complexity can persist in nature.

Thomas Ings et al., 2009: 254

Ecological networks attempt to summarize the multitude of potential interactions between species within a community by representing those species as nodes in the network and using links between nodes to represent interactions between species (**Figure 10.1**). Traditionally, ecologists have approached the description of ecological networks from the perspective of consumers and their resources—basically, by constructing a diagram showing who eats whom within a community, which is known as a food web. More recently, ecologists have expanded their studies of ecological networks to include other types of species interactions, such as mutualistic webs (e.g., Pascual and Dunne 2006; Ings et al. 2009; Thébault and Fontaine 2010) and host–parasitoid webs (e.g., Hawkins 1992). Food webs and host–parasitoid webs describe antagonistic interactions, in which one species benefits and another loses. Mutualistic webs describe interactions that benefit both species (e.g., plant–pollinator networks, plant–seed disperser networks).

Food webs focus on "typical" predator–prey interactions, where consumers are generally larger than their prey. Host–parasitoid webs concentrate on

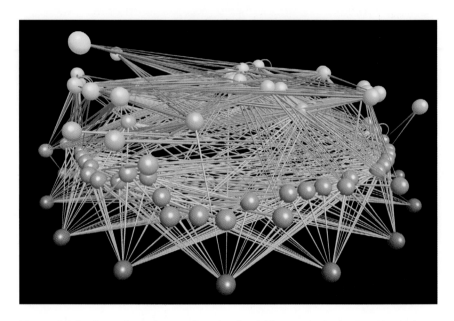

Figure 10.1 An ecological network, or food web, for species in the East River Valley, near Crested Butte, Colorado (based on the research of Neo Martinez and Brett Harvey). Node colors represent trophic levels: red nodes represent basal species, such as plants and detritus; orange nodes represent intermediate consumer species; yellow nodes represent top consumers (predators). Links characterize the interaction between nodes, with the link being thicker at the consumer end and thinner at the resource end. (Image produced with FoodWeb3D, written by R. J. Williams and provided by the Pacific Ecoinformatics and Computational Ecology Lab: www.foodwebs.org.; Yoon et al. 2004.)

a special type of "predator–prey" feeding relationship, in which parasitoids lay eggs within their host; the larvae that hatch from those eggs then consume and kill the host. Parasites, which feed on but may not kill their hosts, are conspicuously absent from most studies of ecological networks, although the need to better incorporate parasites more fully into webs that describe trophic relations is well recognized (Marcogliese and Cone 1997; Lafferty et al. 2006, 2008). Similarly, interactions between plants and herbivores (which often result in the partial consumption but not the death of an individual plant) do not fit neatly within the traditional food web approach, and the systematic study of herbivory networks is relatively recent (e.g., Melián et al. 2009; Thébault and Fontaine 2010; Fontaine et al. 2011).

The study of ecological networks is an active and fast-moving field, fueled in part by analytical techniques developed in the larger field of "network science," which includes social, biological, and technological networks (e.g., the Internet), and by the ability to access information on ecological networks through online databases (e.g., the Interaction Web Database at http://www.nceas.ucsb.edu/interactionweb/index.html). In this chapter, we will explore a variety of different ecological networks and their properties, beginning with the best-studied networks—food webs.

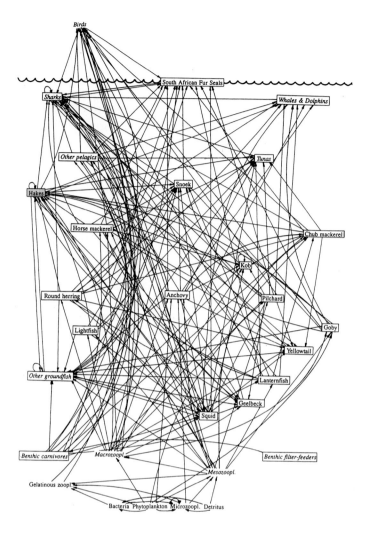

Figure 10.2 A food web for the Benguela marine ecosystem off the coast of South Africa. Arrows point from a resource to a consumer. Hake, the commercial fish of interest in this food web, are located near the upper left. (From Yodzis 2001.)

Food Webs

Although simple food web diagrams appeared as early as the late 1800s (e.g., Camerano 1880; Shelford 1913), it was Elton (1927) who introduced the concept of the food web to the mainstream of ecology (see discussion in Leibold and Wootton 2001). **Figure 10.2** is a typical food web diagram—in this case, for the Benguela marine ecosystem off the southwestern coast of Africa (Yodzis 2001). This diagram is an example of a connectedness web (Paine 1980) or structural web, in which the trophic links between resources and consumers (i.e., who eats whom) are illustrated by arrows pointing toward the consumer.

Connectedness webs

Connectedness webs show the presence of an interaction between species, but they do not specify the strength of the interaction. As in most ecosystems, the predators in Figure 10.2 feed on multiple species of prey, and the prey species likewise have many predators. Note too that consumers near the top of the food web are identified by species or feeding guilds, whereas consumers and prey lower in the food web tend to be aggregated into broad functional groups (e.g., "benthic filter-feeders," or even "bacteria" and "phytoplankton" at the base of the food web in Figure 10.2). This is typical of most food webs, which tend to be better resolved at the top than at the bottom, in part because species richness is greater at lower trophic levels and in part because species at lower trophic levels tend to be small, difficult to identify, and have feeding relationships that are hard to quantify (Dunne 2006).

Ecologists often separate food webs into those in which the basal trophic level is composed of primary producers ("green food webs") and those in which the basal trophic level is composed of detritus ("brown food webs"). Both primary production and detritus are sources of energy in most communities, and there have been some efforts to link the "green" and "brown" portions of food webs (e.g., Rooney et al. 2006; Butler et al. 2008), although this approach is still uncommon. Rooney and McCann (2012) suggest that the "green" and "brown" portions of food webs may represent "fast" and "slow" energy channels, respectively, and that these different energy channels play an important role in determining the stability of food webs. We will return to the question of network structure and stability in more detail later in this chapter.

The Benguela food web shown in Figure 10.2 was analyzed extensively by the late Peter Yodzis, a theoretical ecologist at Guelph University. Yodzis was asked by the South African government to address whether culling (i.e., killing) Cape fur seals (*Arctocephalus pusillus*), which can consume up to their own body weight daily in hake (*Merluccius capensis*; a delicacy in Europe), would increase the fish harvest for humans. The annual consumption of commercial fish by seals is about 2 million tons, which equals the annual catch by fishermen. When we look at the food web in Figure 10.2, it is clear that seals have a direct, negative effect on the abundance of hake. However, seals also eat hake predators and hake competitors. Therefore, as Yodzis (2001) noted, the combined impact of the direct and indirect effects of seals on hake is not obvious. Yodzis (1998) calculated that there are over 28 million potential interaction pathways between seals and hake. His tentative conclusion was that culling seals was unlikely to benefit the hake fishery because of the multitude of indirect pathways between seals and hake—some of which are positive and some negative. We will return to the question of compounding indirect effects later in this chapter.

Ecologists have many examples of connectedness webs because the data needed to generate them (i.e., presence or absence of interactions) are relatively easy to collect (although there are significant differences in the quality of published food webs and in whether they are based on observations of individuals or species; Woodward et al. 2010). Thus, ecologists have spent considerable time and effort in analyzing the properties of connectedness webs. Initial work on connectedness food webs suggested some general topological patterns, and the structural properties of networks are known as **network topology**. (For reviews of the early literature, see Cohen et al. 1990; Pimm et al. 1991; Martinez 1992.) For example, it was once thought that the proportions of species found at different trophic levels (i.e., top predators, consumers, producers) remained relatively constant across webs of different species richnesses, and that the ratio of the total number of links (L) to the total number of species (S) was roughly constant at a value of about 2. Hence, the average number of species with which a given species interacts, $n = 2L/S$, was estimated to be about 4 and appeared to be independent of the total number of species in the web (Cohen et al. 1990; Polis 1991). More recent work, however, using more fully resolved food webs, suggests that the above generalizations are not all that general (Pascual and Dunne 2006; Ings et al. 2009). Other topological properties of food webs (and of other ecological networks), however, appear more robust. We will discuss three of these properties—connectance, nestedness, and modularity—in detail later in this chapter when we compare the structures of trophic and mutualistic networks.

Current interest in network topology is becoming focused on the *mechanisms that produce the structural patterns* of food webs as well as on those patterns themselves (Pascual and Dunne 2006). Williams and Martinez (2000), for example, showed that many of the structural properties of empirical food webs (i.e., those based on actual observations)—including generality (the number of species that are prey for a particular species), vulnerability (the number of predators that prey on a particular species), and average food chain length—can be predicted from a simple model specifying that predators consume prey species falling within a contiguous range of "niche values" (e.g., body sizes). In other words, large species consume smaller species that are within some particular size range. Although this "niche model" and related approaches can predict many of the structural properties of real food webs, these models are still largely phenomenological because the rules specifying consumer diets are not based on explicit ecological or evolutionary processes (Stouffer 2010). More mechanistic approaches to understanding food web structure have been proposed. For example, Loeuille and Loreau (2005) and Rossberg et al. (2006) examined the evolutionary development of food webs, and Beckerman et al. (2006) and Petchey et al. (2008) used optimal foraging theory to predict predator diets and trophic links. Despite their differences, the phenomenological and mechanistic approaches to understanding food web structure share some common features, most notably a strong focus on body size (mass) as a key species trait (Stouffer 2010).

Although relatively easy to construct, connectedness webs miss a lot of biological reality. In particular, they ignore the amount of energy (or biomass) transferred between nodes within the food web, and they are mute about the strength of interactions between species. Paine (1980) suggested two other ways to construct food webs that addressed these deficiencies: energy flow webs and functional webs (**Figure 10.3**).

Energy flow webs

An energy flow web measures the amount of energy (biomass) moving between species within a food web (**Figure 10.4**). Constructing an energy flow web for a community involves much more effort than building a connectedness web, and as a consequence, high-quality energy flow webs for complex communities are relatively rare (see Baird and Milne 1981; Baird and Ulanowicz 1989; Raffaelli and Hall 1996; and Benke et al. 2001 for examples). Implicit in the energy flow approach to food webs is the idea

(A)

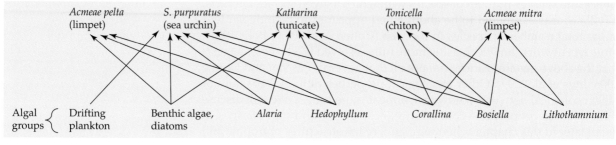

(B)

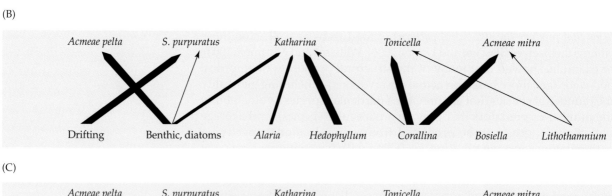

(C)

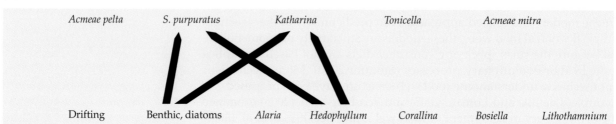

Figure 10.3   Three distinct approaches to constructing food webs, illustrated for the same set of species/functional groups found in the marine intertidal zone of Tatoosh Island, off the coast of Washington State. Arrows point from algal groups to consumers of algae (grazers). (A) A connectedness web, based on observations of who eats whom. (B) An energy flow web, based on estimates of biomass consumption (plus values from the literature). Arrow thicknesses correspond to different amounts of energy flow. (C) A functional web, based on species removal experiments. Arrows connect strongly interacting species. (From Paine 1980.)

that there is a relationship between the amount of energy flowing through a pathway and the importance of that pathway to community dynamics. This is true for some community properties, such as the bioaccumulation of pollutants like PCBs or mercury, which closely mirrors the flow of energy between species or trophic levels (Rasmussen et al. 1990; Vander Zanden and Rasmussen 1996). However, energy flow has been shown to be a surprisingly poor predictor of the strength of interactions between species or of the impact of removing a particular species from a community (e.g., Paine 1980; Berlow 1999).

When I was a graduate student in Earl Werner's lab in the late 1970s, the limnologists in Robert Wetzel's lab (located one floor below us at the Kellogg Biological Station) would sometimes needle us about the relevance of our research, suggesting that the fishes we studied (bluegill and bass) might be interesting, but probably didn't contribute much to the functioning of lake ecosystems because only 1% of the primary production of lakes ended up in the bass population. Ironically, less than 10 years later, pioneering studies by Power et al. (1985) and Carpenter et al. (1985) showed that bass may function as keystone species in streams and lakes, and that the strong

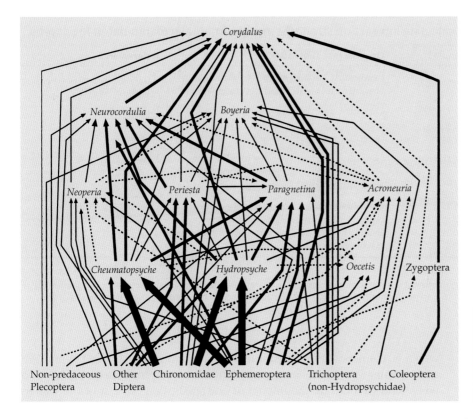

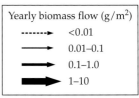

Figure 10.4 An energy flow web for the major predaceous invertebrates living on woody debris in the Ogeechee River, Georgia. Arrows point from prey to predators; arrow thicknesses indicate the amount of energy (biomass) flow. Primary consumers (non-predators) are in the bottom row. Predators are arranged in a hierarchy according to their distance, in food chain links, from the primary consumers. *Cheumatopsyche* and *Hydropsyche* are omnivores, and their ingestion of basal food resources is not shown. (From Benke et al. 2001.)

cascading effects of these top predators on species at lower trophic levels could have major impacts on community composition, primary production, and even water quality (see also Mittelbach et al. 1995, 2006). We will talk much more about keystone species and cascading effects later in this chapter and in the next, but for now the take-home message is this: energy flow is often a poor predictor of the impact that a particular species or group of species may have within a food web. This recognition leads us to functional webs, the third type of food web suggested by Paine (1980) .

Functional webs

A functional web (or interaction web) shows the strength of the interactions between species within a community, implicitly recognizing that not all species and interactions are equally important. Paine (1980) proposed that we measure interaction strengths by experimentally removing species from the community and looking at the responses of the remaining species. He did this for a portion of the algal/grazer community found in the rocky intertidal of Tatoosh Island, on the outer coast of Washington State, and found "few strong interactions embedded in a majority of negligible effects" (Paine 1992; see also Figure 10.3).

Since Paine's pioneering work, ecologists have developed a variety of ways to measure interaction strengths in theoretical and real food webs (Berlow et al. 2004). Measures of interaction strength in model food webs tend to focus on the individual interactions between species pairs within a

TABLE 10.1 Indices used to calculate interaction strengths in experimentally manipulated food webs

Index	Formula[a]	Original intent of index	References
Raw difference	$(N - D)/Y$	Commonly used in studies to show absolute, untransformed treatment effects	
Paine's Index (PI)	$(N - D)/DY$	Quantify effect of consumer on competitive dominant resource that has potential to form monoculture	Paine 1992
Community Importance (CI)	$(N - D)/Np_Y$	Quantify effect of a species relative to its abundance (i.e., distinguish "keystones" from "dominants"	Power et al. 1996
Dynamic Index	$\ln(N/D)/Yt$	Quantify an effect size that is theoretically equivalent to the coefficient of interaction strength in the discrete-time version of the Lotka–Volterra model	Osenberg & Mittelbach 1996; Wootton 1997

Source: From Berlow et al. 1999, *Ecology* 80: 2206–2224.

[a] Where N is the abundance of prey in the treatment with predators present (i.e., Normal condition); D is the abundance of prey in the treatment where predators are "Deleted" (i.e., experimentally removed); Y is the abundance of the predator; p_Y is the proportional abundance of the predator; and t is time.

web (i.e., elements of the community or interaction matrix; see Levins 1968 and Chapter 7), whereas measures of interaction strength in empirical food webs tend to focus on the impact of one species on the rest of the web as measured by removal experiments (Laska and Wootton 1998; O'Gorman and Emmerson 2009). As noted by Berlow et al. (2004: 587), "In the first category individual interaction strengths are independent of the network, in the second category they are in theory inseparable from their network context" (see also Emmerson and Raffaelli 2004).

Table 10.1 lists four indices commonly used to measure interaction strength in empirical food webs, including the metric used by Paine (1992). All of these metrics require that responses be measured over relatively short time intervals; otherwise, indirect effects and density-dependent feedbacks may occur and influence the estimates of interaction strength between directly interacting species (Bender et al. 1987; Laska and Wootton 1998). One of the measures listed in Table 10.1, the "dynamic index" (also called the "log response ratio" or "log-ratio metric"), compares the log of the ratio of prey abundance "with" versus "without" predators; it is equivalent to the coefficients of interaction between species in the Lotka–Volterra predator–prey model (Navarrete and Menge 1996; Osenberg et al. 1997; Laska and Wootton 1998). This metric provides a useful link between theoretical and empirical estimates of interaction strength in food webs; however, it does so only for the simplest predator–prey model, assuming a linear functional response and no predator interference (Berlow et al. 2004). Bascompte et al. (2005) proposed an additional measure of interaction strength in food webs using observational rather than experimental data, where interaction strength is measured as

$$IS = \frac{(Q/B)_j \times DC_{ij}}{B_i} \qquad \text{Equation 10.1}$$

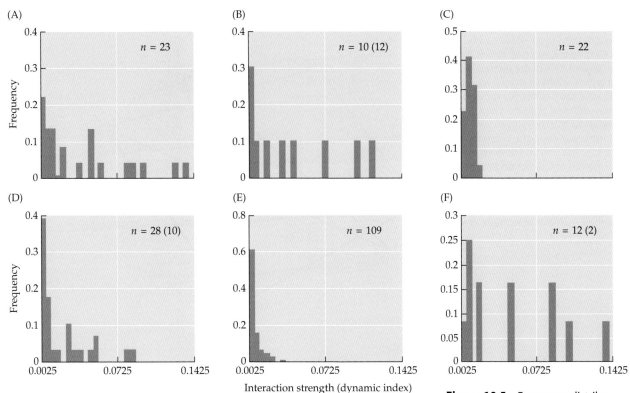

Figure 10.5 Frequency distributions of the absolute values of per capita effects of consumers on their prey. Interaction strengths between consumers and prey were calculated as ln(C/E × P), where C and P are the abundances of prey and consumers, respectively, in control treatments and E is the prey abundance in consumer removal treatments (Osenberg et al. 1997; Wootton 1997). The distributions were calculated using raw data obtained from the studies detailed in (A) Paine 1992, (B) Raffaelli and Hall 1996, (C) Sala and Graham 2002, (D) Wootton 1997, (E) Fagan and Hurd 1994, and (F) Levitan 1987. All distributions are skewed toward many weak interactions and few strong interactions, a pattern that is consistent across all systems studied. Numbers on each graph represent sample size (number of interactions); numbers in parentheses indicate the number of interaction strengths found above the scale of the graphs. (After Wootton and Emmerson 2005.)

where $(Q/B)_j$ is the number of times a population of predator j consumes its own weight per day, DC_{ij} is the proportion of prey i in the diet of predator j, and B_i is biomass of prey i.

Despite the challenges of measuring interaction strengths, a consistent feature has emerged from many different studies of both theoretical and empirical food webs: *most food webs contain a few strong and many weak links*. For example, Wootton and Emmerson (2005) analyzed the frequency distributions of the absolute value of per capita effects of consumers on their prey (using a form of the dynamic index; see Table 10.1) using data collected from experimental manipulations of a number of different food webs. They found that interaction strengths in the communities studied were not normally distributed; rather, strong interactions occurred between relatively few species and weak interactions occurred between most species (**Figure 10.5**). A very similar result was found by Bascompte et al. (2005; **Figure 10.6**) and Rooney and McCann (2012). Theoretical studies show that weak interactions, which are overrepresented in nature, can be a powerful stabilizing force in food webs (McCann et al. 1998; Emmerson and Yearsley 2004; Rooney and McCann 2012). We will explore this idea further in our discussion of complexity and stability later in this chapter.

Keystone Species

The skewed distribution of interaction strengths in real food webs is also consistent with the observation that some species play extraordinarily large

Figure 10.6 Variation of interaction strengths in a real food web. (A) A random sample of 30% of the species and 11% of the interactions in a large Caribbean marine food web (249 total species/trophic groups). Arrows connect predators and their prey; arrow thickness is proportional to interaction strength. Loops represent cannibalism. (B) The frequency distribution of interaction strengths in the food web in A, as calculated by Equation 10.1. The solid line represents the best fit to a lognormal distribution. (After Bascompte et al. 2005.)

(A)

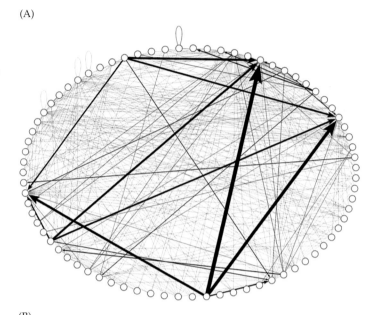

(B)

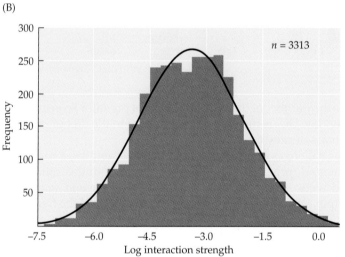

roles within communities. Paine (1969) called these keystone species—species whose effect on the community is disproportionately large relative to their abundance (**Figure 10.7**; Power et al. 1996). Classic examples of keystone species and their impacts include:

- Predation by *Pisaster* starfish increases species diversity by preventing the monopolization of space by mussels in the Pacific intertidal (Paine 1966, 1969).
- Sea otters limit the abundance of grazing sea urchins, thereby allowing kelp forests and their associated species to flourish (Estes and Palmisano 1974; also see Chapter 11).

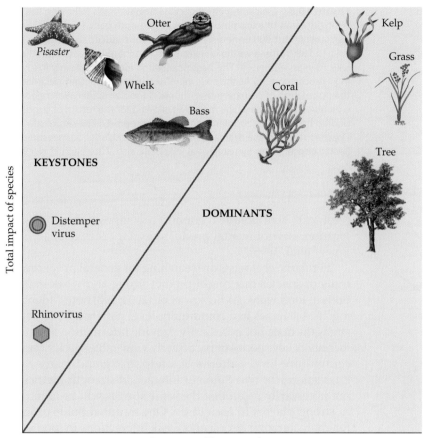

Figure 10.7 Keystone species (upper left) are defined as species whose impacts on the community are large relative to their biomass. Dominant species (upper right) are those that constitute a large fraction of a community's biomass and whose impacts are large but not disproportionate to their abundance. (After Power et al. 1996.)

- Piscivorous bass (*Micropterus*) control the abundance of small fishes in lakes (**Figure 10.8**), affecting a trophic cascade down through the grazer and algal trophic levels and ultimately affecting water clarity (Carpenter et al. 1985; Power et al. 1985; Mittelbach et al. 1995, 2006).

A longer list of likely keystone species and their mechanisms of action can be found in Power et al. (1996).

Identifying a keystone species is relatively easy after the species is lost or purposefully removed from an ecosystem; identifying keystones *a priori* is much harder. Ecologists have struggled to list the traits that might characterize keystone species. For example, a high feeding rate and a preference for consuming prey that are competitive dominants typify some keystone predators (Power et al. 1996). However, in many cases, whether a species acts as a keystone species or not depends on the context of the interaction

(A)

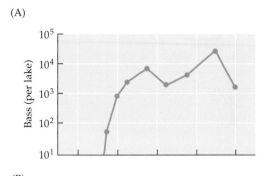

(B)

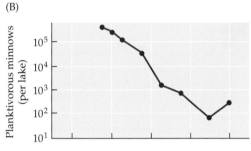

(C)

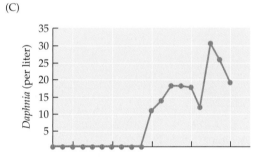

(D)

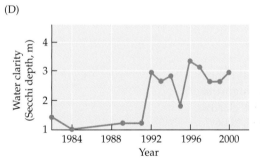

Figure 10.8 Cascading effects across trophic levels following the reintroduction of a top predator, the largemouth bass, to a small Michigan lake. About 600 fingerling bass were reintroduced to Wintergreen Lake in 1987, after a winterkill event had eliminated all bass from the lake 10 years earlier. (A) The bass population increased rapidly following introduction. (B) As the bass increased in numbers, they decimated the population of planktivorous minnows (golden shiner, *Notemigonus crysoleucas*). (C) Loss of the minnows allowed for the return of large-bodied herbivorous zooplankton (*Daphnia* spp.). (D) The return of *Daphnia* resulted in a dramatic increase in water clarity, as measured by Secchi depth. (After Symstad et al. 2003; see also Mittelbach et al. 1995.)

("context" here includes community composition, geographic location, environmental productivity, etc.; Menge et al. 1994; Power et al. 1996).

Similarly, ecologists are searching for general properties or traits of species that might predict strong and weak interactions in food webs. As Berlow et al. (2004: 590) note, "Identifying consistencies in a community-level pattern of interaction strengths does not necessarily provide information about the identity of key species or particularly vulnerable species. Similar community-level patterns of a few strong and many weak interactions for two different interaction strength metrics do not necessarily mean that the same species will be identified as strong players in each case." One trait that holds promise for characterizing strong and weak interactions in food webs is body size.

Body Size, Foraging Models, and Food Web Structure

Not surprisingly, Charles Elton (1927: 59) was one of the first to recognize the importance of predator and prey body sizes in structuring food web interactions: "A little consideration will show that size is the main reason underlying the existence of these foodchains, and that it [body size] explains many of the phenomena connected with the food-cycle [food web]." Species attributes that influence their interactions with other species often scale with body size; examples include metabolic rate, movement speed, rate of encounter, handling time, and feeding rate (Mittelbach 1981; Peters 1983; Brown et al. 2004; Woodward et al. 2005; McGill and Mittelbach 2006; see also Chapter 6). Thus, we might expect the relative sizes of predator and prey to go a long way toward determining who eats whom within food webs (the trophic links) and thus, potentially, the strength of ecological interactions.

Beckerman et al. (2006) and Petchey et al. (2008) incorporated size-specific handling times for predators, along with the size-based energy content

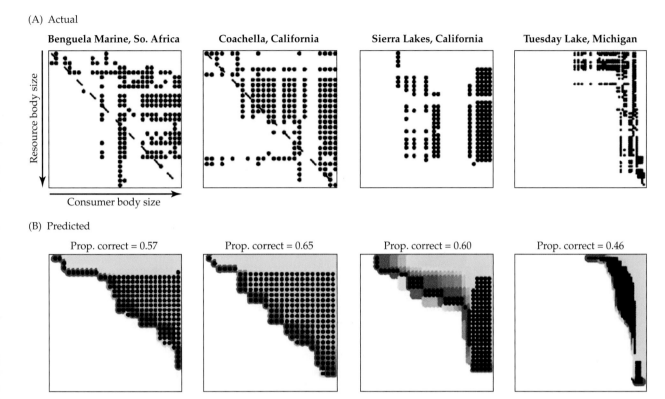

Figure 10.9   Size-based foraging relationships predict food web structure. (A) Trophic links from four real food webs in matrix format, with resources in rows and consumers in columns. Body size increases from left to right and top to bottom. A black dot indicates that the consumer in that column feeds on the resource in that row. (B) Trophic links predicted by a model of optimal foraging with parameters tuned to the data in A. Yellow to red indicates low to high resource profitability. Consumer diets always include the darker red (most profitable) resources and extend by different amounts into the yellow (less profitable) resources. (From Petchey et al. 2008.)

of prey, into a foraging model to predict trophic links within a food web. By modeling handling time as an increasing function of the ratio of prey size to predator size (assuming a minimum handling time; see Chapter 6), and by modeling prey energy content as a positive function of prey size, they correctly predicted up to 65% of the trophic links in four real-world food webs, based on the assumption that predators prefer to eat the most energetically rewarding prey (**Figure 10.9**). Their study thus makes an important stride toward linking foraging theory and species interactions. The link, however, is still incomplete, as Petchey et al. (2008) were not able to parameterize all components of the optimal foraging model independent of the data used to test the model, and Allesina (2011) showed that a phenomenological food web model performed as well as Petchey et al.'s more mechanistic model. However, this does not diminish the value of applying optimal foraging theory to the study of food webs (Petchey et al. 2011), and

Figure 10.10 Per capita interaction strength increases with the ratio of prey to predator body weight (fourth-root transformed data). This conclusion is based on data from avian rocky intertidal predators, subtidal grazers, and estuarine predators. The regression line shown is for avian intertidal predators only. (After Wootton and Emmerson 2005.)

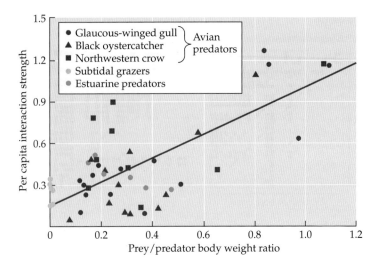

I suspect it won't be long before parameters such as prey handling times and encounter rates are directly incorporated into a food web model that distinguishes consumers and resources by body size. Recently, Woodward et al. (2010) showed that an extension of Petchey et al.'s 2008 model quite accurately predicted the trophic links of consumers classified by their body size, but not by their species.

We might wonder, then, is there a connection between the relative sizes of predators and prey and the strength of their interactions as well? Perhaps. In a preliminary analysis, Wootton and Emmerson (2005) showed that per capita interaction strength (measured by the dynamic index) was positively related to the ratio of prey weight to predator weight in a collection of four food web studies (**Figure 10.10**). They noted, however, that this relationship showed a great deal of variation between studies, and they suggested that the underlying relationship might be unimodal rather than linear. Emmerson and Raffaelli (2004) also found that the predator/prey size ratio was correlated with the strength of trophic interactions in

Figure 10.11 Predatory beetles (A) and spiders (B) feeding on large and small prey items in the laboratory show feeding rates that are a hump-shaped function of the ratio of predator mass to prey mass. Observations suggest that predation was limited by the relatively long time needed to subdue and handle prey at low predator–prey body mass ratios. At high predator–prey body mass ratios, predation was limited by the high escape efficiencies of prey species due to fast reaction times and their use of small refuges. (After Brose et al. 2008.)

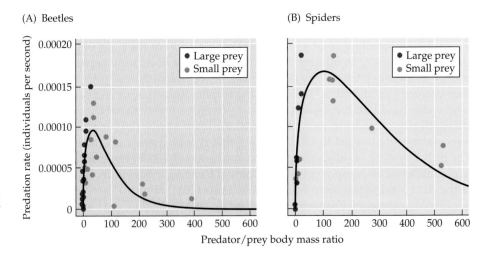

experimental food webs, but that the form of this relationship varied with predator species. Finally, a unimodal relationship between predation rate (a measure of energy flux) and the predator/prey body mass ratio was found in a laboratory study using predaceous ground-dwelling beetles and spiders feeding on collembolans, crickets, or fruit flies (Brose et al. 2008). In this experiment, the observed predation rate for predators and their different sizes of prey was accurately predicted by a foraging model predicated on the size-specific scaling of handling times, prey detection, and prey escape, but did not conform well to the predictions of a model based on the metabolic scaling of predation rates **Figure 10.11**). On the great plains of Africa, adult body size has been shown to be a strong predictor of predator-driven mortality rates (**Figure 10.12**; Sinclair et al. 2003).

In sum, it appears that Elton's intuition was correct: body size relationships play a major role in determining the pattern and strength of trophic interactions within food webs (e.g., Zook et al. 2011). In addition, by developing general foraging models in which important parameters such as encounter rates and handling times are allometric functions of predator

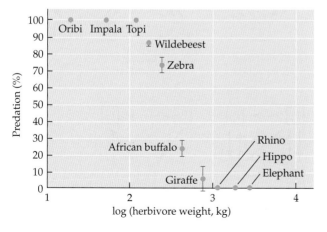

Figure 10.12 Predator-related adult mortality in non-migratory African ungulates drops off sharply at a body size of about 150 kg. Error bars are 95% confidence limits. (After Sinclair et al. 2003; photo © Images of Africa/Alamy.

and prey sizes (e.g., McGill and Mittelbach 2006), ecologists may begin to forge important links between individual behaviors and the structure and dynamics of ecological networks (Berlow et al. 2008; Petchey et al. 2010; Stouffer 2010).

Indirect Effects

Looking at a food web, or any other diagram of an ecological network, emphasizes the potential importance of indirect effects in communities. An indirect effect occurs when one species affects a second species through a "change" in an intermediate (third) species (Abrams et al. 1996). This "change" may be an alteration in the abundance of a species, referred to as a density-mediated effect, or an alteration in a species' phenotype (e.g., behavior, habitat use, morphology), referred to as a trait-mediated effect.

There are many types of indirect effects in food webs; a few are modeled in **Figure 10.13**. While all are potentially important, four types of indirect effects stand out as being particularly common and significant: exploitative competition, apparent competition, trophic cascades, and keystone predation. We discussed exploitative competition and apparent competition in Chapters 7 and 8. In Chapter 11, we will examine trophic cascades and keystone predation in relation to top-down versus bottom-up control of trophic-level biomass and species richness.

Because indirect effects involve the influence of one species on another via an intermediate species, the pathway for an indirect effect is necessarily longer than that for a direct effect. Because of this, we might expect (1) that indirect effects will take longer to develop than direct effects, and (2) that indirect effects will be weaker than direct effects (because of attenuation of effect size with each trophic link involved). An early literature review by Schoener (1993) found some support for these predictions. Schoener summarized the detailed findings of six experimental studies of community regulation in terrestrial and aquatic habitats. He concluded that (1) direct

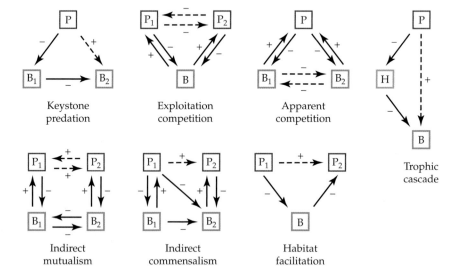

Figure 10.13 Models of seven types of indirect effect sequences. Solid arrows represent direct effects; indirect effects are shown by dashed arrows. Plus and minus signs indicate positive and negative effects, respectively. B, basal species; P, predator; H, herbivore. (After Menge 1995.)

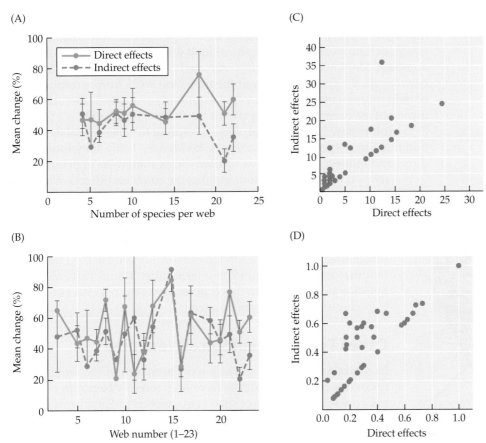

Figure 10.14 Importance of direct and indirect effects in marine intertidal food webs. (A) Mean percentage change in the abundance of organisms (absolute value) caused by experimentally manipulated direct and indirect effects plotted against the number of species in each food web. Webs with equal numbers of species are lumped. (B) The same data plotted for each food web individually (webs are not lumped). Error bars are ±1 SE. (C) The number of months required to demonstrate significant direct and indirect effects. (D) The same data expressed as the proportion of the duration of each experiment, in order to adjust for variation in experiment duration. (After Abrams et al. 1996; data from Menge 1995.)

effects were usually stronger than indirect effects, and (2) short-chain indirect effects were stronger than long-chain indirect effects. Schoener's analysis was insightful, but it was limited in scope because of the few studies available at the time.

In a later analysis, Menge (1995) examined the results of perturbation experiments in 23 marine rocky intertidal habitats around the globe. Although the mean change in species densities due to direct effects was greater than that due to indirect effects in 12 of 18 food webs, this difference was not significant (**Figure 10.14A,B**). On average, 40%–50% of the change in species densities after an experimental perturbation was due to indirect effects, with the remainder due to direct effects. Thus, direct and indirect effects were comparable in magnitude in the communities studied. Menge (1995) also found no difference in the time it took to observe significant direct and indirect effects (**Figure 10.14C,D**). Thus, based on Menge's review, we may conclude that indirect effects are comparable in magnitude to direct effects, and that direct and indirect effects take place at similar rates. However, we need to temper these conclusions somewhat because the results are based on a single well-studied habitat—the marine intertidal zone.

To my knowledge, no other comprehensive reviews comparing indirect and direct interactions have been published since Schoener (1993) and Menge

(1995) conducted their studies, which is surprising. Recently, however, Montoya et al. (2009) evaluated the relative importance of direct and indirect effects in nine well-studied empirical food webs using a nonexperimental approach. Montoya and colleagues estimated per capita interaction strengths between all the species in each of the nine food webs from a combination of field measurements and prey/predator body size ratios, then used the inverse of the community matrix (Bender et al. 1984; Yodzis 1988; see Chapter 7) to explore how each food web should respond to a theoretical perturbation that alters species' abundances. Like Menge (1995), they found that direct and indirect effects played similarly important roles in determining a response to perturbation, and that the combined effects of indirect interactions could often counteract the direct effects. For example, they found that for 40% of predator–prey links, predators had a positive net effect on the abundance of their prey due to a predominance of indirect effects (see also Abrams 1992). Thus, the evidence suggests that indirect effects are of fundamental importance in food webs and that they may often lead to counterintuitive results (e.g., recall the question posed earlier in this chapter about the net effect of culling Cape fur seals on hake abundance).

Other Types of Ecological Networks

In contrast to food webs, which have been recognized and studied for over a century, ecologists have only recently begun to examine networks of other types of species interactions. Mutualisms, in particular, have become the focus of much recent work on ecological networks.

Mutualistic networks

Two of the most common and important mutualistic interactions occur between plants and animals: pollination and seed dispersal. Over 90% of angiosperm species in the tropics, for example, are thought to depend on animals for pollination (Bawa 1990) and dispersal (Jordano 2000). We discussed many examples of plant–animal mutualisms in Chapter 9. Here, we will consider some of the patterns of interaction observed between plant and animal mutualists within communities and their potential consequences.

Like trophic interactions, mutualistic interactions can be diagrammed as webs of links between species (or functional groups). However, unlike food webs, in which species are often represented by nodes of a single type (one-mode networks), plant–animal mutualisms have two well-defined types of nodes (e.g., plants and pollinators), and interactions occur between, but not within, node types (Bascompte and Jordano 2007). These two-mode networks may be represented by bipartite interaction webs. **Figure 10.15** shows an example of such an interaction web for plants and their bee pollinators (Bezerra et al. 2009). In this example, the lines connect plant species (circles) to the bee species (squares) that pollinate them. The strength of the interaction is represented by the thickness

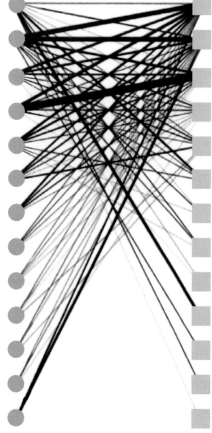

Figure 10.15 A plant–pollinator interaction network. Boxes represent pollinating bee species; circles represent plant species. The thickness of the line connecting bee and plant indicates the number of visits by a bee species to a plant species (thicker lines indicate more visits). There are an equal number of bee and plant species in this example simply due to chance. (From Bezerra et al. 2009.)

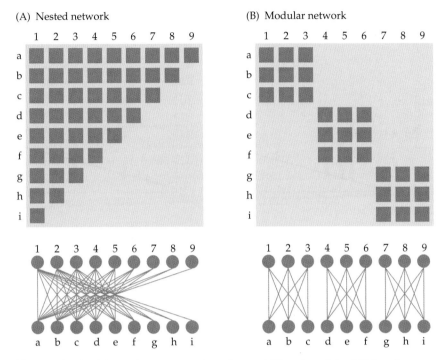

Figure 10.16   Schematic representation of nested (A) and modular (B) bipartite networks. In matrix representations (top), each row and column corresponds to a species; squares represent species interactions. In web representations (bottom), each node represents a species, and interacting species are connected by lines. (From Fontaine et al. 2011.)

of the connecting line, which indicates the relative number of visits by a bee species to a plant species.

Some important topological properties of this and other mutualistic webs are (1) a high level of connectance, (2) a high degree of nestedness, and (3) relatively low modularity (Thébault and Fontaine 2010). Connectance refers to the observed number of links in the network, expressed as a proportion of the total number of possible links in the network. Nestedness refers to a specific type of interaction structure in which species with many interactions (generalists) form a core of interacting species and species with few interactions (specialists) interact mostly with generalists (Bascompte et al. 2003; Melián and Bascompte 2004; Ings et al. 2009). Modularity exists in a network when groups of species interact more among themselves than with species from other groups (**Figure 10.16**).

Nonrandom patterns of nestedness have been found in ant–plant networks (Guimarães et al. 2006a), pollination and seed dispersal networks (Bascompte et al. 2003; Bezerra et al. 2009), and marine cleaning networks (Guimarães et al. 2006b). These observations suggest that mutualisms may evolve into a predictable community structure (Lewinsohn et al. 2006). Thébault and Fontaine (2010) compared the structures of 34 mutualistic (pollination) networks with those of 23 trophic (herbivory) networks. They

Figure 10.17 The relative importance of connectance, nestedness, and modularity in the stability and architecture of mutualistic (plant–pollinator in this study) and trophic (plant–herbivore) model networks. Final network structure is plotted against initial network structure for a series of model simulations in which mutualistic (A) and trophic (B) networks were allowed to develop over time. During the simulations, some species became extinct before equilibrium was reached, thus altering network structure. These extinctions caused mutualistic networks to become more connected and more nested over time, whereas trophic networks became more modular over time. (After Thébault and Fontaine 2010.)

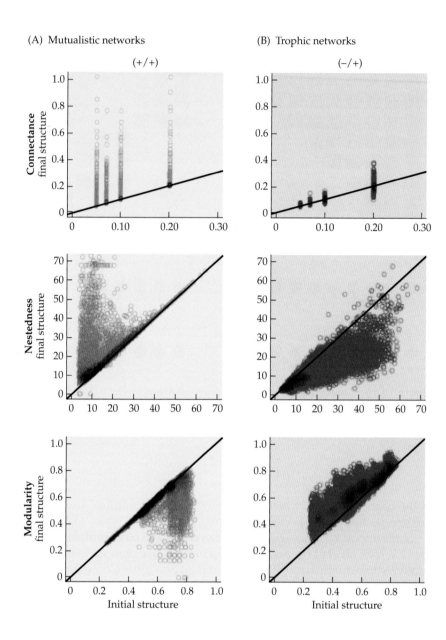

also analyzed models of population dynamics to assess how differences in network topology (nestedness, modularity, and connectance) affected the persistence and resilience of mutualistic and trophic networks. They found that increased nestedness and connectance promoted stability in model mutualistic networks, whereas increased modularity promoted stability in model trophic networks (**Figure 10.17**). These same properties were found to differ between real-world mutualistic and trophic networks: mutualistic networks had higher connectance and higher nestedness than did trophic networks, whereas trophic networks tended to have higher modularity (**Figure 10.18**). Thus, their analyses suggest that food webs and mutualistic webs differ systematically in their topologies, and that these differences dif-

(A)

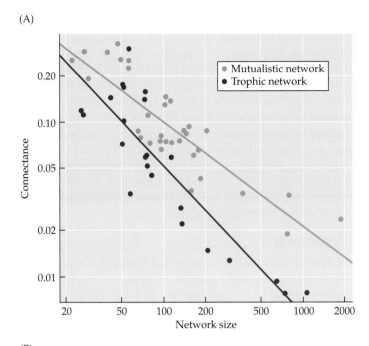

Figure 10.18 The degrees of connectance, nestedness, and modularity differ among real mutualistic and trophic networks. Each dot represents an empirical network involving either pollination (green) or herbivory (red). (A) Connectance plotted as a function of network size. Connectance is greater in pollination networks than in herbivory networks. (B) Relationship between network nestedness and modularity. Box plots of relative nestedness (below) and relative modularity (left) show that nestedness is greater in pollination (mutualisitic) networks and modularity is greater in herbivory (trophic) networks. (After Thébault and Fontaine 2010.)

(B)

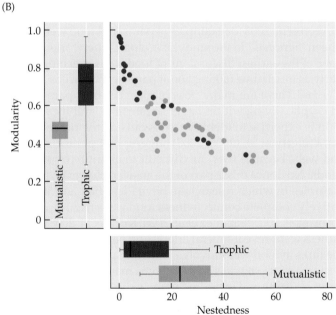

ferentially affect network stability such that each type of network develops a structure that tends to stabilize that network.

The study of mutualistic networks to date has focused largely on describing patterns of interaction and predicting their consequences for network functioning. Future steps toward understanding these networks should be experimental. Just as the study of food webs has moved from the analysis of patterns in connectedness webs to the construction of functional webs, experimental manipulations of mutualistic webs (e.g., species removals) are needed to judge whether the patterns observed in these networks reflect their functional behaviors. For example, interaction strength in pollination webs has been estimated by the observed number of visits by a pollinator species to a plant species (e.g., Bezerra et al. 2009), yet we do not know whether the number of pollinator visits in fact reflects the degree to which the reproductive success of the plant species depends on a particular pollinator species (but see Vázquez et al. 2005). Likewise, we do not know how the extinction of a particular pollinator (or plant species) would influence the persistence of other pollinator or plant species within the network if pollinators are flexible in their behaviors and therefore shift the structure of the interaction web in response to the extinction (or removal) of a particular plant or pollinator species (Rezende et al. 2007).

Parasites and parasitoids

Most of the world's species are probably parasitic (Price 1980), especially if we take a broad view of parasites that includes bacteria, viruses, fungi, and other symbiotic organisms as well as protozoans and metazoans that feed on their hosts without killing them. Indeed, Lafferty et al. (2008) argue that parasitism is the most common consumer strategy among organisms. Yet parasites are rarely included in the study of food webs (insect parasitoids, which *do* kill their hosts and thus act like "standard" predators, are an exception; see van Veen et al. 2008 for examples of parasitoid food webs). Why have parasites been "ignored" in the study of ecological networks, despite the fact that in one of the earliest discussions of food webs in ecology, Elton (1927) spent nearly as much time talking about parasite–host webs as about predator–prey webs? There are many possibilities. For one, parasites are often small and cryptic (living within larger organisms), and they require a level of taxonomic expertise to identify that is outside the range of most ecologists. Thus, Lafferty et al. (2006, 2008) suggest that parasites are left out of most food webs because they are difficult to quantify by standard ecological methods. No doubt this is part of the answer. However, there are a few recent examples in which researchers have identified and included parasites in relatively complete connectedness food webs (e.g., Huxham et al. 1995; Thompson et al. 2005; Vázquez et al. 2005; Kuris et al. 2008). Analyses of connectedness webs reveal several interesting patterns. For example, about 75% of the links in food webs involve parasitic species (Lafferty et al. 2006); the total biomass of parasites in food webs may exceed that of top predators (Kuris et al. 2008); and including parasites in food webs increases food chain length and food web connectance (Lafferty et al. 2006).

These studies suggest some general patterns that may emerge from including parasites in food webs. However, as we noted earlier in this chapter, connectedness webs (which define most of the current attempts to include parasites in food webs) have their limitations. Moreover, other traits of parasites—beyond their size and crypticity—present significant challenges to incorporating these consumers into food webs. First, the presence of complex life cycles and multiple hosts makes it difficult to position a parasite species within a food web. (This problem is similar to the one raised by ontogenetic niche shifts in free-living organisms that change their diet as they grow; see, for example, Rudolf and Lafferty 2011.) Second, parasites feed on, but rarely kill, their hosts. This trait makes them similar to many herbivores, but measuring the energy transfer from hosts to parasites is more difficult than measuring energy transfer from plants to herbivores. And finally, parasites affect the behaviors of their hosts, often making them more susceptible to predators (Moore 2002) and thereby influencing host mortality rates and the rate of energy transfer between trophic levels. This trait makes their effects similar to (but opposite in direction from) the nonlethal (nonconsumptive) or trait-mediated effects of predators on their prey (see Chapter 6). Thus, while the challenges of incorporating parasites into food webs are qualitatively no different from the challenges incurred with other types of consumers (Lafferty 2008), the sum total of these challenges is significant and explains why so few webs to date include parasites, despite their recognized importance in nature.

Complexity and Stability

From their very beginning, studies of food webs and other ecological networks have wrestled with the question of whether more diverse communities are more stable (McCann 2000). Charles Elton, a pioneer in the study of food webs, was a strong proponent of the idea that simple communities showed greater fluctuations in species abundance, and were more prone to species invasions, than were diverse communities (Elton 1927). Elton's ideas were echoed by Odum (1953) and MacArthur (1955), and the general consensus among ecologists at the time was that more complex communities, with their greater numbers of consumer and resource species, were better buffered from the impacts of species loss and environmental fluctuations. This idea was severely challenged by Robert May (1973a), whose theoretical analysis of model ecosystems showed that more diverse communities tended to be less stable.

May constructed his theoretical communities by randomly assigning interaction strengths to the different species. We now know that the random assignment of interaction strengths in May's model was critical to his finding that diversity tended to destabilize community dynamics and that the distribution of interaction strengths in real communities is distinctly nonrandom. As discussed earlier in this chapter, natural communities show a pronounced skew in the distribution of interaction strengths: interactions between a few species are very strong, but most species interact only weakly (Bascompte et al. 2005; Wootton and Emmerson 2005). Recent theoretical studies of food webs show that this skewed distribution of interaction strengths strongly promotes stability.

McCann et al. (1998) and McCann (2000) showed how weak interaction strengths can promote stability in food webs by damping population oscillations associated with strongly interacting species. Weak interactions can limit runaway consumption along strong consumer–resource pathways, and they can generate negative covariances between resources that share a consumer, thus helping to ensure that consumers have weak effects on a resource when that resource is at low density. Diversity may therefore promote stability in food webs if the number of weak trophic interactions (and therefore the importance of the weak interaction effect) increases with diversity (McCann et al. 1998; Jiang et al. 2009). Jiang and Pu (2009) found, in a recent meta-analysis of experiments manipulating species diversity, that diversity had a strong, positive effect on temporal stability in food webs with multiple trophic levels. Empirical food webs also differ from the randomly constructed webs studied by May (1973a) by being significantly more modular or compartmentalized (e.g., subsets of species interact more often amongst themselves than with other species; Krause et al. 2003; Rezende et al. 2009; Guimerà et al. 2010; Thébault and Fontaine 2010; but see Kondoh et al. 2010). Modularity has been shown to increase food web persistence by buffering the propagation of species extinctions through the community (Melián and Bascompte 2002; Stouffer and Bascompte 2011; Thébault and Fontaine 2010).

The stabilizing effects of modularity and weak interactions may be promoted in food webs by the segregation of energy flow into "fast" and "slow" channels. Energy channels can be defined as groups of highly interacting species at lower trophic levels that derive their energy largely from the same basal resource. Fast energy channels tend to have smaller, faster-growing populations with higher biomass turnover rates than slow energy channels (Rooney and McCann 2012). In marine food webs, for example, the flow of energy originating from phytoplankton production moves through a "fast channel" and the energy flow originating from detritus moves through a "slow channel." Mobile consumers at the top of the food web couple the energy flow from these two channels. Species diversity tends to be greater in the slow, detritus-based channel than in the fast, plankton-based channel, and Rooney and McCann argue that the many weak interactions observed within the slow channel are critical to stabilizing the overall food web.

Conclusion

The boom in network analysis in the natural and social sciences during the past decade has spawned a variety of analytical and graphic tools that are helping to power the next generation of studies of ecological networks (Watts and Strogatz 1998; Barabasi 2009; Bascompte 2009; Fontaine et al. 2011). These tools make it possible to better visualize the structure of complex ecological networks, characterize their properties, and develop testable models. Food webs continue to be the most studied and best understood of all ecological networks. Our current knowledge of food webs shows that many of the patterns observed in the early analysis of connectedness webs are less general than once thought, but that some properties, such as modularity, are common to most food webs. Studies of interaction webs have increased our understanding of the roles of individual species and of particular types of interactions within food webs. While species remain the principal focus in most studies of ecological networks, ecologists are beginning to broaden their scope and focus on more general properties of organisms (e.g., body size) that may determine metabolic rates or rates of consumer–resource interactions, thus allowing greater insight into the mechanisms underlying the structure and functioning of food webs (Beckerman et al. 2010).

Although the new tools of network analysis have greatly increased the ability of ecologists to visualize ecological networks and analyze their structural properties, fundamental challenges remain. Perhaps the greatest of these challenges is to incorporate multiple types of species interactions into the same ecological network (Allesina and Pascual 2008; Melián et al. 2009; Fontaine et al. 2011). Ecologists have significantly broadened their focus to include networks that include species interactions beyond the traditional predator–prey relationships considered in food webs. However, analyses of these different types of ecological networks are mostly proceeding in parallel (i.e., ecologists study either pollination networks or plant–herbivore networks or host–parasite networks), whereas real communities are composed of multiple species interacting in multiple ways (e.g., mutualism, facilitation, predation). Experimental studies of interaction webs in nature have demonstrated the importance of multiple modes of interaction (e.g., Menge 1995; see Figure 10.3). Thus, we should expect that new insights and a far greater understanding of how network structure affects the diversity and stability of communities will result from the integration of multiple interaction types within a single network.

Summary

1. Interactions among the species in a community can be diagrammed as ecological networks. Food webs focus on "typical" predator–prey interactions, in which consumers are larger than their prey. Less commonly studied are mutualistic webs, host–parasitoid webs, and the roles of parasites and herbivores.

2. Food webs tend to be better resolved at the top than at the bottom, in part because species richness is greater at lower trophic levels and in part because species at lower trophic levels tend to be small, difficult to identify, and have feeding relationships that are hard to quantify. Ecologists often separate food webs into those in which the basal trophic level is made up of primary producers ("green food webs") and those in which the basal trophic level is detritus ("brown food webs").

3. Connectedness, or structural, webs show the presence of an interaction between species but do not specify the strength of that interaction.

4. Energy flow webs measure the amount of energy (biomass) moving between species in a food web. Implicit in this approach is the idea that there is a relationship between the amount of energy flowing through a pathway and the importance of that pathway to community dynamics. However, energy flow has been shown to be a surprisingly poor predictor of the strength of interactions between species or of the impact of removing a particular species from a community.

5. Functional, or interaction, webs measure the strength of the interactions between species within a community, implicitly recognizing that not all species and interactions are equally important. Measures of interaction strength in model food webs tend to focus on the individual interactions between species pairs, whereas measures of interaction

strength in empirical food webs tend to focus on the impact of one species on the rest of the web, as measured by removal experiments.

6. Studies of theoretical and empirical food webs have shown that most food webs contain a few strong links and many weak links. A keystone species is one whose effect on the community is disproportionately large relative to its abundance. Body size relationships play a major role in determining the pattern and strength of trophic interactions within food webs.

7. Evidence suggests that the net effects of indirect interactions—when the actions of one species influences a second species via a third species—are important in food webs. Such indirect interactions include exploitative competition, apparent competition, cascading effects, and keystone predation.

8. Mutualistic interactions can be diagrammed as webs of links between two well-defined types of nodes, in which interactions occur between, but not within, node types. Some important properties of mutualistic webs are a high level of connectance, a high degree of nestedness, and relatively low modularity.

9 Parasites are left out of most food webs because their interactions and impacts on other species are difficult to quantify by standard ecological methods. Ecologists need to find ways to better incorporate parasites and other infective agents into ecological networks.

10. Food webs that are more diverse tend to have more weak interactions and greater modularity than simpler communities, and thus tend to be more stable.

11 Food Chains and Food Webs

Controlling Factors and Cascading Effects

Why the sky is blue is a matter of basic physics, but why land is green is a much trickier question.

Shahid Naeem, 2008: 913

[P]opulations in different trophic levels are expected to differ in their method of control.

N. Hairston, F. E. Smith, and L. B. Slobodkin, 1960: 421

Everything should be made as simple as possible, but not simpler.

Albert Einstein

In the previous chapter, we saw how the multitude of links between species in a food web can generate a complex network of potential interactions and indirect effects (see, for example, Figure 10.2). In this chapter, we will see how these complex networks can be simplified into chains of linked consumer–resource populations in order to address the fundamental question of what controls the abundances of organisms at different trophic levels. Are these abundances controlled from the bottom up, by resource availability, or from the top down, by predation? Are simple food chain models, which lump together all the organisms at each trophic level, too simple? We'll begin here by looking at two classic papers that launched a vigorous exploration of the factors controlling the abundances of producers, herbivores, and carnivores in ecosystems.

Why Is the World Green?

Given the complexity of natural food webs, it's surprising that one of the most influential and highly cited papers in community ecology (more than 1,400 citations as of December 2011) takes as its premise that communities can be simplified into just three interacting trophic levels: producers, herbivores, and carnivores. In 1960, three ecologists from the University of Michigan, Nelson Hairston, Frederick E. Smith, and Lawrence B. Slobodkin, stirred up a hornet's nest when they published a short note to the *American Naturalist* entitled "Community structure, population control, and competition." Hairston et al. (1960; hereafter HSS) were interested in what controlled the abundance of populations and by extension, the total abundances of organisms at different trophic levels. They reasoned as follows:

1. In the absence of higher-level predation, carnivores should be limited by competition for their food (herbivores).

2. Consequently, herbivore populations should be held below their carrying capacity by carnivores and have little impact on their food (plants).

3. In the absence of control by herbivores, plants should be dense and limited by competition.

Thus, they concluded that "populations in different trophic levels are expected to differ in their methods of control" (HSS 1960: 421)—in other words, that carnivores and plants are limited by resource competition and herbivores are limited by predation. Their argument quickly became known as the "world is green hypothesis" because the mechanisms it proposed would lead to a world rich in plants; they cited as evidence for their reasoning that "obvious depletion of green plants by herbivores are exceptions to the general picture" (HSS 1960: 423–424). However, HSS's logic did not sit well with a number of ecologists, who pointed out that many plants had chemical and other defenses to prevent them from being eaten, that populations were limited by factors other than resources and predators, and that simple deductive arguments are not proof of underlying processes (Murdoch 1966; Ehrlich and Birch 1967). After some spirited debate, the ideas of HSS faded from view, only to be resurrected 20 years later in a landmark study by Oksanen et al. (1981).

The papers by HSS and Oksanen et al. launched a vigorous exploration of the factors controlling the abundances of plants, herbivores, and carnivores in ecosystems. Studies of trophic-level regulation and related topics (e.g., top-down versus bottom-up control, trophic cascades, and determinants of food chain length) have played a major role in the study of ecological networks. Summarizing this wealth of information is a challenge. I have chosen to follow (albeit roughly) the historical development of these ideas and their empirical tests. The older papers (e.g., HSS, Rosenzweig 1977; Oksanen et al. 1981) provide a conceptual foundation for much that follows, and this historical thread illustrates the progress that can be made when theory gives rise to testable predictions and empirical tests in turn guide the development of theory. As we will see, viewing food chains and food webs as coupled consumer–resource models yields some very interesting and testable predictions, and the outcome of testing these predictions has greatly increased our understanding of the factors controlling the abundances of producers and consumers in ecosystems.

What Determines Abundance at Different Trophic Levels?

Oksanen et al. (1981), building on the mathematical framework of Rosenzweig (1973), formalized and extended the ideas of HSS. Their approach was to apply consumer–resource (predator–prey) equations to interacting trophic levels. Like HSS, Oksanen and colleagues considered trophic levels to be ecological units that could be modeled as linked consumer or resource populations, and they treated each trophic level as a single, homogeneous population. This approach simplifies nature to an extreme, but it turns out to be quite insightful.

Oksanen et al. focused on how the biomass at each trophic level should change with a change in the potential primary productivity of the ecosystem. "Potential primary productivity" (G) can be thought of as the maximum gross primary productivity allowed by the environment if there were no consumers present; for simplicity, we will sometimes refer to this as "potential productivity." In many systems, G is correlated with the supply of a limiting factor, such as phosphorus or nitrogen in lakes (Elser et al. 2007) or precipitation in grasslands (Scurlock and Olson 2002). The Oksanen et al. model predicts that ecosystems with low potential primary productivity will support fewer trophic levels than ecosystems with high potential productivity. We will examine whether food chain length increases with productivity in natural systems later in this chapter. For now, we will take as a given that more productive systems can support more trophic levels, and we will use a simplified version of Oksanen et al.'s analysis to examine HSS's prediction that "different trophic levels are expected to differ in their methods of control."

Using three-dimensional isocline plots (such as **Figure 11.1**) based on the Rosenzweig and MacArthur predator–prey model (see Chapter 5), Oksanen et al. showed how the abundances of producers, herbivores, and carnivores should change with potential primary productivity. Recall that in Chapter 5 we used two-dimensional isocline plots to visualize the responses of predator and prey populations to an increase in the prey's carrying capacity (see Figure 5.9). Rather than using complex three-dimensional isocline plots here, however, we will focus instead on a simplified picture of Oksanen et al.'s main result.

Figure 11.2 shows how the number of trophic levels is predicted to increase as potential productivity (G) increases and how the equilibrium biomass of a trophic level is expected to change with an

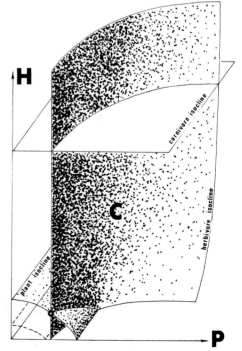

Figure 11.1   An example of a three-dimensional isocline plot showing the equilibrium biomass of three trophic levels (plant, herbivore, and carnivore abundances) in a system of linked consumer–resource populations. (From Oksanen et al. 1981.)

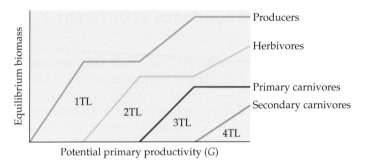

Figure 11.2 The predicted change in equilibrium biomass of different trophic levels in response to a change in potential primary productivity (G). These predicted responses are based on the model of Oksanen et al. (1981), which assumes that increases in potential productivity permit the addition of higher trophic levels (TLs). Complexities arising due to unstable consumer–resource interactions are ignored in this simplified treatment. (After Mittelbach et al. 1988 and Leibold 1989.)

increase in potential primary productivity. Note the interesting stair-stepped pattern. Whether a trophic level responds to an increase in potential primary productivity depends on the number of trophic levels in the system, with adjacent trophic levels showing an alternating pattern of response. That is, in food chains with an odd number of trophic levels (e.g., three), the abundances of producers and of primary carnivores are predicted to increase with potential primary productivity and the abundance of herbivores is predicted to stay constant. In food chains with an even number of trophic levels (e.g., four), producer and primary carnivore abundances stay constant and herbivore and secondary carnivore abundances increase with potential productivity.

To better understand what drives this stepwise response to potential productivity, let's focus on a system with three trophic levels (as did HSS). It is important to remember that all trophic-level abundances in Figure 11.2 represent equilibrium biomasses. Therefore, one can view Figure 11.2 as representing different ecosystems with different G's or, alternatively, as a single ecosystem that has undergone a change in G in response to which the various trophic levels have reached a new equilibrium.

In **Box 11.1**, we can see how solving for the density of each trophic level at equilibrium yields the alternating pattern of trophic-level response shown in region 3TL of Figure 11.2. The equilibrium abundances of the top and bottom trophic levels (primary carnivores and producers, respectively) are functions of resources available in the environment, whereas the equilibrium abundance of the middle trophic level (herbivores) is not (Chase et al. 2000). The model described in Box 11.1 is a linear version of the consumer–resource model analyzed by Oksanen et al. (1981). Nonlinear consumer–resource models with saturating functional responses can lead to unstable dynamics and patterns different from those shown in Figure 11.2 (Abrams and Roth 1994; Oksanen et al. 2001), as can other modifications to the model (Mittelbach et al. 1988). For now, however, we will ignore these complications and ask how well the predictions in Figure 11.2 correspond to HSS's intuition about the factors controlling trophic levels.

BOX 11.1

A linear version of the three-species consumer–resource model analyzed by Oksanen et al. (1981) can be written as

$$\frac{dR}{dt} = S - a'NR - cR \qquad \frac{dN}{dt} = N\left(a'b'R - a''P - c'\right) \qquad \frac{dP}{dt} = P\left(a''b''N - c''\right)$$

where R, N, and P represent the abundances of the resource, consumer, and predator populations, respectively, S is the total supply of R's resource, a's represent attack rates, b's convert food eaten into new consumers, and c's are density-independent (nonconsumptive) loss rates (Holt et al. 1994; Chase et al. 2000). Parameters without primes (a, b, c) are those of the resource population R, parameters marked with primes (a', b', c') are those of the consumer N, and parameters with double primes (a'', b'', c'') are those of the predator P. As shown by Chase et al. (2000), the equilibrium numbers of resources, consumers, and predators in this system can be written as

$$R^* = \frac{Sa''b''}{a'c'' + ca''b''} \qquad N^* = \frac{c''}{a''b''} \qquad P^* = \frac{a'b'R - c'}{a''}$$

Note that the abundance of R increases with S and the abundance of P increases with R, but the abundance of N is a function of three constants and therefore does not vary with the environment (Chase et al. 2000). Thus, in this three-trophic-level system, the bottom and top trophic levels increase with productivity, whereas the middle trophic level remains constant (i.e., the pattern shown in region 3TL of Figure 11.2).

The above model assumes Type I (linear) functional responses, whereas the model of Oksanen et al. (1981) assumed a Type II (nonlinear) functional response in the herbivore and carnivore populations. Such nonlinear functional responses can lead to unstable dynamics and patterns of trophic level responses different from those shown in Figure 11.2 (Abrams and Roth 1994). However, these unstable dynamics were not the focus of Oksanen et al.'s analysis, and their main conclusions can be reasonably represented by the linear consumer–resource model described here.

A simple thought experiment illustrates the duality of top-down and bottom-up control

In a three-trophic-level system, herbivore biomass does not change with increasing G because all of the increased production of herbivores goes into carnivores, so herbivore abundance appears to be controlled by predation (Figure 11.2). Increasing G increases the equilibrium biomass of carnivores and producers; thus, the abundances of producers and carnivores appear to be controlled by competition (i.e., both producer and carnivore equilibrium biomasses increase with an increase in the productivity of their resources). Therefore, Figure 11.2 seems to support HSS's argument that the different trophic levels are controlled by different factors. However, in a more complete sense, the equilibrium biomass at each level below the top trophic level in Figure 11.2 represents the balance between the effects of predation and competition. To see this, consider the following "thought experiment." What would happen to the herbivore population in a three-trophic-level system if we were to remove the potentially limiting effects of predation and competition—say, by adding unlimited food or removing all predators? How would we expect the herbivore population to respond in the 3TL region of Figure 11.2? In the *short term*, herbivores should respond to unlimited food through an increase in per capita birth rates and an increase in herbivore biomass. Likewise, the removal of predators should lead to a *short-term* reduction in herbivore death rates and an increase in herbivore biomass. Experimentally, we could compare these short-term responses to food addition and predator removal with responses in control treatments in which the potential sources of limitation (resources, predators) are kept at natural levels (**Figure 11.3**).

As this simple thought experiment shows, the herbivore population in region 3TL is simultaneously limited in abundance by *both* competition (resource availability) and predation. It is not an either/or dichotomy. However, the degree to which resources and predators limit trophic-level biomass may not be equal, and we can estimate the relative degree of predator limitation and resource limitation by looking at the magnitude of population response to the removal of these limiting factors in the short term (see Figure 11.3). We will look more closely at the degree of resource (bottom-up) limitation and predator (top-down) limitation in real food webs in the next section of this chapter.

Some conclusions

What can we conclude from these models about the factors limiting trophic-level biomass? First, linear consumer–resource models predict that trophic levels should alternate in their responses to increases in potential productivity (resulting in the stepped pattern in Figure 11.2). Second, although this stepped response pattern appears to support HSS's hypothesis that competition and predation alternate in importance in controlling trophic levels, a more accurate assessment is that each trophic level (below the top level) is simultaneously limited by both competition and predation. Third, the relative strengths of predator limitation and resource limitation can vary with trophic level and ecosystem productivity, and the strengths of those limiting factors can be estimated experimentally and compared with predictions.

Testing the Predictions

In this section we explore tests of three important predictions of consumer–resource theory as applied to trophic levels:

1. In a food chain of a given length, an increase in potential primary productivity will increase the abundances of populations at the top trophic level and at alternating trophic levels below the top level; however, the abundances of populations at intervening trophic levels should not increase with potential productivity.

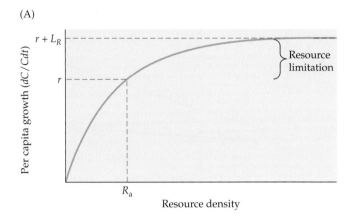

(A)

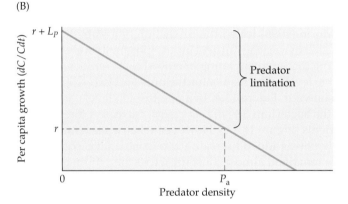

(B)

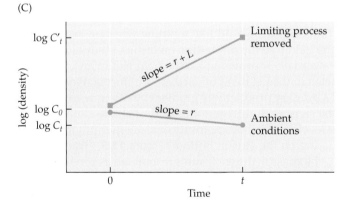

(C)

Figure 11.3 Graphic representation of the concepts of (A) resource limitation, (B) predator limitation, and (C) the assessment of limitation. (A) Per capita growth of the consumer population increases with resource density and is equal to r under ambient conditions (i.e., when $R = R_a$) and to $r + L_R$ in the absence of resource limitation (L_R). The difference between per capita consumer growth rates r and $r + L_R$ is a measure of resource limitation. (B) Per capita growth of the consumer population declines as predator density increases. The difference between per capita consumer growth rate in the absence of predators ($r + L_P$) and per capita consumer growth rate at ambient predator density (P_a) is a measure of predator limitation. (C) Limitation can be estimated with short-term field experiments as the difference between per capita growth under natural conditions (r) and under the conditions in which the focal (limiting) process has been removed ($r + L$)—that is, the difference between the slopes of the two lines in the bottom panel. (After Osenberg and Mittelbach 1996.)

2. A reduction in the abundances of populations at the top trophic level will lead to an alternating increase and decrease in the abundances of populations at sequentially lower trophic levels. This prediction focuses on the transmission of effects down the food chain, which is often referred to as a "trophic cascade" (Paine 1980; Terborgh and Estes 2010).

3. An increase in potential productivity should lead to an increase in the number of trophic levels that can be supported in an ecosystem (i.e., an increase in food chain length).

(A)

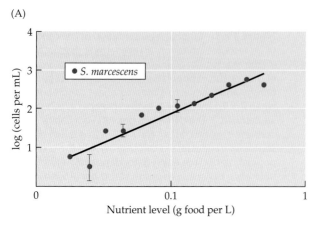

(B)

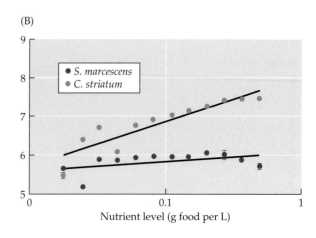

Figure 11.4 One- and two-trophic-level food chains respond differently to an increase in resources in a laboratory microbial community. (A) In the absence of predators, the abundance of the bacterium *Serratia marcescens* increases directly with an increase in the amount of resources. (B) In a two-trophic-level system, an increase in resources leads to an increase in the abundance of the bacterium's predator (the ciliate protozoan *Colpidium striatum*), but no change in the abundance of *S. marcescens*. (After Kaunzinger and Morin 1998.)

Effects of productivity on trophic-level abundances

Kaunzinger and Morin (1998) tested the prediction of a stair-stepped response (see Figure 11.2) by varying nutrient inputs (potential productivity) in simple laboratory microcosms containing microbial communities with one, two, or three trophic levels. The results for the one- and two-trophic-level communities are shown in **Figure 11.4**. In a one-trophic-level community, the abundance of the bacterium *Serratia marcescens* increased directly with an increase in nutrients (Figure 11.4A). However, when a predator (the ciliate *Colpidium striatum*) was added to create a two-trophic-level community, *Serratia* abundance remained almost constant across the productivity gradient while the abundance of the predator increased (Figure 11.4B). Thus the results of Kaunzinger and Morin's experiment match the predictions of simple consumer–resource theory applied to food chains (regions 1TL and 2TL in Figure 11.2).

As ecologists, however, we would like to know whether the same results hold true in the field (see the discussion on testing theory in Chapter 8). Wootton and Power (1993) provided such a field test in an innovative experiment conducted in the Eel River of northern California. Wootton and Power established a primary productivity gradient by covering experimental enclosures in the river with different amounts of shade cloth, thereby manipulating light levels and the productivity of algae (periphyton) growing on the stream bottom (**Figure 11.5**). Trophic-level biomass in the enclosures responded as predicted. The biomass of the top trophic level (predators, including small fish and predatory invertebrates such as dragonfly and damselfly larvae) increased with increasing light, as did the biomass of the basal trophic level (i.e., attached algae) (**Figure 11.6**). However, the biomass of the middle trophic level (herbivores, including mayfly

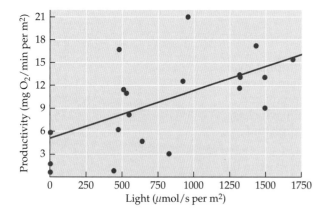

Figure 11.5 Shading of experimental enclosures in northern California's Eel River limited the productivity of algae on the stream bottom. Primary productivity in these systems was directly related to the amount of light transmitted to the stream. (After Wootton and Power 1993.)

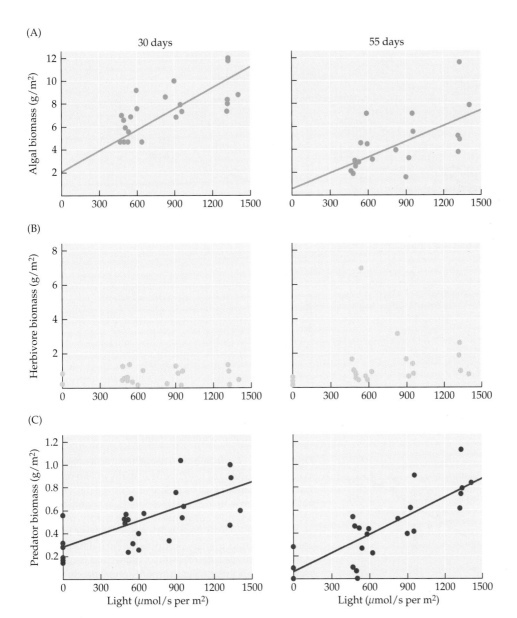

Figure 11.6 Trophic levels in a three-trophic-level food chain in the Eel River responded differently to increased primary productivity brought about by changes in resource (light) availability. The biomass of the bottom and top trophic levels (algae and predators, respectively; A and C) increased with increasing light availability, but the abundance of the middle trophic level (herbivores; B) did not change. This differential trophic level response is consistent with predictions of Oksanen et al.'s 1981 model (see Figure 11.2). The stability of the patterns after 30 and 55 days indicates that the system is in steady-state, although there is a suggestion of an increase in herbivore abundance with increasing resources after 55 days. Regression lines show significant relationships. (After Wootton and Power 1993.)

nymphs, caddisfly larvae, midge larvae, and snails) showed no significant response to the productivity manipulation after 30 and 55 days. These experiments show a clear pattern of alternating trophic level response to increased productivity, supporting perhaps the strongest prediction of

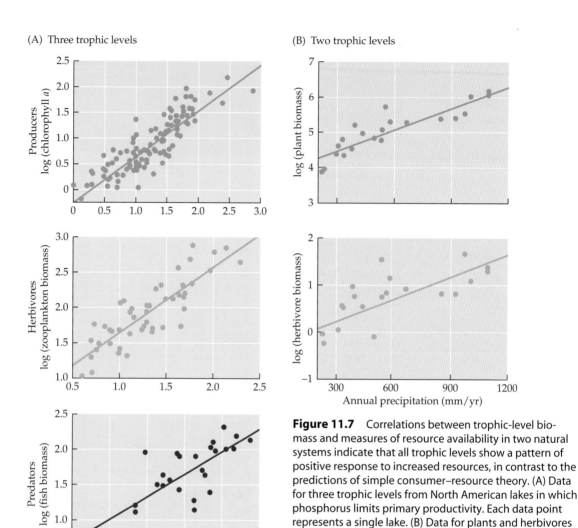

(A) Three trophic levels

(B) Two trophic levels

Figure 11.7 Correlations between trophic-level biomass and measures of resource availability in two natural systems indicate that all trophic levels show a pattern of positive response to increased resources, in contrast to the predictions of simple consumer–resource theory. (A) Data for three trophic levels from North American lakes in which phosphorus limits primary productivity. Each data point represents a single lake. (B) Data for plants and herbivores on North American grasslands where annual precipitation limits primary productivity. Each data point represents a grassland. (A after Ginzburg and Akçakaya 1992; B after Chase et al. 2000.)

consumer–resource models applied to food chains. (See Aunapuu et al. 2008 for another example showing that Arctic food webs may also respond as predicted by the food chain model.)

Observations of trophic-level biomasses in many natural systems, however, are at odds with this pattern. Ginzburg and Akçakaya (1992) found that the abundances of all trophic levels (producers, herbivores, and carnivores) increased with potential productivity in North American lakes (**Figure 11.7A**). Chase et al. (2000) observed a similar pattern in grasslands, where the abundance of plants and herbivores both increased with increasing rainfall (which is correlated with productivity; **Figure 11.7B**).

If natural systems show positive responses by all trophic levels to increases in potential productivity, how do we reconcile these results with the predictions of consumer–resource theory and with the experimental results of Kaunzinger and Morin (1998) and Wootton and Power (1993)? One potential clue comes from looking closely at the types of species responding to increased productivity in natural systems. McCauley et al. (1988) and Watson et al. (1992) showed that increases in algal biomass with lake productivity were strongly correlated with an increase in the abundance of grazer-resistant algal species and that edible algae showed no significant increase in more phosphorus-rich lakes (**Figure 11.8**). These findings suggest that shifts in species composition within trophic levels (e.g., from relatively edible to relatively inedible species) may affect the pattern of trophic-level responses.

We can see how such shifts might work in a simple food web model. Consider a food web in which the middle trophic level is composed of two prey types (e.g., two herbivore species) that share a common resource (e.g., plants) and a common predator (**Figure 11.9**). The properties of this "diamond-shaped" food web have been well studied by theoretical ecologists (e.g., Holt et al. 1994; Leibold 1996). Figure 11.9A illustrates a case of **keystone predation**, where the better resource competitor (N_1) is also more vulnerable to predation (Paine 1966; Leibold 1996). Under these conditions, and assuming there is no interference competition between prey types N_1 and N_2, an increase in the potential productivity of resource R (due to increased nutrient input) leads to a shift in

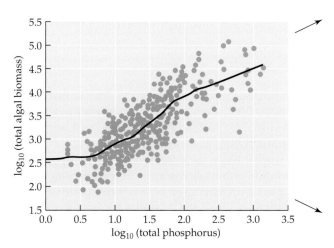

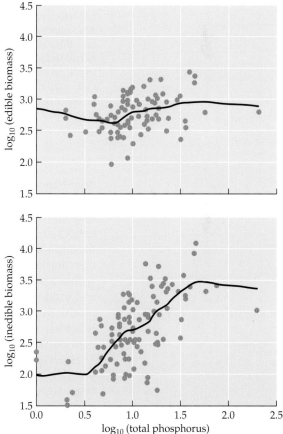

Figure 11.8   The overall positive relationship between phosphorus availability and total algal biomass in lakes (left-hand graph) has two separate components. Algal species small enough to be consumed by herbivores ("edible biomass"; above right) show no increase in abundance with increasing phosphorus input, whereas large, grazer-resistant algae ("inedible biomass"; below right) increase in abundance as phosphorus levels increase. All units are µg/L; the curve is fitted by LOWESS ("locally weighted smoothing") regression algorithms. (After Watson et al. 1992.)

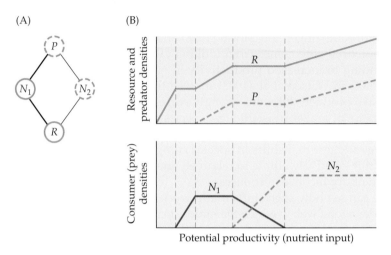

Figure 11.9 (A) An example of keystone predation, in which two prey types (N_1 and N_2) share a resource (R) and a predator (P) and there is a trade-off in the prey species' competitive ability and vulnerability to predation. The thick lines linking N_1 to R and P indicate that N_1 is a better competitor for resources and is more vulnerable to predation, whereas the thin lines linking N_2 to R and P indicate that N_2 is a poorer competitor but is less vulnerable to predation. (B) The predicted effect of an increase in potential productivity on the abundances of resources (R), prey (N_1, N_2), and predators (P), based on a theoretical model of the diamond-shaped food web. An increase in productivity (increased input of resource R) leads to a shift in dominance from the better competitor, N_1, to the more predator-resistant species, N_2. A region of intermediate productivity may allow for the coexistence of N_1 and N_2. Looking across the entire range of resource productivities, we see that the abundance of all three trophic levels increases with nutrient input. (A after Bohannan and Lenski 2000; B after Leibold 1996.)

dominance from the better resource competitor, N_1, to dominance by the more predator-resistant species, N_2. A region of intermediate productivity may allow for the coexistence of N_1 and N_2. Further, if we look across the entire range of resource productivities modeled in Figure 11.9B, we see that the abundance of all three trophic levels increases with nutrient input. Thus, this simple food web model shows how heterogeneity in species composition within a trophic level, coupled with a trade-off between competitive ability and vulnerability to predation, can lead to (1) species replacements along productivity gradients, from good resource competitors to good predator resisters; and (2) increases in the abundances of all trophic levels with an increase in the potential productivity of the bottom trophic level.

Such species turnover along productivity gradients has been observed in ponds, lakes, and streams (Rosemond et al. 1993; Osenberg and Mittelbach 1996; Leibold et al. 1997; Darcy-Hall 2006) and in grasslands (Milchunas and Lauenroth 1993; Chase et al. 2000), providing some evidence that more predator-resistant species predominate in more productive systems. However, laboratory experiments provide the clearest examples of how an increase in potential productivity can favor more predator-resistant taxa and lead to an increase in biomass at all trophic levels.

(A) Food chain

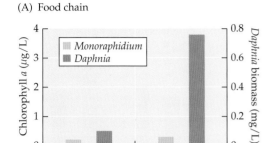

(B) Food web

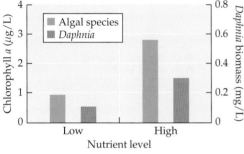

Nutrient level

Figure 11.10 Trophic-level biomass responds differently to nutrient inputs in food chains and in food webs. In Steiner's laboratory microcosms, food chains included only a single species of algae (*Monoraphidium*) and a single species of herbivorous zooplankton (*Daphnia*), whereas a food web comprised *Daphnia* and a mixture of multiple algal species, some edible and some relatively inedible. (A) The food chain system responded as predicted by Oksanen et al.'s (1981) model: nutrient addition had little effect on average algal biomass at the end of the experiment, but resulted in a dramatic increase in average herbivore biomass. (B) In the food web, both the algae and the herbivore increased in average biomass with increased nutrient input. Note that the size of *Daphnia*'s response to nutrient addition was less in the food web than in the food chain, which is expected if the composition of the producer trophic level is shifted towards less edible species at high nutrient concentrations (After Steiner 2001.)

Bohannan and Lenski (1997, 1999, 2000) and Steiner (2001) tested whether food chains (in which, recall, each trophic level is homogeneous) and food webs responded differently to increases in productivity by constructing simple microbial communities in the laboratory. Steiner's (2001) experiment contained two trophic levels: an herbivore trophic level (the grazing zooplankton *Daphnia*) and a producer trophic level. The producer level was composed of either a single species of edible algae (*Monoraphidium*) or a mixture of algal species, some of which were edible and some of which were relatively inedible. We will refer to these two experimental systems as a food chain and a food web, respectively. The food chain system responded as predicted by the model illustrated in Figure 11.2: nutrient addition had little effect on producer biomass, but caused a dramatic increase in herbivore biomass. The food web, on the other hand, showed a very different response: both producer and herbivore trophic levels increased significantly with nutrient input (**Figure 11.10**). Note that the size of the *Daphnia* response to nutrient addition was less in the food web than in the food chain, which would be expected if the composition of the producer trophic level is shifted toward less edible species at high nutrient concentrations (Leibold 1996).

Bohannan and Lenski (1997, 1999, 2000) found a very similar result, but with an interesting twist. They grew bacteria (*E. coli*) in continuous-culture chemostats with glucose as the resource and the bacteriophage T4 as the predator. The system responded to an increase in glucose as predicted by a two-link food chain model: predator abundance increased and prey (*E. coli*) abundance remained (nearly) constant (**Figure 11.11A**). However, under these experimental conditions, phage-resistant mutants of *E. coli* arise spontaneously (and predictably) after a small number of bacterial generations.

Figure 11.11 Bohannan et al. performed experiments with a simple microbial food chain maintained in continuous-culture chemostats, with glucose as the resource for the bacterium *E. coli* (the prey) and the bacteriophage T4 as the predator. (A) The experimental system responded to an increase in glucose as predicted by a two-link food chain model—that is, predator abundance increased while prey abundance remained nearly constant. (B) When phage-resistant mutants of *E. coli* arose spontaneously, bacterial abundance increased and phage abundance decreased. (C) Over time, phage-resistant *E. coli* came to dominate the bacterial population, at which point *both* trophic levels responded positively to increased glucose. The invasion of the phage-resistant bacteria thus changed the dynamics of the system from that of a food chain to that of a food web. (After Bohannan and Lenski 1997, 1999.)

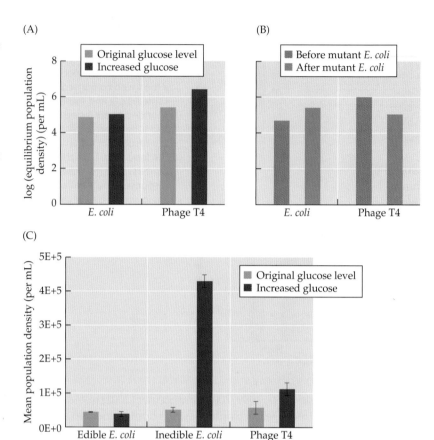

When predator-resistant *E. coli* entered the community, they changed the dynamics of the system from that of a food chain to that of a food web; *E. coli* abundance increased and phage abundance decreased (**Figure 11.11B**). Over time, phage-resistant *E. coli* came to dominate, and both trophic levels responded positively to increased glucose (**Figure 11.11C**).

These results support the predictions of the food web model described above, in which species occupying an intermediate trophic level (between a resource and a predator) experience a trade-off between competitive ability and resistance to predation. If species at an intermediate trophic level are completely invulnerable to predation, this will prevent any bottom-up flow of biomass, as all increases in potential productivity will go into increasing the biomass of inedible species (Phillips 1974; Abrams 1993; Leibold et al. 1997). Davis et al. (2010) recently observed a similar response during a 5-year enrichment experiment in a forest stream. Nutrient enrichment led to a positive biomass response by predators and prey in the first 2 years, but predator biomass eventually declined to pretreatment levels as the prey population became dominated by large-bodied, predator-resistant species Adaptive foraging behavior, in which foragers respond to increased resource availability by reducing their foraging time in favor of increased safety, and thus make resources less available to the next trophic level up, can also limit the bottom-up response of trophic levels to increased productivity (Abrams

1993), as can inducible defenses in species occupying intermediate trophic levels (Vos et al. 2004).

We have seen that the responses of natural systems to nutrient additions and increases in potential productivity are often very different from those predicted by simple consumer–resource models applied to food chains (Figures 11.2 versus Figure 11.7). Shifting species composition within a trophic level can prevent predator control and maintain the importance of resource limitation, even in high-nutrient systems. However, other factors may also lead to positive responses by adjacent trophic levels to increasing productivity, including direct density dependence (interference competition) within a trophic level (Wollkind 1976), complex life histories and invulnerable life stages (Mittelbach et al. 1988; de Roos et al. 2003; Darcy-Hall and Hall 2008), ratio-dependent predation (Arditi and Ginzburg 1989; see Chapter 5), and adaptive foraging and omnivory (Abrams 1993). Thus, we cannot infer a particular mechanism or model of trophic interaction based on patterns observed in nature. What we *can* say is that resource limitation and predator limitation should often act in concert to limit the abundances of organisms in ecosystems.

Thus, initial attempts to characterize ecosystem regulation as top-down versus bottom-up presented a false dichotomy. However, the relative strengths of resource and predator limitation at any trophic level may differ depending on the nature of interactions between consumers and resources or on ecosystem productivity. Recall that we employed a "simple thought experiment" earlier in this chapter to illustrate how short-term trophic-level responses to resource additions or predator removals might estimate the relative strengths of limitation by these two factors. In fact, ecologists have conducted many such experiments in the natural world, and in a variety of ecosystems. Recent meta-analyses of these experiments reveal some very interesting features of predator limitation and resource limitation in different ecosystems (these analyses are often placed within the context of top-down and bottom-up control). We will examine the results from these meta-analyses shortly. But first, let's look at some examples of strong top-down control in nature, focusing on **trophic cascades** and the impact of top predators on the abundance and distribution of species at lower trophic levels.

Trophic cascades and the relative importance of predator and resource limitation

Hairston, Smith, and Slobodkin (1960) sparked the study of trophic cascades by proposing that the consumption of herbivores by predators keeps the terrestrial world green. The first experimental demonstrations of trophic cascades came not from terrestrial ecosystems, however, but from studies of marine and freshwater communities. Paine's (1966) pioneering experiment showing the consequences of removing a predatory starfish (*Pisaster*) for the structure of an intertidal community is often cited as the first experimental demonstration of a trophic cascade (a term Paine himself coined in 1980). Other examples from freshwater systems are equally dramatic and well known; impacts of piscivorous fish, for example, cascade down to the producer trophic level, affecting algal abundance and water quality (e.g.,

Figure 11.12   A sea otter consuming a sea urchin, one of its favorite prey. (Photograph © Tom Mangelsen/Naturepl.com.)

Carpenter and Kitchell 1993; see Shurin et al. 2010 for a review and Figure 10.8 for an example).

Perhaps the most remarkable of all trophic cascades, both in terms of the geographic scope of the interaction and the charismatic nature of the players, is the effect of sea otters on marine kelp forests in the Pacific Northwest. Sea otters (*Enhydra lutris*) ranged widely across the North Pacific Rim until the late 1700s, when their discovery by European explorers sparked a fur trade that led to their being hunted to near extinction. Pockets of sea otter populations survived the decimation in Russia, southwestern Alaska, and central California. Since an international ban on their hunting was imposed in 1911, otter populations have become successfully reestablished in many areas, and sea otters are again found from Russia to Alaska to California. Jim Estes and his colleagues (e.g., Estes and Palmisano 1974; Estes and Duggins 1995) compared islands in the Aleutian archipelago where sea otter populations had recovered to islands where they had not. They found that islands with abundant otter populations had low densities of herbivorous sea urchins (the otter's favored prey; **Figure 11.12**) and extensive stands of kelp (*Macrocystis*). However, in the absence of otters, urchin populations boomed, and their grazing resulted in the creation of "urchin barrens" that were devoid of kelp and other attached algae. Thus, the very existence of the kelp forest along these islands, as well as the diverse fauna of fish and invertebrates found therein, was shown to be linked to the presence of sea otters—a dramatic example of a cascading effect (**Figure 11.13**, left). Studies in other parts of the otter's range, however, found somewhat different results. For example, in the kelp forests off Southern California, a diverse array of predators (including lobsters and fish) appears able to control urchin populations even in the absence of sea otters (Estes et al. 1989; Tegner and Dayton 2000; Steneck et al. 2002). Recent studies have shown that increasing predator diversity may strengthen trophic cascades

1972–1990

1991–1997

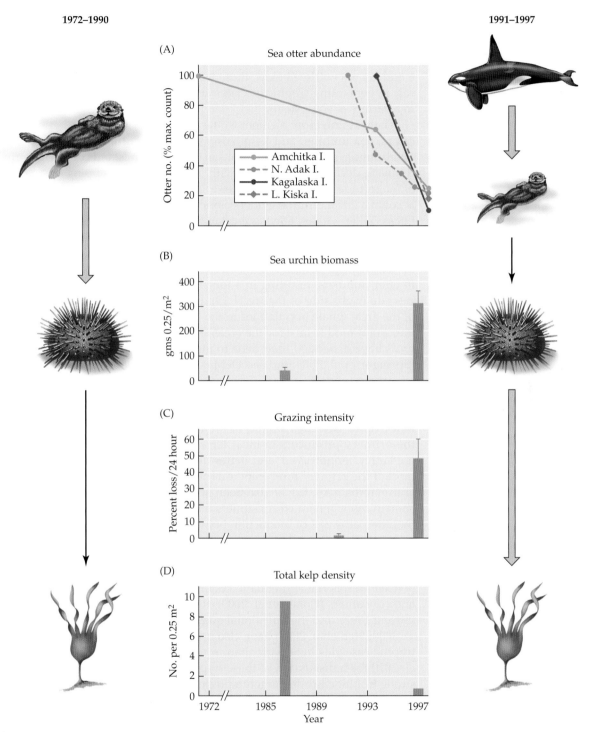

(A) Sea otter abundance

Otter no. (% max. count)

Legend:
— Amchitka I.
- - ◆ - - N. Adak I.
—●— Kagalaska I.
- - ◆ - - L. Kiska I.

(B) Sea urchin biomass

gms 0.25 / m²

(C) Grazing intensity

Percent loss / 24 hour

(D) Total kelp density

No. per 0.25 m²

Year
1972   1985   1989   1993   1997

Figure 11.13 Examples of strong trophic cascades in a nearshore marine ecosystem. In the 1970s, abundant sea otter populations along the Aleutian Islands limited the numbers of sea urchins, reducing urchin grazing intensity and promoting dense stands of kelp (left). However, what appears to be a shift in the foraging behavior of a small number of orcas (killer whales) in the 1990s led to a dramatic decline in sea otter numbers, an increase in urchin biomass and urchin grazing, and a decline in kelp (right). (After Estes et al. 1998.)

in these kelp forests by modifying herbivore behavior (Byrnes et al. 2006). We discussed nonlethal (nonconsumptive) effects of predators in Chapter 6 and, as we will see below, these trait-mediated effects can play a large role in generating trophic cascades and controlling trophic-level biomass in a variety of ecosystems.

There is one more element to the fascinating interaction between sea otters, sea urchins, and kelp. After nearly a century of recovery from overhunting, sea otter populations in the Aleutian Islands began suffering dramatic declines in the 1990s, with otter densities dropping by an order of magnitude in some areas (Estes et al. 1998). The cause of these declines is controversial, but the evidence once again points to a cascading effect. Orcas (killer whales, *Orcinus orca*) in the Aleutians appear to have shifted their foraging behavior to include sea otters in their diet. Calculations by Williams et al. (2004) suggest that such a change in feeding behavior by even a very small number of orcas could have caused the observed decline in otter numbers, resulting in a four-level instead of a three-level trophic cascade (right side of Figure 11.13). Industrial whaling and the collapse of the great whale populations in the North Pacific—which eliminated an important food source for the orcas—has been proposed as a reason why orcas have expanded their diets to include sea otters, seals, and sea lions (Springer et al. 2003; but see DeMaster et al. 2006; Wade et al. 2009 for counterarguments).

Strong trophic cascades have been suggested to be less common in terrestrial than in aquatic ecosystems. For example, Strong (1992) argued that trophic cascades should be weak in terrestrial ecosystems because the primary producers in most terrestrial ecosystems (vascular plants) are larger, longer-lived, and better defended against herbivory than are most of the primary producers in aquatic ecosystems (e.g., phytoplankton and macroalgae). It has also been suggested that terrestrial food webs contain more generalist consumers than aquatic food webs, and therefore that the pathways of interaction between trophic levels are more diverse and reticulate (branching) in terrestrial than aquatic ecosystems, again potentially weakening the propagation of top-down effects (Polis and Strong 1996; Hillebrand and Shurin 2005; Frank et al. 2006). Differences between terrestrial and aquatic systems in the relative sizes of producers and consumers (Cohen et al. 2003), the degree of omnivory in food webs (Strong 1992), and the nutritional quality of plants (stoichiometry; Elser et al. 2000; Hall et al. 2007) have also been hypothesized to lead to stronger trophic cascades in aquatic than in terrestrial ecosystems (see Shurin et al. 2010 for a discussion of these hypotheses). Are top-down effects and strong trophic cascades largely a property of aquatic systems—or, as Strong (1992) quipped, "Are trophic cascades all wet"? A number of meta-analyses have tested this idea by comparing the strengths of trophic cascades in experiments that manipulated consumer densities in freshwater, marine, and terrestrial ecosystems. The results of these meta-analyses have recently been synthesized and discussed in an excellent paper by Shurin et al. (2010), whose findings I summarize below.

HOW COMMON ARE TROPHIC CASCADES IN DIFFERENT ECOSYSTEMS? Consistent cascading effects of predators on plants were reported by Schmitz et al. (2000) in their analysis of more than 40 terrestrial predator removal studies. They concluded that the magnitude of these cascading effects in terrestrial ecosystems was similar to that observed by others in aquatic ecosystems. Furthermore, they concluded that the indirect effects of predators on plant damage were much greater than their indirect effects on plant biomass and reproduction (Schmitz et al. 2000). Shurin et al. (2002) focused solely on the effects of predator removal on plant biomass and compared effect sizes across terrestrial, marine, and freshwater ecosystems. They found that predator removal had a greater impact on plant community biomass in aquatic than in terrestrial ecosystems; on freshwater than on marine plankton; and in marine benthic (bottom) habitats than in marine open-water habitats. Gruner et al. (2008), in their meta-analysis of 191 factorial (cross-classified) manipulations of herbivores and nutrients, also found evidence that aquatic systems showed stronger trophic cascades. Herbivore removal generally increased producer biomass in both freshwater and marine systems, but its effects were inconsistent on land. A recent meta-analysis restricted to effects of birds on plants showed that birds had significant positive effects on plant survival and plant leaf damage (Mäntylä et al. 2011).

Although the weight of the evidence suggests that top-down effects are more pronounced in aquatic than in terrestrial ecosystems, it is clear that trophic cascades are commonly found in both habitats. Moreover, Shurin et al. (2010) note that some of the best examples of terrestrial cascades (e.g., Schoener and Spiller 1999; Terborgh et al. 2001; Ripple and Beschta 2007) were not included in the meta-analyses discussed above because they did not report plant community biomass. Top-down effects on ecosystem properties other than plant biomass and plant leaf damage have not been well quantified. Schmitz (2006) showed that spiders had little effect on plant biomass in an old-field community, but significantly affected the relative abundances of plant species. It is unknown whether species turnover in response to changing predator densities (Leibold 1996) is greater in some ecosystems than in others.

Although trophic cascades provide strong evidence for the importance of top-down processes (Schoener 1989; Terborgh and Estes 2010), and although they have been viewed as supporting the predictions of HSS (e.g., Power 1990; Spiller and Schoener 1990), the existence of a trophic cascade says little about the *relative* importance of limitation imposed by predators versus resource limitation. As noted earlier in this chapter, the relative importance of predator and resource limitation can be measured experimentally through short-term removals of these limiting factors (see Figure 11.3). In one of the most comprehensive meta-analyses to date, Gruner et al. (2008) compared 191 factorial (cross-classified) manipulations of herbivores and nutrients from freshwater ($n = 116$), marine ($n = 60$), and terrestrial ($n = 15$) ecosystems in a quantitative assessment of the relative effects of fertilization and herbivory on plant biomass (**Figure 11.14**). Because

Figure 11.14 Results of a meta-analysis of 191 factorial manipulations (cross-classified for all possible combinations of fertilization and herbivore removal) show that herbivores and nutrients control plant community biomass to similar degrees across freshwater, marine, and terrestrial systems, and that these two processes act together in an additive fashion (i.e., interactive effects were minimal). LRR (natural log response ratio) measures effect size of fertilization or herbivore removal or their interaction on plant biomass. (After Gruner et al. 2008.)

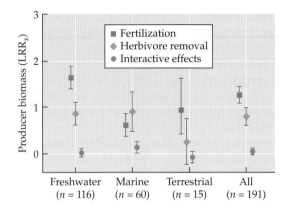

the experimental treatments were cross-classified, Gruner et al. could also assess whether the effects of herbivores and nutrients acted independently (in an additive fashion) or if there was a positive or negative synergy (a significant interaction between treatments). They concluded that herbivores and nutrient supply control plant community biomass to similar degrees across freshwater, marine, and terrestrial systems, and that these two processes act together in an additive fashion (i.e., that interactive effects are minimal). They then concluded (Gruner et al. 2008: 748) that "the near equivalence and marked independence of fertilization and herbivore effect sizes … provides additional justification to retire the antediluvian notion that *either* top-down or bottom-up forces predominantly control plant biomass within major ecosystem types."

Hillebrand et al. (2007) similarly analyzed the effects of herbivores and nutrients on plant species richness and plant species evenness in different ecosystems. They found that fertilization consistently reduced plant species richness and evenness in terrestrial systems, but tended to increase species richness in aquatic systems while decreasing evenness. Herbivory enhanced plant species evenness across ecosystems and promoted species richness in terrestrial, but not aquatic, ecosystems.

TROPHIC CASCADES AND NONCONSUMPTIVE (TRAIT-MEDIATED) EFFECTS As we learned in Chapter 6, predators have both lethal (consumptive) and nonlethal (nonconsumptive) effects on their prey. Predator-induced changes in prey behavior, habitat use, and morphological and chemical defenses can dramatically alter the nature of the interaction between a prey population and the prey's own resources. Thus, although HSS and subsequent studies (Paine 1980; Oksanen et al. 1981) focused on top-down effects in food chains that were driven by changes in predator and prey densities (or, more generally, consumer and resource densities), it is now clear that trophic cascades often result from a combination of consumptive (also called density-mediated) and nonconsumptive (also called trait-mediated) effects (Abrams 1995; Schmitz 2010).

Some 20 years ago, my graduate student Andy Turner and I discovered an early example of nonconsumptive trophic cascade when we stocked bluegill sunfish into two ponds that had been partitioned into quadrants with and without the bluegill's predator, the largemouth bass (Turner and Mittelbach 1990). The bass had no significant effect on bluegill mortality, as the bluegill retreated to the protection of the nearshore vegetation when bass where present. However, this predator-induced habitat shift significantly reduced bluegill predation on herbivorous zooplankton found in the open water (e.g., *Daphnia*), leading to a trophic cascade in which the bass had positive indirect effects on the abundance of herbivorous zooplankton (Turner and Mittelbach 1990). Similarly, He and Kitchell (1990) documented a nonconsumptive trophic cascade in a whole-lake experiment, in which adding piscivorous northern pike (*Esox lucius*) caused a rapid habitat shift in the prey fish population: 50%–90% of the total change in prey fish numbers resulted from the emigration of prey fish through an outlet stream.

These and other early studies of behaviorally induced trophic cascades in fresh waters (e.g., Power et al. 1985) were followed by a wealth of empirical and theoretical studies examining the nonconsumptive effects of predators in terrestrial and aquatic food webs (reviews in Werner and Peacor 2003; Abrams 2010; Schmitz 2010; Terborgh and Estes 2010). Attempts to measure the relative impacts of density-mediated and trait-mediated cascading effects across studies have concluded that trait-mediated indirect effects are as important as density-mediated indirect effects (e.g., Bolnick and Preisser 2005; Preisser et al. 2005), and that a simple foraging gain versus predation risk trade-off can explain much of the variation in the strength of trophic cascades in terrestrial ecosystems (Schmitz et al. 2004). However, as discussed in Chapter 6, estimating the relative importance of consumptive and nonconsumptive effects of predators in natural food webs is difficult. Abrams (2010) has concluded that the experimental and analytical methods employed to date probably overestimate the relative importance of trait-mediated effects. Still, as Abrams (2010: 15) notes, "There is little doubt that adaptively flexible foraging and defense behaviors have the potential to alter the population, community, and evolutionary dynamics of species that comprise food webs."

Trait-mediated trophic cascades need not affect plant biomass to have a strong and significant impact on plant community structure and ecosystem functioning. Schmitz (2003, 2010) provides an excellent example of this in his experimental studies of the impacts of predatory spiders in old-field food webs. Schmitz's (2010) study system focuses on strong interactions between spiders (predators), grasshoppers (herbivores), and plants (grasses and forbs). The spider *Pisaurina mira* is a sit-and-wait predator whose presence causes grasshoppers (*Melanoplus femurrubrum*) to switch from feeding on their preferred plant species (a grass, *Poa pratensis*) to feeding on a less nutritious (but safer) plant species, the goldenrod *Solidago rugosa*. Thus, these spiders exert most of their top-down control by altering grasshopper foraging behavior rather than grasshopper population density. Because *S. rugosa* is a competitively superior plant species in this system, the net result of the predator-induced shift in herbivore foraging behavior is a change in plant diversity and productivity. In the absence of *P. mira*, *S. rugosa* dominates the plant community. But where this predatory spider is present, herbivores suppress the growth of *S. rugosa*, reducing its competitive effects on other plants species. Thus, the spider's presence increases plant diversity (measured as species evenness), but decreases the overall productivity of the plant community (**Figure 11.15A**).

Schmitz (2008) went on to show how the impact of this trait-mediated trophic cascade depended on the hunting mode of the predator. When an actively roaming spider (*Phidippus rimator*) was substituted for the sit-and-wait spider *P. mira*, grasshoppers did not switch to feeding on *Solidago*—instead, they fed on their preferred resource, the grass *P. pratensis*. As a result, there was a 14% reduction in plant species evenness, a 33% increase in nitrogen mineralization (a measure of nutrient recycling), and a 163% increase in aboveground productivity, all due to a simple difference in the foraging mode of the top predator and the behavioral response of its prey (**Figure 11.15B**). More work is needed to determine whether such dramatic cascading effects of predator foraging mode occur in more complex communities where a variety of species (with differing traits) are found at each trophic level.

There are many other examples of trait-mediated cascading effects of predators in terrestrial and aquatic ecosystems (Werner and Peacor 2003; Schmitz 2010). For example, in subtidal marine systems, an increase in predator species richness has been shown to decrease herbivore foraging activity or herbivore per capita feeding rate, leading to a positive indirect effect of predators on plants (e.g., Byrnes et al. 2006; Stachowicz et al. 2007). Adaptive foraging behaviors and other trait-mediated responses can also change the bottom-up effects of resource enrichment on trophic-level abundances. For example, if consumers at intermediate trophic levels face a foraging gain–predation risk trade-off, they may reduce their foraging effort in favor of increased safety when food becomes more abundant. Such a behavioral response will limit the bottom-up flow of resources to upper trophic levels (Abrams 1984, 2010). However, unlike top-down interactions, in which the trait-mediated effects of predators on their prey's resources are always expected to be positive, the bottom-up effect of resources on predator abundance may (in theory) be either positive or negative, depending on the behavioral response of the prey (Abrams 1984; Werner and Anholt 1993; Abrams 2010). To date, there has been little empirical work on this question, and we cannot say which outcome is more likely (Abrams 2010).

What determines food chain length?

It is rare to find natural ecosystems with fewer than three trophic levels (producers, herbivores, and carnivores), and those ecosystems occur only at the very lowest productivity levels, such as near the limits of vegetation in deserts or the high Arctic (Aunapuu et al. 2008; Terborgh et al. 2010). However, food chains with four, five, six, or more trophic levels exist in a number of places. Ecologists have long questioned what limits the length of food chains in nature (e.g., Elton 1927). As we have seen in this chapter, the number of food chain links can have important effects on the pattern of trophic-level regulation (e.g., see Figure 11.4) and the nature of cascading effects. In addition, the extent of bioaccumulation of contaminants in the food chain depends on the number of trophic links from a basal species to a top predator (e.g., Cabana and Rasmussen 1994).

(A)

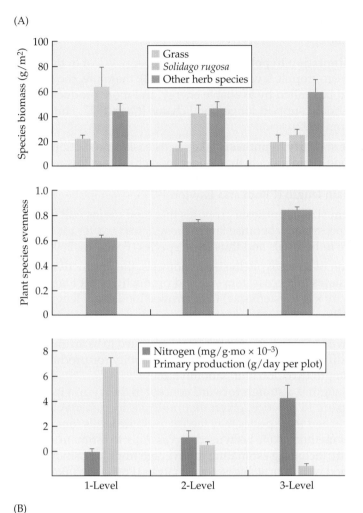

(B)

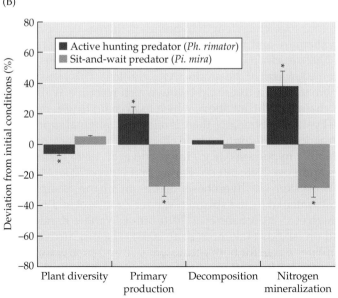

Figure 11.15 Behavioral responses to predators can lead to strong cascading effects. The spider *Pisaurina mira* is a sit-and-wait predator whose presence causes grasshoppers to switch from feeding on their preferred plant species (a grass, *Poa pratensis*) to feeding on goldenrod (*Solidago rugosa*). Because *S. rugosa* is competitively dominant, the net result of the predator-induced shift in herbivore foraging behavior is a change in plant diversity and productivity. (A) In the absence of *P. mira* (a two-trophic-level system), *S. rugosa* is abundant in the plant community. But in the presence of the spider (a three-level system), grasshoppers suppress the growth of *S. rugosa*, reducing its competitive effect on other plant species, and thus increasing plant diversity (measured as species evenness), but decreasing the overall productivity of the plant community. (B) This trait-mediated trophic cascade depends on the hunting mode of the predator. When an actively roaming spider (*Phidippus rimator*) was substituted for the sit-and-wait spider *P. mira*, grasshoppers did not switch to feeding on *Solidago*; instead, they fed on *P. pratensis*. As a result, there was a 14% reduction in plant diversity in the actively hunting predator treatments compared with the sit-and-wait predator treatments; a 163% increase in aboveground productivity in the active predator treatments compared with the sit-and-wait treatments; no difference in plant matter decomposition rates between treatments; and a 33% increase in nitrogen mineralization in the active as opposed to the sit-and-wait predator treatments. Values are mean ±1 SD. Asterisks indicate significant treatment effects ($p < 0.05$) by *t*-test. (After Schmitz 2008, 2010.)

Four major hypotheses have been proposed to explain why food chain length differs among ecosystems (Post 2002):

1. The **energy limitation hypothesis** notes that energy is lost in the transfer between trophic levels (Lindeman 1942); therefore, food chain length should be limited by available energy, and more productive environments should support longer food chains (Elton 1927; Hutchinson 1959; Pimm 1992).

2. The **dynamic stability hypothesis** takes as its basis a prediction from theoretical food chain models that longer food chains are less resilient to disturbance; therefore, disturbance frequency or intensity should limit food chain length (Pimm and Lawton 1977).

3. The **ecosystem size hypothesis** proposes that food chains will be longer in larger ecosystems because larger ecosystems (greater total area) support more individuals and thus more species (Post et al. 2000).

4. A modification of the ecosystem size hypothesis, the **productive space hypothesis**, states that ecosystem size and productivity act together (productivity × ecosystem size) to determine food chain length (Schoener 1989).

Recently, adaptive foraging behavior has been suggested to be an important component modifying the impact of energy availability (or productivity) on food chain length (Kondoh and Ninomiya 2009).

Food chain length is a straightforward concept in theory, but it is complicated to measure in practice, given that multiple species may occupy each trophic level and that species may feed at multiple levels (omnivory). Still, a variety of methods have been used successfully to estimate average food chain length, including estimating the average trophic position of top predators in a food web from stable isotope analysis (Post 2002; Takimoto et al. 2008). Recent comparative studies from aquatic and terrestrial environments allow us to evaluate the four major hypotheses listed above.

Although the energy limitation hypothesis is the oldest and most widely discussed of the four food chain length hypotheses, there is little evidence that resource availability and food chain length are positively correlated in nature (Post 2002), aside from the observation that food chains may be short at the extreme low end of the productivity gradient. This observation suggests that there may be a threshold below which resource availability constrains food chain length, but above which food chain length depends on other factors (Post 2002). Similarly, there is relatively minimal support for the role of disturbance in determining food chain length in natural systems (the dynamic stability hypothesis) (e.g., Townsend et al. 1998; Takimoto et al. 2008; Walters and Post 2008; but see Parker and Huryn 2006; McHugh et al. 2010). The ecosystem size hypothesis, on the other hand, is supported by studies in freshwater lakes and streams (Vander Zanden et al. 1999; Post et al. 2000; McHugh et al. 2010; Sabo et al. 2010; but see Vander Zanden and Fetzer 2007) and on islands (Takimoto et al. 2008). The productive space hypothesis postulates a role for both ecosystem size and productivity per unit size in determining food chain length (Schoener 1989). To date, the ecosystem size component of the hypothesis has been supported, with relatively less support for the influence of productivity per unit size (Post 2002). Thus, although the question of what determines food chain length in ecosystems is far from settled, the evidence suggests that ecosystem size, perhaps in combination with ecosystem productivity, is a primary driver of food chain length in aquatic and terrestrial ecosystems.

Conclusion

We started this chapter by asking whether complex networks of species interactions might be simplified into chains of linked consumer–resource populations in order to address the question of what controls the abundances of plants, herbivores, and carnivores in ecosystems. Not surprisingly, Hairston, Smith, and Slobodkin's 1960 explanation for our "green world" is deficient in many ways. However, their fundamental insight—that we can apply our knowledge of density-dependent processes and species interactions to ecosystem-level questions—opened up a whole new way to look at food webs and trophic dynamics. We have seen how top-down and bottom-up forces combine to determine the abundances of organisms at different trophic levels, and how the strength of these forces may vary across gradients in ecosystem productivity. Trophic cascades are evidence of the importance of top-down control in many ecosystems, and they highlight the consequences of the loss of top predators from ecosystems worldwide (Estes et al. 2011). Moreover, simplifying complex food webs into model food chains has brought to light the important role that species diversity within a trophic level plays in determining a community's responses to resource enhancement or to the addition or removal of top predators. Particularly important are species' traits related to competitive ability and predator avoidance and whether species at intermediate trophic levels may respond behaviorally to changes in resource or predator abundance (trait-mediated effects).

One important aspect of food webs that we have not addressed in this chapter is the question of system stability in the face of environmental change. In Chapter 14, we will see how top-down and bottom-up forces may lead to dramatic changes in the structure and functioning of communities and ecosystems (e.g., regime shifts and alternate stable states), and we will look at the consequences of variable environments for species interactions, species diversity, and community stability.

Summary

1. The "world is green hypothesis" of Hairston, Smith, and Slobodkin ("HSS") proposed that (1) in the absence of higher level predation, carnivores should be limited by competition for their food (herbivores) and, consequently, (2) herbivore populations should be held below their carrying capacity and have little impact on plants; and finally (3) in the absence of control by herbivores, plants should be dense and limited by competition.

2. Oksanen et al. applied consumer–resource equations to interacting trophic levels. They modeled food chains as linked consumer–resource interactions. This model predicts that whether a trophic level responds to an increase in potential productivity depends on the number of trophic levels in the system, and that adjacent trophic levels will show an alternating pattern of response.

3. Although the stepped response pattern predicted by Oksanen et al. appears to support HSS's hypothesis that competition and predation alternate in importance in controlling trophic levels, a more accurate assessment is that each trophic level below the top one is simultaneously limited by competition and predation. Initial attempts to characterize regulation as top-down versus bottom-up presented a false dichotomy; the relative strengths of predator limitation and resource limitation vary with trophic level and ecosystem productivity.

4. Experiments show that organisms arrayed in food chains and food webs respond to changes in potential productivity as predicted by consumer–resource models.

5. Most natural systems show increases in the abundances of all trophic levels with an increase in potential productivity. A simple diamond-shaped food web model illustrates how heterogeneity in species composition within a trophic level, coupled with a trade-off between competitive ability and vulnerability to predation, can lead to this result, as well as to species replacements along productivity gradients, from good resource competitors to good predator resisters.

6. Shifting species composition within a trophic level can prevent predator control and maintain the importance of resource limitation, even in high-nutrient systems.

7. Although the weight of the evidence suggests that top-down effects are more pronounced in aquatic than in terrestrial ecosystems, trophic cascades are commonly found in both habitats. Trophic cascades provide strong evidence for the importance of top-down processes, but the existence of a trophic cascade says little about the relative importance of predator limitation versus resource limitation.

8. Strong cascading effects result from both the consumptive (density-mediated) and nonconsumptive (trait-mediated) effects of predators.

9. Although natural ecosystems with fewer than three trophic levels are found only at very low productivity levels, there is little evidence that resource availability and food chain length are positively correlated in nature (the energy limitation hypothesis). Similarly, there is minimal support for the role of disturbance in determining food chain length in natural systems (the dynamic stability hypothesis). The ecosystem size hypothesis, which proposes that food chains will be longer in larger ecosystems (because larger systems support more individuals and thus more species) is the best supported, although productivity per unit size may also play a role (the productive space hypothesis).

Spatial Ecology

Metapopulations and Metacommunities

12 Patchy Environments, Metapopulations, and Fugitive Species

Species interactions take place in a spatial context.

Martha Hoopes et al., 2005: 35

It is conceivable that species ... may actually be able to exist primarily by having good dispersal mechanisms, even though they inevitably succumb to competition with any other species capable of entering the same niche. Such species will be termed fugitive species.

G. Evelyn Hutchinson, 1951: 575

Up to this point, we have focused on populations and species without considering their spatial distribution in the environment. By assuming that individuals mix freely in a homogeneous environment, we have been able to use simple models to understand the fundamental machinery of predation, competition, and facilitation, and we have been able to combine these species interactions into modules to explore the functioning of food webs and other ecological networks. However, it is time for us to relax the assumption of environmental homogeneity, as there are important ecological phenomena that depend critically on the fact that populations and species may be distributed heterogeneously across the landscape. We begin our exploration of these phenomena by considering the case of a single species in a patchy environment.

Metapopulations

H. G. Andrewartha and Charles Birch, in their landmark 1954 text *The Distribution and Abundance of Animals*, were the first to advocate thinking about populations in a spatial context. Their experience with studying insects suggested that local populations experience large fluctuations in abundance, potentially disappearing for years only to be reestablished at a later time by colonists from other populations. Their premise—that populations occur in **patches** of suitable habitat surrounded by areas of unsuitable habitat, that such local populations are connected by the migration of individuals, and that local populations may be reestablished by colonists from other populations after going extinct—describes what we now call a **metapopulation,** or "population of populations." Unfortunately, Andrewartha and Birch's early insights into metapopulations appear to have been lost in an ongoing debate over density dependence and density independence (Hanski and Gilpin 1991; see Chapter 1), and 15 years passed before Richard Levins (1969, 1970) formally introduced the concept of the metapopulation and developed a mathematical model to describe its dynamics.

The classic Levins metapopulation model

The **Levins metapopulation model** is simple in form (Levins 1969, 1970) and serves as a foundation for all other metapopulation models (Hanski 1997). Like the logistic equation, it can be viewed as a paradigm of population growth, but in this case, population growth in a patchy environment (Hanski and Gilpin 1991). The basic assumptions of the Levins model are as follows:

1. The environment is composed of a large number of discrete patches, all identical and all connected to each other via migration (i.e., dispersal is global).

2. Patches are either occupied or not (actual sizes of populations within patches are ignored, and we assume that each colonized patch quickly reaches its carrying capacity). Species abundance is described by the fraction of total patches occupied (P).

3. Populations within patches have a constant (per patch) rate of extinction (m).

4. The rate of patch colonization is proportional to the per patch colonization rate (c) times the fraction of currently occupied patches (P) times the fraction of currently empty patches ($1 - P$), which are the targets of colonization.

Given the above assumptions, the rate of population growth can be described as

$$dP/dt = cP(1-P) - mP \qquad \text{Equation 12.1}$$

Rearranging this equation yields

$$dP/dt = (c - m)P\left(1 - \frac{P}{1 - m/c}\right) \qquad \text{Equation 12.2}$$

Equation 12.2 shows that the Levins model is directly analogous to the well-known logistic equation (see Chapter 4). Like the logistic model, the Levins model has a globally stable equilibrium point, and the population size at equilibrium (i.e., the fraction of patches occupied when $dP/dt = 0$) is equal to

$$\hat{P} = 1 - m/c \qquad \text{Equation 12.3}$$

The major conclusions (Nee and May 1992) of the Levins metapopulation model are:

1. At equilibrium (steady state), there will always be some unoccupied patches in the environment. The fraction of occupied patches at equilibrium is $\hat{P} = 1 - m/c$, and the fraction of unoccupied patches is $1 - \hat{P} = m/c$.

2. Decreasing the number of patches in the environment (for example, if habitat destruction permanently removes a fraction D of the patches from the system) will decrease the rate of population growth as

$$dP/dt = cP(1 - D - P) - mP \qquad \text{Equation 12.4}$$

3. The population reaches an "extinction threshold" (where the equilibrium fraction of occupied patches is zero) when the fraction of patches permanently removed from the system is $D = 1 - m/c$.

The Levins model is simple in form, but it captures the essence of metapopulation dynamics and is very general in its application. Like other general models we have studied (e.g., the logistic growth model, the Rosenzweig–MacArthur predator–prey model), the Levins model greatly simplifies the real world. It is useful, however, in that it offers important insights into the roles of dispersal and local extinction in determining the dynamics and persistence of patchily distributed populations and species. As we will see when we discuss metapopulations and metacommunities (Chapter 13), **dispersal limitation** occurs when a species is unable to occupy all suitable patches in the environment, and the degree to which species are dispersal-limited has important consequences for population dynamics, species coexistence, and community structure.

Implications of the metapopulation model for conservation biology

Metapopulation theory has largely replaced the theory of island biogeography (MacArthur and Wilson 1967; see Chapter 2) as a conceptual underpinning for the field of conservation biology (Merriam 1991; Hanski and Simberloff 1997). In particular, the Levins model helps us understand how habitat loss may affect species persistence. Its first important insight is that we should expect to find unoccupied patches of suitable habitat in all metapopulations. In the Levins model, the fraction of unoccupied patches at equilibrium is $1 - \hat{P} = m/c$. Thus, a significant fraction of unoccupied patches should be found in metapopulations of species that have low rates of colonization (c) relative to their rate of extinction within patches (m). Moreover, the model shows that the presence of unoccupied patches in the landscape should not be viewed as an indication that there are "spare" or "extra" habitats into which a species might expand, or that some of these unoccupied habitats could be eliminated with no consequence to the species. Instead, decreasing the number of patches decreases the rate of population growth (Equation 12.4). Further, if the number of patches is decreased sufficiently (that is, to the point where $D = 1 - m/c$) the species will disappear from the landscape.

The second important insight from the Levins model is that the disappearance of a metapopulation due to habitat loss does not happen immediately, but may occur generations after the critical loss of habitat. Tilman et al. (1994) referred to this delay as the **extinction debt**; Hanski (1997) conjures up an even more chilling description by describing the members of metapopulations facing slow extinction as the "living dead." Therefore, as Hanski (1997: 88) notes, "Conservationists should dismiss the false belief that protecting the landscape in which a species now occurs is necessarily sufficient for long-term survival of the species."

In other words, preserving existing habitat may not be enough to save a species whose dynamics are those of a metapopulation. For example, Hanski and Kuussaari (1995) suggest that about 10% of the resident butterfly species in Finland may be facing extinction debt. On the bright side, however, is the fact that the decline toward extinction can be slow, giving conservationists time to intervene and potentially save a species by increasing the number of available habitat patches, or by increasing the rate of dispersal among patches (e.g., by establishing habitat corridors or assisting the migration of individuals between patches). As Equation 12.3 shows, species abundance in a metapopulation increases as the rate of dispersal between patches increases. If dispersal between habitat patches is restricted by a surrounding landscape of unfavorable or impenetrable habitat (e.g., roads, agricultural fields, urbanized areas), then the establishment of "movement corridors" that allow individuals to more readily disperse between patches should increase the overall population size.

Even though ecologists and conservation biologists were quick to recognize the potential for corridors of favorable habitat to help maintain species in fragmented landscapes, scientific evidence supporting this idea was slow to materialize (Simberloff et al. 1992; Van Der Windt and Swart 2008). Simberloff et al. (1992) were particularly critical of the early, unbridled acceptance of movement corridors as a conservation tool, noting that corridors also have potential negative consequences, such as increasing the spread of disease or of competitor and predator species, or enabling the invasion of exotic species (Simberloff and Cox 1987).

In the decades since Simberloff et al.'s 1992 paper, there have been a number of experimental tests of the efficacy of corridors in promoting species dispersal and persistence. For example, Gillies and St. Clair (2008) found that corridors facilitated the movement of tropical bird species between isolated forest patches. They began their study by capturing territorial individuals of barred antshrikes (*Thamnophilus doliatus*) and rufous-naped wrens (*Campylorhynchus rufinucha*) in a highly fragmented tropical dry forest of Costa Rica. They translocated the birds 0.7–1.9 km from their "home" territories, placing them in several different habitat types. Then, using radiotelemetry and GPS, they follow the movements of the individual translocated birds. Barred antshrikes, which are forest specialists (normally found in only the most intact forest in this region), returned to their territories faster, and with greater success, in treatments containing corridors of forest trees. However, return rates of rufous-naped wrens, which are forest generalists (found in both intact and degraded forests), were about the same whether forest corridors were available or not (**Figure 12.1**). A meta-analysis (Gilbert-Norton et al. 2010) of 78 corridor experiments from 35 studies concluded that corridors increase movement between habitat patches by approximately 50%

(A)

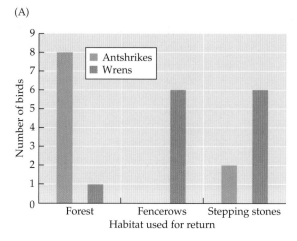

(B)

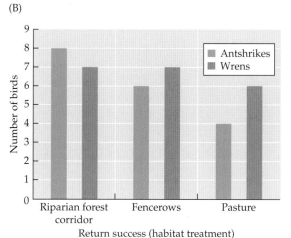

Figure 12.1 The dispersal of tropical forest-dwelling birds is enhanced by the existence of habitat corridors. Individual barred antshrikes (*Thamnophilus doliatus*; photo) and rufous-naped wrens (*Campylorhynchus rufinucha*) were removed from their territories in a tropical dry forest in Costa Rica and transported 0.7–1.9 km distant into three habitat treatments: (1) riparian forest corridors, (2) fencerows, and (3) pastures. Ten birds of each species were released in each treatment; the birds were followed using radiotelemetry and GPS. (A) Antshrikes, which are forest specialists, primarily used riparian forest corridors to return to their territories. Wrens, which are forest generalists, used a combination of habitat types (including fencerows and small patch "stepping stones") to return to their territories. (B) The return success of antshrikes was significantly greater when riparian forest corridors were available than in the other two treatments. Wrens showed no difference in return success by treatment type. (After Gillies and St. Clair 2008; photograph courtesy of Brian Gratwicke.)

compared with patches that are not connected by corridors. Gilbert-Norton et al. also found that natural corridors (i.e., those existing in landscapes prior to the studies) facilitated movement more than corridors created and maintained for the studies.

Movement corridors also may serve to increase species richness. In a large-scale experiment, Damschen et al. (2006) examined the long-term effect of corridors on plant species richness by experimentally creating patches of longleaf pine habitat at the Savannah River Ecological Laboratory in South Carolina (**Figure 12.2A**). They found that habitat patches connected by corridors retained more native plant species than did isolated patches, and that this difference increased over time (**Figure 12.2B**). Thus, the available evidence indicates that corridors can be a significant tool in the conservation of populations in fragmented landscapes.

The assisted dispersal of species, which has been proposed as a way of ameliorating the effects of climate change, is a much more controversial measure. Global climate change is occurring at a rapid rate, and there is clear evidence that species are shifting their ranges in response to changing

Figure 12.2 The presence of movement corridors can affect species richness. (A) One of six experimental study landscapes where Damschen et al. constructed large (100 m × 100 m) patches of longleaf pine habitat within a surrounding matrix of pine plantation. The patches mimic native longleaf pine habitat, which is characterized by an open and species-rich understory maintained by frequent low-intensity fires. Some patches were connected by corridors to other patches; others were not. (B) Patches connected by corridors were found to be richer in plant species native to longleaf pine habitat than unconnected patches, and that this difference in native species richness increased over time. The number of exotic plant species was unaffected by the presence of corridors. Data were unavailable for 2004, when patches were burned by the U.S. Forest Service as part of restoration management. (A courtesy of USDA Forest Service; B after Damschen et al. 2006.)

(A)

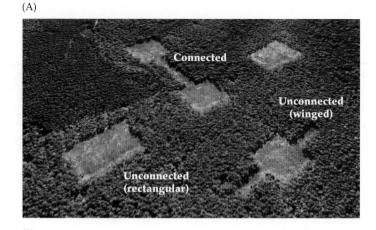

(B)

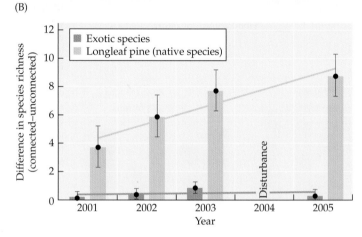

environments. The ranges of a variety of organisms have shifted poleward at average rates of 6–17 km per decade over the past 50 years (e.g., Parmesan and Yohe 2003; Chen et al. 2011). For some species, these rates of range shift have kept pace with changing environments. However, there is concern that less mobile species may not be able to keep up with climate change. Studies of past extinctions (such as those of the Pleistocene) show that species with small ranges are particularly vulnerable to rapid climate change (Sandel et al. 2011). Moreover, past rates of climate change generally have been far slower than those taking place today. Therefore, some ecologists and conservation biologists have advocated the active translocation of at-risk species to areas where future climates are projected to be more suitable (see discussions in Hunter 2007; McLachlan et al. 2007; Hoegh-Guldberg et al. 2008). Such **assisted dispersal** (also called assisted colonization or assisted migration) carries with it many risks and uncertainties, however, as evidenced by the well documented negative impacts of unintended species introductions on native communities. Thus, Ricciardi and Simberloff (2009: 252) suggest that "assisted colonization is tantamount to ecology roulette." Assessing the pros and cons of assisted dispersal and developing sound conservation strategies to reduce the impacts of global climate change on species with limited dispersal ability is a major challenge facing ecologists and conservation biologists in the very near future.

Parallels between metapopulation models and epidemiology

There are many parallels between the spread of infectious organisms (viruses, bacteria, metazoan parasites) and the dynamics of a metapopulation. For example, the host can be seen as analogous to a patch, infection as akin to colonization, and host recovery, death, or immunization as akin to patch extinction (Lawton et al. 1994). The threshold theory of epidemiology has long recognized that a minimum number of susceptible host individuals is required to sustain a disease outbreak (Kermack and McKendrick 1927; Anderson and May 1991). This number is directly analogous to the minimum number of patches required to maintain a metapopulation in the Levins model (Lawton et al. 1994). Further, such metapopulation thinking demonstrates that eradicating a disease does not require immunizing all the hosts (or eliminating all disease-transmitting vectors, such as mosquitoes for malaria). Instead, one only needs to reduce the number of susceptible hosts (or the density of vectors) to a level at which the disease is no longer able to persist (i.e., that specified by $D = 1 - m/c$ in the Levins model). Thus, it has been possible to eradicate smallpox without vaccinating every person, and to eliminate malaria from some regions of the world without eliminating every mosquito (Nee et al. 1997). The extinction threshold in the Levins metapopulation model is analogous to the eradication threshold in epidemiology models that assume weak homogeneous mixing (i.e., that new infections are proportional to the number of susceptible hosts; Anderson and May 1991; Nee 1994). In both cases, the threshold for extinction/eradication occurs when the fraction of patches destroyed (or the fraction of hosts vaccinated) equals $1 - m/c$ (Lande 1988a; Nee 1994; Bascompte and Rodríguez-Trelles 1998). Note that if the system is at equilibrium, the quantity $1 - m/c$ is equal to the fraction of patches occupied (Equation 12.3).

Empirical examples of metapopulation dynamics

Are there real-world populations that behave as predicted by the simple Levins metapopulation model? The answer is yes, although their number seems limited. Studies of frogs (Sjögren 1988; Sjögren Gulve 1991), butterflies (Hanski et al. 1994; Hanski and Kuussaari 1995), birds (Verboom et al. 1991), zooplankton (Pajunen and Pajunen 2003), and plants (Menges 1990) all show the signatures of a classic metapopulation: populations exist in a network of occupied and unoccupied patches, and local populations undergo colonization and extinction events in each generation. However, as we might expect, nature is more varied than our simple models. As ecologists have delved further into the study of metapopulations, they have uncovered a range of different spatial structures beyond those assumed by the classic Levins model. Harrison (1991) categorized these spatial structures into four main types (**Figure 12.3**):

1. **Classic metapopulations**, as characterized by the Levins model, in which all local populations are similar in size and type. All populations may go extinct, but have the potential to be reestablished by colonization.

2. **Mainland–island metapopulations**, in which a large population or patch (the "mainland") exists without significant risk of extinction and smaller surrounding populations are supported by immigrants from that mainland population.

Figure 12.3 Stylistic representation of four different kinds of metapopulations. Circles represent habitat patches: filled circles represent occupied patches, open circles represent vacant patches. Arrows indicate migration, and dashed lines indicate the boundaries of local populations. (A) The classic Levins metapopulation. (B) Mainland–island (or source–sink) metapopulation. (C) Patchy population. (D) Nonequilibrium population; this differs from (A) in that there is no recolonization. (After Harrison 1991.)

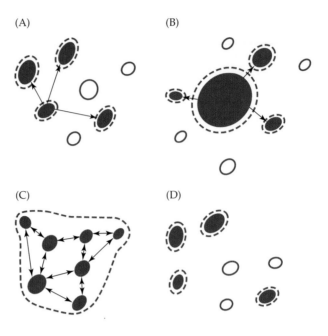

3. **Patchy populations**, in which individuals within a single interbreeding population are clumped in space, but the clumps do not exist as separate populations (i.e., the degree of movement and gene flow within the population is high).

4. **Nonequilibrium populations**, in which extinction is not balanced by recolonization. If local populations are effectively isolated from one another, then the extinction of local populations eventually leads to regional species extinction (as in the case of boreal mammals isolated on desert mountaintops following post-Pleistocene climate change; Brown 1971).

In reality, these different classifications grade into one another, and we can think of them as points along a continuum of spatially structured populations that are connected by dispersal (Driscoll 2007; **Figure 12.4**). The Levins

Figure 12.4 The four metapopulation types shown in Figure 12.3 can be viewed as points along a continuum characterized by varying degrees of extinction risk due to variation in patch size and/or quality; degree of dispersal between patches; and extinction rate relative to colonization rate. (After Driscoll 2007.)

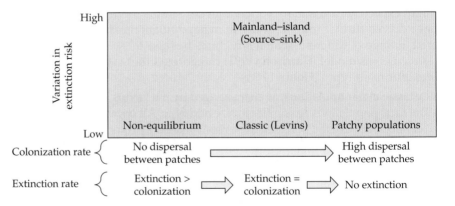

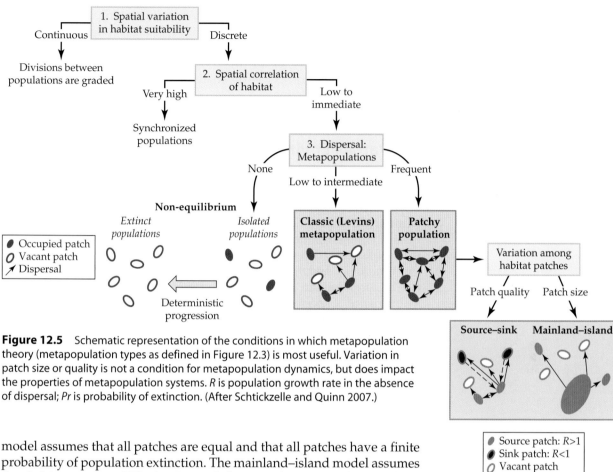

Figure 12.5 Schematic representation of the conditions in which metapopulation theory (metapopulation types as defined in Figure 12.3) is most useful. Variation in patch size or quality is not a condition for metapopulation dynamics, but does impact the properties of metapopulation systems. *R* is population growth rate in the absence of dispersal; *Pr* is probability of extinction. (After Schtickzelle and Quinn 2007.)

model assumes that all patches are equal and that all patches have a finite probability of population extinction. The mainland–island model assumes that habitat patches differ in size and that populations on the mainland have essentially zero probability of extinction. It is conceptually similar to models of **source–sink dynamics** (Pulliam 1988), which assume that habitat patches differ in quality. High-quality habitats (**source habitats**) support populations where birth rates exceed death rates, leading to an export of individuals to lower-quality habitats (**sink habitats**), where birth rates do not keep up with death rates. Populations in source habitats have a very low probability of going extinct, whereas populations in sink habitats are maintained only by the flow of individuals dispersing from productive source populations [the "rescue effect" of Brown and Kodric-Brown (1977) or the "mass effect" of Shmida and Ellner (1984)].

The challenge facing ecologists and conservation biologists today is to determine where a particular population falls on this continuum of meta-population types (**Figure 12.5**). That categorization can guide us in applying the knowledge gleaned from simple metapopulation theory, as well as from more sophisticated metapopulation models (e.g., Hanski 1994; Keymer et al. 2000; Hastings 2003), to predict the dynamics and persistence of a species (e.g., the classic work of Lande 1987, 1988a,b on the dynamics of the Northern spotted owl). Even though relatively few populations in nature exactly match the assumptions of the classic Levins model or the mainland–island model, the concept of the metapopulation is of great importance because it focuses our attention on the flow of individuals between populations (Baguette 2004). This exchange of individuals between habitat patches has consequences not only for the preservation of a metapopulation, but also for the maintenance of species diversity within a landscape.

Fugitive Species: Competition and Coexistence in a Patchy Environment

G. Evelyn Hutchinson, more than anyone else, directed our attention to the problem of how species diversity might be maintained in the face of interspecific competition (Hutchinson 1959). His suggestion that niche partitioning could ameliorate the effects of competition and allow species to sidestep the competitive exclusion principle laid the foundation for a theory of species diversity that dominated ecological thought for decades (see Chapters 1 and 8). Hutchinson also realized that the dispersal of species between habitat patches could provide another mechanism for the maintenance of species diversity.

The competition/colonization trade-off

Hutchinson published his insight in a short note with the curious title "Copepodology for the Ornithologist." His writing is so clear and inspired that it is worth quoting at length (Hutchinson 1951: 575):

> It is conceivable that species … may actually be able to exist primarily by having good dispersal mechanisms, even though they inevitably succumb to competition with any other species capable of entering the same niche. Such species will be termed *fugitive species*. They are forever on the move, always becoming extinct in one locality as they succumb to competition and always surviving by reestablishing themselves in some other locality as a new niche opens. The temporary opening of a niche need not involve a full formal successional process. Very small, seemingly random changes in the physical environment might produce locally a new niche suitable for the fugitive, or some unfavorable circumstance might cause the temporary disappearance of a competitor from such a suitable niche over a small region. Any such change would give the fugitive species its chance, for a time, until some more slowly moving competitor caught up with it. Later it would inevitably become extinct, but some of its descendants could occupy transitorily a temporarily available niche in some other locality. A species of this sort will enjoy freedom from competition so long as small statistical fluctuations in the environment give it a refuge into which it can run from competitors.

Recall that some patches remain unoccupied when a metapopulation is at equilibrium. Thus, as Hutchinson noted above, it may be possible for an inferior competitor to coexist with a superior competitor if the inferior competitor possesses the superior ability to consistently move on to colonize open patches—that is, it is a **fugitive species**.

Simple models similar to the Levins (1969) metapopulation model show that two (or more) species may coexist in a patchy environment when there is a trade-off between competitive ability and dispersal ability (the **competition/colonization trade-off**; e.g., Skellam 1951; Levins and Culver 1971; Horn and MacArthur 1972; Armstrong 1976; Hastings 1980; Tilman 1994). For example, we can easily modify the Levins model to consider two competing species. Let us assume that the environment consists of a large number of habitat patches, all of the same type, and that the superior competitor (species 1 in this case) always displaces the inferior competitor (species 2) from a habitat patch. Thus, species 1's population dynamics is unaffected by species 2. The inferior competitor (species 2), however, can colonize only those patches where species 1 is absent. If the two species have equal mortality rates (m), then species 2 will persist in the system only if it is a sufficiently better colonist than species 1. The relationship describing the condition for two species' coexistence when the species have equal mortality rates is

$$c_2 > c_1{}^2/m$$

where c_1 and c_2 are the colonization rates for species 1 and species 2, respectively (Hastings 1980). Thus, even though coexistence within a patch is impossible, species 1 and 2 can coexist regionally in a system of patches due to the fact that species 2 is the better colonizer. Tilman (1994) shows how this simple two-species model can be generalized to include multiple coexisting species that differ in their colonization and mortality rates.

The competition/colonization trade-off is a potentially strong stabilizing mechanism for maintaining biodiversity in a patchy environment (Chesson 2000a; Muller-Landau 2008). Theoretically, a large number of species could coexist via this mechanism if (1) species exhibit a competitive hierarchy, (2) the better competitor always wins within a patch, and (3) there is a strict trade-off between the ability to compete within a patch and the ability to colonize a patch (Tilman and Pacala 1993; Tilman 1994). Do natural systems fit these expectations?

A number of studies have proposed that species diversity in terrestrial plant communities may be maintained via a competition/colonization trade-off (e.g., Platt and Weiss 1977; Tilman 1994; Rees 1995; Rees et al. 1996). For example, Tilman and colleagues documented what appears to be a competition/colonization trade-off in five grass species in abandoned agricultural fields (**Figure 12.6**) as a result of differential allocation of biomass to roots and reproductive structures. A competition/colonization trade-off in plants may also result from differential allocation of resources to seed number and seed size. That is, for a given level of reproductive investment, plants may either make many small seeds or a few large seeds each time they reproduce. A **seed size/seed number trade-off** is well documented in species with similar adult sizes and life histories, as seen in **Figure 12.7** (e.g., Shipley and Dion 1992; Greene and Johnson 1994; Rees 1995; Turnbull et al. 1999; Jakobsson and Eriksson 2003; Moles et al. 2004). Note, however, that the relationship between seed size and lifetime fecundity can be different if large-seeded plant species are bigger or longer lived. Small-seeded species are expected to have a greater probability of colonizing open patches, both because small seeds disperse farther and because plants that produce small seeds have higher fecundity (Harper et al. 1970; Howe and Westley 1986;

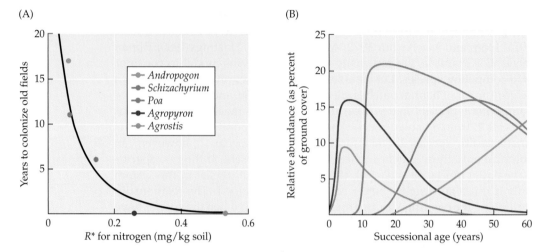

Figure 12.6 Grass species growing in abandoned agricultural fields ("old fields") in Minnesota exhibit an apparent trade-off between their ability to compete for soil nitrogen and their ability to colonize old fields (measured as the number of years required to colonize the fields). (A) Species competitive ability is expressed as the observed R^* for soil nitrogen (determined for each species grown in monoculture). Species that are better competitors (i.e., have lower R^* values) are slower to colonize old fields. (B) The approximate dynamics of succession for these five grass species at the Cedar Creek Ecosystem Science Reserve, based on a chronological sequence of old fields. (After Tilman 1994.)

Greene and Johnson 1993). Large-seeded species, on the other hand, are expected to be better competitors because they have higher germination success and higher survival through the seedling stage (Gross and Werner 1982; Westoby et al. 1996; Eriksson 1997; Turnbull et al. 1999).

The above arguments make a good case for the existence of competition/colonization trade-offs in plants, and thus we might expect to find evidence for this mode of species coexistence in a variety of systems. Surprisingly, however, the evidence is mixed. Support for a competition/colonization trade-off in plant communities has come from studies by Platt and Weiss (1977)

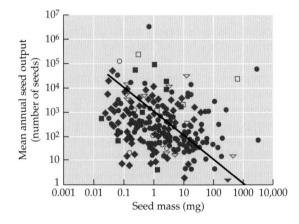

Figure 12.7 A trade-off between seed size and seed number exists for a wide variety of plant species that are similar in adult size and life history. Each data point represents a species; different symbols refer to species from different habitat types. (After Moles et al. 2004.)

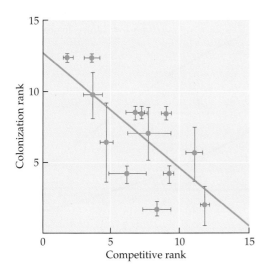

Figure 12.8 Microbial species (protozoans and rotifers) coexisting in laboratory microcosms show a competition/colonization trade-off. Data points are means ±95% confidence intervals. (After Cadotte et al. 2006b.)

on a native prairie, Tilman and Wedin (1991) on a grassland/savanna, Rees et al. (1996) on sand-dune annuals, and Turnbull et al. (1999) on a limestone grassland. On the other hand, studies by Thompson et al. (2002), Jakobsson and Eriksson (2003), Leigh et al. (2004), and Muller-Landau (2008) have found little support for the hypothesis in other systems, including tropical forests. For animal communities the results are similar; only a handful of studies document species coexistence via what appears to be a competition/colonization trade-off (e.g., Hanski and Ranta 1983; Hanski 1990a; Marshall et al. 2000). Interestingly, one of the best-documented examples of a competition/colonization trade-off comes from an artificial assemblage of protozoan and rotifer species used in laboratory experiments (**Figure 12.8**; Cadotte et al. 2006b). In this case, the experimental system closely matched the assumptions of simple theory, where all patches are assumed to be identical.

Why do we have so few real-world examples of a species coexistence mechanism that was suggested by Hutchinson (1951) more than 60 years ago—a mechanism that has been rigorously examined in theoretical models and that continues to hold a prominent place in ecological thinking? This question has puzzled me for years. I believe Hutchinson's basic insight about fugitive species is correct: some species are "forever on the move … becoming extinct in one locality as they succumb to competition and always surviving by reestablishing themselves in some other locality." But it is also likely that the "patchiness" of real-world communities is much more varied than envisioned in models focused on a single patch type, where species coexistence occurs via a competition/colonization trade-off. As we will see below, when the environment contains a variety of patch types (e.g., differences in soil type or nutrient availability for plants), additional mechanisms of species coexistence come into play.

Consequences of patch heterogeneity

If we allow the landscape to vary in the spatial arrangement of patches, or in the types of environment in the patches available, we open up many other avenues for species coexistence via colonization-related trade-offs and

(A) *Azteca*

(B) *Allomerus*

(C)

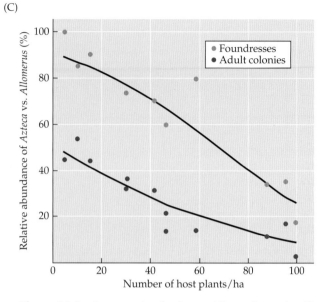

Figure 12.9   An example of a dispersal/fecundity trade-off in a community of Neotropical ants living symbiotically on a single plant species (*Cordia nodosa*). *Azteca* ants are able to disperse over longer distances, whereas *Allomerus* is more likely to colonize nearby patches successfully. (A) An *Azteca* queen with a brood inside a plant domatium (cutaway view). (B) *Allomerus* workers enter a domatium. (C) The relative abundance of *Azteca* foundresses and adult colonies increases as host plant density decreases, as would be predicted when there is a dispersal/fecundity trade-off. (Photographs © Alexander Wild; C after Yu et al. 2004.)

niche partitioning (Pacala and Tilman 1994; Levine and Rees 2002; Muller-Landau 2008). In a heterogeneous environment, species may coexist via combinations of different colonization and competitive abilities, such that each species does best in a different place or time (Chesson 2002; Muller-Landau 2008). For example, if we modify an environment composed of identical patches and add spatial heterogeneity, such that some patches are close together and some are far apart, then species may coexist via a **dispersal/fecundity trade-off**. The more fecund species will be more successful in areas of high patch density, and the better dispersers will be more successful in areas of low patch density. Yu et al. (2001, 2004) provide a nice example of species coexistence via this mechanism in a community of Neotropical ants that live symbiotically on a single plant species, *Cordia nodosa*. The plants produce structures called domatia ("domiciles") that house and protect the ants (**Figure 12.9A,B**), while the ants are thought to help protect the plant from herbivory. Ants of the genus *Azteca* are stronger fliers, and their foundresses are able to disperse over longer distances, whereas the ant species *Allomerus* cf. *demerarae* has a higher per capita fecundity and is more successful at colonizing nearby patches. Looking across a gradient in host plant (i.e., patch) density, Yu et al. (2004) found that the relative abundance of *Azteca* increases as host plant density decreases (**Figure 12.9C**), as would be predicted via a dispersal/fecundity trade-off.

If we expand our thinking to consider an environment containing different patch types, then we open up the opportunity for species to differ in their tolerance of or performance in a patch type (i.e., for species to exhibit niche differences), as well as in their probability of reaching a patch (dispersal ability). This situation describes a more complex competition/colonization trade-off, which Muller-Landau (2008) refers to as the **tolerance/fecundity trade-off**. In this case, species coexistence is promoted via niche differences (by means of habitat partitioning); however, differences in colonizing ability (fecundity) can modify the conditions for coexistence and strongly affect the relative abundances of species within the community (Levine and Rees 2002). As Muller-Landau (2008) notes, there is an extensive literature (empirical and theoretical) on species coexistence via habitat partitioning (see Chapter 8) and a parallel literature focusing on species coexistence via competition/colonization traits in a homogeneous environment. What we need next is a combination of theory and empirical work exploring how differences in colonization ability among species may interact with habitat heterogeneity to contribute to species coexistence and determine the relative abundances of species within communities (e.g., Pacala and Tilman 1994; Hurtt and Pacala 1995; Levine and Rees 2002; Muller-Landau 2008).

Conclusion

By considering population dynamics and species interactions within a spatial context, we have opened up a rich array of mechanisms by which populations may be maintained across a landscape (metapopulations) and by which species may interact and coexist (patch dynamics and fugitive species). Metapopulation theory provides the conceptual foundation for much of conservation biology, and in addition is fundamentally linked to models of epidemiology. Metapopulation theory continues to develop at a rapid pace. And, as we will see in the next chapter, it is natural to extend our thinking about two species interacting as metapopulations in a patchy landscape to thinking about local communities linked by the dispersal of species into metacommunities. By incorporating the effects of species dispersal into community dynamics, we begin to come full circle, back to a question introduced at the beginning of this book: How do local and regional processes interact to produce and maintain the diversity of life?

Summary

1. A metapopulation is a set of local populations, each of which occupies a patch of suitable habitat surrounded by unsuitable habitat, that are connected by the migration of individuals. Local populations that go extinct may be reestablished by colonists from other populations.

2. The Levins metapopulation model is a simple model that offers insights into the roles of dispersal and local extinction in determining the dynamics and persistence of patchily distributed populations and species. In this model, decreasing the number of patches decreases

the rate of population growth and may lead to the extinction of the species. Thus, the presence of unoccupied patches is to be expected and should not be viewed as an indication that these habitats could be eliminated with no consequence to a species.

3. The Levins model shows that the extinction of a species with metapopulation dynamics does not happen immediately, but may occur generations after a critical habitat patch is lost. For this reason, the preservation of existing habitat may not be enough to save the species. Such a delay in extinction is referred to as extinction debt.

4. The Levins model also shows that abundance in a metapopulation increases as the rate of dispersal between patches increases. Recent evidence supports the effectiveness of movement corridors between patches in promoting dispersal, and therefore species persistence, in fragmented landscapes.

5. The extinction threshold in the Levins metapopulation model is analogous to the eradication threshold in epidemiology models.

6. Metapopulations can be characterized into four main types: (1) classic metapopulations, as characterized by the Levins model; (2) mainland–island metapopulations, in which a large population that has a low risk of extinction supports smaller surrounding populations by emigration; (3) patchy populations, within which movement and gene flow are high; and (4) nonequilibrium populations, in which extinction is not balanced by recolonization. In reality, these four types grade into one another.

7. It may be possible for an inferior competitor to coexist with a superior competitor if the inferior competitor is better at colonizing open patches—a "fugitive species". This competition/colonization trade-off is a potentially strong stabilizing mechanism for maintaining biodiversity in a patchy environment.

8. Evidence for the competition/colonization trade-off as a mechanism of species coexistence in natural communities is mixed, probably because models of this trade-off focus on a single patch type. When the environment contains a variety of patch types, additional mechanisms of species coexistence come into play.

9. Where some habitat patches are close together and some are far apart, species may coexist in those patches via a dispersal/fecundity trade-off: the more fecund species will be more successful in areas of high patch density, and the better dispersers will be more successful in areas of low patch density.

10. If species differ in their tolerance of or performance in a patch type (i.e., in their niches) as well as in their probability of reaching a patch (i.e., in dispersal ability), they may coexist via a tolerance/fecundity trade-off. In this case, species coexistence is promoted via habitat partitioning, but the difference in colonizing ability can modify the conditions for coexistence and strongly affect the relative abundances of species within the community.

13 Metacommunities and the Neutral Theory

Metacommunity ecology ... involves an expansion and enrichment of traditional community ecology, not a replacement for it.

Marcel Holyoak et al., 2005: 466

I examine the consequences of assuming that population and community change arises only through ecological drift, stochastic but limited dispersal, and random speciation.

Stephen Hubbell, 2001: 7

The concept of the metacommunity follows closely from the study of metapopulations. Just as the dispersal of individuals may link the dynamics of populations separated in space, the dispersal of species among communities may link local communities into a metacommunity. A **metacommunity** thus may be defined as *a set of local communities linked by the dispersal of one or more of their constituent species* (Wilson 1992; Leibold et al. 2004). The local community, in this framework, includes all species that potentially interact in a single locality. (In this chapter, we will use "locality" and "patch" interchangeably to describe the space occupied by a local community.)

The study of metacommunities is a young but rapidly expanding field, extending many of the ideas developed from the study of metapopulations to the study of multispecies communities. In this chapter, we will examine ways in which the dispersal of species between communities may have significant impacts on local and regional diversity as well as affecting the ability of communities to track environmental change.

Leibold et al. (2004) and Holyoak et al. (2005) outline four perspectives that have been used to approach the study metacommunities:

1. The **patch dynamics perspective** is an extension of metapopulation models to more than two species.
2. The **mass effects perspective** extends the principles of source–sink dynamics and rescue effects to multiple species.

Species attributes	Spatial environment	
	Homogenous	Heterogeneous
Different niches	1. Patch dynamics (see Chapter 12)	2. Mass effects/source–sink dynamics 3. Species sorting
Demographically equivalent	4. Neutral theory	

Figure 13.1 Matrix classification of the four perspectives for approaching the study of metacommunities described by Leibold et al. (2004) and Holyoak et al. (2005). These perspectives apply to different types of environments (homogeneous or heterogeneous localities) and to different species attributes (niche differences or demographic equivalence). The fourth quadrant of the matrix is empty because species that are demographically equivalent are assumed to experience all environments equally.

3. The **species-sorting perspective** emphasizes differences in species abilities to utilize different patch types in a heterogeneous environment.

4. The **neutral perspective** assumes that species are functionally identical and that niche differences are unimportant (Bell 2001; Hubbell 2001).

A more comprehensive and synthetic theory may someday replace these different perspectives. For now, however, they provide a useful organizing framework for the study of metacommunities. We can begin by considering how each perspective fits into a matrix that partitions environment and species along two major axes (**Figure 13.1**). The first axis distinguishes the spatial environment of a metacommunity as being either homogeneous or heterogeneous (Amarasekare 2003). In a homogeneous environment, a species' performance (e.g., its competitive success) is the same in all patches; in a heterogeneous environment, a species' performance differs between patch types (see Chapter 12). The second axis characterizes the species occupying a metacommunity as either having niche differences that affect their tolerance of or performance in different patch types (or their ability to disperse between patches), or as being ecologically or demographically equivalent (Hubbell 2006).

We have already encountered the patch dynamics perspective; it was introduced in the latter half of Chapter 12 when we extended metapopulation models to include multiple competing species. There we saw that multiple species can coexist in an environment composed of homogeneous habitat patches when there is a trade-off between dispersal ability and competitive ability (the competition–colonization trade-off). We saw that species can also coexist in a homogeneous environment if there is spatial variation in the density of patches and there is a trade-off between dispersal ability and fecundity (Yu et al. 2004). Thus, as Hutchinson (1951) hypothesized, and as Skellam (1951), Levins and Culver (1971), and others showed theoretically, the addition of spatial patchiness to an otherwise homogeneous environment could act to increase biodiversity. However, as we also saw in Chapter 12, there are few documented cases of species coexisting via a strict competition/colonization trade-off or a dispersal/fecundity trade-off.

The notion of trade-offs—that is, that no species can be the best at everything (no species can be a "superspecies," *sensu* Tilman 1982)—is fundamental to all branches of ecology, from life history evolution to functional morphology to nutrient cycling and ecosystem processes. When considered in a community context, trade-offs correspond to niche differences among species (Chase and Leibold 2003a; Kneitel and Chase 2004). For example, a root/shoot trade-off implies that a plant species that is good at competing for soil nutrients may be poor at competing for light (see Chapter 8). Likewise, an activity level/mortality risk trade-off may make a species that is good at avoiding predators poor at finding resources or mates (see Chapter 6). Trade-offs play major roles in species coexistence; we have explored some of those roles in nonspatial models of competition and predation (see Chapters 7 and 8). The mass effects and species-sorting perspectives also recognize the importance of niche differences and trade-offs in promoting species coexistence in spatially heterogeneous environments.

The neutral perspective stands in stark contrast to the first three approaches, as it assumes that all individuals, regardless of species, have equal fitness (i.e., equal probabilities of giving birth, dying, migrating, or giving rise to new species). Thus, under the neutral perspective, environmental heterogeneity is irrelevant to the question of species coexistence, since all species experience the environment equally (see Figure 13.1). The idea that species within a community are ecologically equivalent may seem far-fetched, but as we will see, the neutral perspective yields surprisingly good fits to some real-world patterns of biodiversity (Hubbell 2001).

Metacommunities in Heterogeneous Environments

Environmental heterogeneity is common in nature, and most species experience a world that is both patchy *and* heterogeneous. The mass effects and species-sorting perspectives on metacommunities, both of which incorporate environmental heterogeneity, open up a variety of mechanisms for species coexistence via niche differences and ecological trade-offs.

The mass effects perspective: Diversity patterns in source–sink metacommunities

The dispersal of individuals from high-quality patches (source habitats, where populations have positive growth rates) to low-quality patches (sink habitats, where population growth rates are negative; see Chapter 12) can have significant effects on population dynamics and species diversity in local communities ("mass effects," *sensu* Smida and Elner 1984), as we saw in Chapter 12. Here we extend these ideas to consider how the rate of movement between local communities may affect diversity at multiple spatial scales.

A mathematical model developed by Mouquet and Loreau (2003) nicely shows how the amount of dispersal between local communities may affect the three components of species diversity: alpha (within-community) diversity, beta (between-community) diversity, and gamma (regional) diversity (see also Loreau 2010). In their model, Mouquet and Loreau make several assumptions:

1. Each local community contains S species, which compete for a limited proportion of vacant patches.

2. The metacommunity consists of N local communities that differ in their environmental conditions. Species are differentially adapted to these local conditions (i.e., due to niche differences), resulting in higher reproductive output in those local communities to which they are best adapted.

3. At the regional scale, a constant proportion (a) of each local population disperses between communities. This proportion represents the fraction of the local reproductive output that emigrates. In the model, the dispersing fraction a is assumed to be equal for all species.

Using this model, Mouquet and Loreau (2003) ran numerical simulations for a system of 20 species competing in 20 local communities until the system reached equilibrium. They then measured alpha, beta, and gamma diversity. **Figure 13.2** demonstrates how these three diversity components vary as a function of the fraction of individuals dispersing between communities (a). At zero dispersal, each local community is dominated by the best local competitor; thus, alpha diversity is at a minimum (one species per local community), while gamma and beta diversities are at their maximum (Figure 13.2).

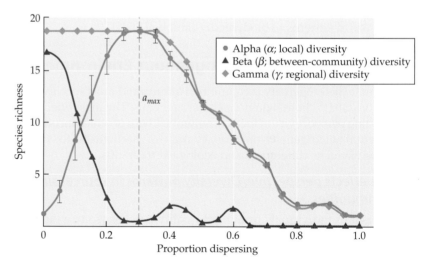

Figure 13.2 Simulation results from Mouquet and Loreau's model of source-sink metacommunities shows how species richness at different spatial scales (alpha, beta, and gamma diversity) should vary as a function of the proportion the local reproductive output that disperses between communities. Local species diversity is maximized at a_{max}. (After Mouquet and Loreau 2003.)

Figure 13.3 The expected relationship between local and regional species richness at different degrees of dispersal between communities in Mouquet and Loreau's metacommunity model. When dispersal between communities is high, local richness is proportional to regional richness (a linear function). As dispersal decreases, the local-to-regional richness relationship becomes curvilinear and saturating due to competitive exclusion within communities. (After Mouquet and Loreau 2003.)

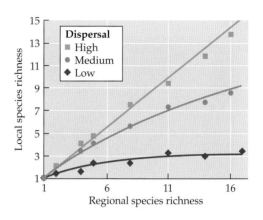

As the proportion of dispersing individuals increases, alpha diversity increases rapidly, the result of species being rescued from competitive exclusion in local communities by immigration from other local communities where they were dominant (a mass effect). Beta diversity decreases with increasing dispersal because local communities become more similar in their composition, whereas gamma diversity remains constant because all species remain in the metacommunity. As the fraction of individuals dispersing increases above a_{max}, local diversity decreases because the best competitor at the scale of the region comes to dominate each community and other species are progressively excluded. Finally, when dispersal is very high, the metacommunity functions as a single large community in which the regionally best competitor excludes all other species and local and regional diversities are at their minimums (Mouquet and Loreau 2003). Note too how variation in the proportion of individuals dispersing from the local community affects the theoretical relationship between regional and local species richness (**Figure 13.3**).

Empirical studies provide general support for Mouquet and Loreau's predictions. For example, the diversity of aquatic invertebrates living in rock pools in South Africa varied with pool isolation (the average distance separating pools; Vanschoenwinkel et al. 2007). The within-pool (alpha) richnesses of all invertebrate taxa and of taxa that disperse passively were highest at intermediate levels of pool isolation, corresponding to intermediate levels of dispersal between pools. However, actively dispersing species, with their potentially high rates of dispersal between pools, showed no relationship between species richness and isolation distance (**Figure 13.4**).

Experiments with invertebrate communities living inside the leaves of pitcher plants also support the hypothesis that local species richness is maximized at intermediate levels of dispersal. Kneitel and Miller (2003) manipulated the dispersal rates of protozoans inhabiting pitcher plant leaves by withdrawing an aliquot of the water from each of five pitchers, pooling the five aliquots, and then adding back a portion of the pooled mixture

Figure 13.4 Studies of invertebrate species richness in South African rock pools provide qualitative support for Mouquet and Loreau's model, showing that local species richness peaks at intermediate levels of pool isolation (measured as the average distance separating pools) for passively dispersing species and for all species combined. (After Vanschoenwinkel et al. 2007).

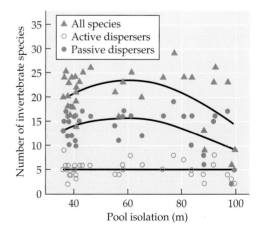

(A)

(B)

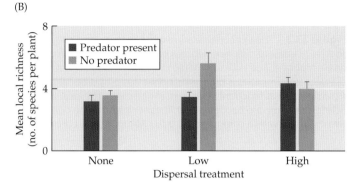

Figure 13.5 (A) The highly modified leaves of *Sarracenia purpurea* contain a characteristic community of invertebrates (e.g., protozoans, rotifers, insect larvae), many of which are obligate pitcher plant dwellers. (B) In the absence of predatory mosquito larvae in the community, mean species richness within a given pitcher plant ("mean local richness") peaked when dispersal was intermediate (the "low dispersal" treatment). When predators were present, however, the frequency of dispersal had no effect on local species richness. Error bars are ±1 SE. (A courtesy of David McIntyre; B after Kneitel and Miller 2003.)

to each pitcher. They performed this procedure weekly (low-dispersal treatment) and biweekly (high-dispersal treatment); a no-dispersal treatment was included as a control. The results show protozoan species richness within pitcher plant leaves to be highest at the low dispersal rate—that is, the intermediate condition, which is consistent with theoretical predictions (see Figure 13.2). They found, however, that dispersal rate had no effect on species richness when predaceous mosquito larvae were included in the community (**Figure 13.5**). Similar results were observed by Howeth and Leibold (2010), who studied zooplankton species diversity in experimental pond metacommunities with and without predation by bluegill sunfish.

Significant effects of dispersal rate on alpha diversity were also found in two laboratory experiments with protozoan metacommunities (Cadotte et al. 2006a; Davies et al. 2009) and in a field mesocosm experiment with pond invertebrates (Pedruski and Arnott 2011). In each of these cases, increasing the rate of dispersal between communities increased alpha diversity. Although a positive effect of dispersal rate on alpha diversity is inconsistent with the prediction of a hump-shaped relationship between those parameters (see Figure 13.2), it is possible that the dispersal rates used in these studies revealed only the left-hand portion of the predicted hump. The authors of the experimental studies attributed the increase in alpha diversity with increasing dispersal rate to mass effects and the maintenance of species in sink habitats via continued dispersal, although more complicated effects of dispersal on species dynamics are possible (Fox 2007). Experiments also show that beta diversity tends to decrease with increased dispersal rate (e.g., Warren 1996; Pedruski and Arnott 2011), as predicted by the model (see Figure 13.2), whereas the effects of dispersal rate on regional (gamma) diversity are variable (positive, negative, or no effect have been observed; see references in Pedruski and Arnott 2011).

The species-sorting perspective and the formation of complex adaptive systems

The species-sorting perspective emphasizes differences in species' abilities to utilize different patch types in a heterogeneous environment. The only significant difference between the species-sorting and mass effects perspectives is in the assumed rate of species dispersal between communities; both perspectives assume a heterogeneous environment and niche differences between species. The species-sorting perspective assumes that dispersal occurs at a relatively low rate, such that the movement of individuals between local communities does not have a direct effect on population abundance within a patch or the outcome of species interactions within local communities (i.e., there are no mass effects). The rate of dispersal, however, is assumed to be high enough so that all species are able to reach every locality where they are capable of persisting (Chase et al. 2005). Thus, the species-sorting perspective encompasses a rather restricted range of dispersal rates, corresponding to the lower end of the scale shown in Figure 13.2.

Future models of metacommunity dynamics in heterogeneous environments are likely to incorporate the species-sorting and mass effects perspectives into a more unified whole, as they clearly represent points along a continuum of dispersal rates (Amarasekare et al. 2004). However, species-sorting, by focusing on the low end of dispersal rates, points out an important and so far unconsidered consequence of dispersal in metacommunities: that a low rate of species input potentially allows communities to track environmental change and respond to changing environments as "complex adaptive systems" (*sensu* Levin 1998; Leibold and Norberg 2004). We can illustrate this effect with an example we encountered back in Chapter 11, when we addressed the question of how trophic-level biomass responds to a change in productivity at the base of the food chain or food web (Leibold 1990; Leibold et al. 2005). Recall that in simple food chains, in which each trophic level is composed of a single species or of a group of species that all respond similarly to food and predators, an environmental change that increases the productivity of the basal resource (e.g., nitrogen deposition in grasslands or nutrient inputs into lakes) is predicted to increase the biomass of the top trophic level and alternating trophic levels below it (see Figure 11.2). Thus, in a three-trophic-level system, the abundances of predators and plants should increase with an increase in productivity, but the abundance of herbivores should remain constant. Simple experimental food chains show this alternating trophic level response, as do short-term manipulations of productivity in natural systems. However, most natural communities show an increase in the biomass of all trophic levels (producers, herbivores, and predators) when potential productivity is increased. As Leibold (1990, 1996) and others have shown, the input of species with differing competitive abilities and differing mortality risks can change the prediction of alternating trophic level responses, leading instead to a positive biomass response by all trophic levels as basal resources are increased (see Figure 11.9B). Thus, the observed positive biomass response of trophic levels in natural communities (see Figure 11.7) may be brought about by changing species composition within a community due to the dispersal of new species from the surrounding metacommunity. Such compositional

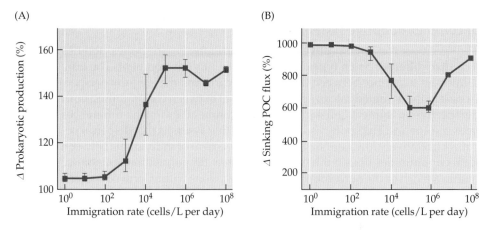

Figure 13.6 The rate of species immigration to the local community may influence ecosystem processes. A model of carbon sequestration by marine heterotrophic prokaryotes shows that the rate of species immigration to the local prokaryote community determines the ability of the community to respond to environmental change (here a change in the supply of particulate organic carbon, or POC). The graphs plot (A) differences in prokaryote community production or (B) the loss of POC by sinking out of the water column [measured as the ratio of values after (numerator) to before (denominator) an increase in POC supply]. Community response to increased POC measured as productivity (A) was greater at a higher rate of species immigration, whereas community response to increased POC measured as POC loss due to sinking (B) was strongest at intermediate rates of species immigration. Points are average values and error bars represents the middle 95% range from 1000 model simulations. (After Miki et al. 2008.)

shifts have been observed in laboratory experiments (Bohannan and Lenski 1997, 1999) and natural systems (Watson et al. 1992; Davis et al. 2010).

As Naeslund and Norberg (2006) note, the metacommunity provides a source of renewal (and new species) to local communities, and its importance for sustaining ecosystem functioning under changing conditions may be large (Elmqvist et al. 2003; Loreau et al. 2003; Leibold and Norberg 2004). For example, a model of carbon sequestration by marine heterotrophic prokaryotes shows that the regional diversity of these prokaryotes and their dispersal among local communities determines the ability of local communities to respond to environmental change (here, a change in carbon supply; **Figure 13.6**). Thus, the process of species-sorting may have a major impact on the rate of carbon sequestration in the deep sea (Miki et al. 2008). Empirically, Naeslund and Norberg (2006) have shown that freshwater communities established from regional species pools show significantly different ecosystem functioning than communities established from local species pools. Here again, the introduction of species from the regional pool allows the local community to more effectively track changing environmental conditions (**Figure 13.7**). In this case, zooplankton (herbivore) abundance and grazing pressure increased with increasing light levels, and this response was greater in the communities established from regional species pools than in those established from local species pools (Figure 13.7C), because "communities that developed from samples of the regional species pool [are] more likely to

Figure 13.7   The introduction of species from the regional pool allows local communities to more effectively track changing environmental conditions. (A–C) Responses of different ecosystem parameters in experimental freshwater communities to a gradient in light flux. The ecosystem response depends on whether the community is drawn from the regional species pool or the local species pool. Note the strong response by zooplankton (panel C); the disparity shows that communities drawn from the regional species pool are more likely to have efficient zooplankton species and/or coexisting zooplankton species. (After Naeslund and Norberg 2006.)

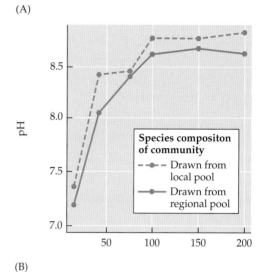

have particularly efficient zooplankton species and/or coexisting species, i.e., a sampling and complementarity effect" (Naeslund and Norberg 2006: 509). These same ideas can be extended to the study of evolutionary dynamics in metacommunities as the result of the dispersal of different genotypes or clones among local communities (Urban et al. 2008; De Meester et al. 2011).

Measuring dispersal in metacommunities

As we have seen, dispersal rates play fundamental roles in metacommunity dynamics: they determine patterns of local and regional diversity (e.g., Figure 13.2; Mouquet and Loreau 2003), and they distinguish the relative impacts of species-sorting versus mass effects on community dynamics and composition. Dispersal, however, is a notoriously difficult process to study. Consequently, we have little information on actual rates of movement of individuals and species between communities. For some taxa, dispersal rates can be estimated by observing their rates of colonization of new habitats. For example, evidence from newly created freshwater ponds shows that the movement of passively dispersed organisms such as zooplankton is surprisingly high, although there is considerable variation among species (Jenkins and Buikema 1998; Louette and De Meester 2007; Louette et al. 2008).

An alternative to directly observing dispersal is to measure some proxy that correlates with dispersal. For example, population geneticists have long used estimates of genetic dissimilarity between spatially separated populations ("isolation by distance") to estimate rates of gene flow (Slatkin 1993). The idea here is that the greater the rate of gene flow between populations, the more similar they should be in genetic composition. Likewise, ecologists have begun to measure the degree of dissimilarity in species composition between local communities to estimate rates of species dispersal (Shurin et al. 2009). Species that are relatively poor dispersers are expected to show more clumped spatial distributions or greater spatial autocorrelation in community composition than good dispersers do. Studies employing measures of community dissimilarity with distance (i.e., the **rate of distance decay of similarity**; Nekola and White 1999; Soininen

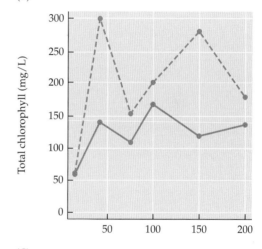

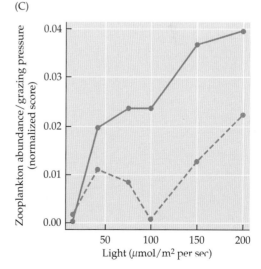

et al. 2007) have documented dispersal limitation in a variety of systems, including temperate and tropical forests (Tuomisto et al. 2003; Gilbert and Lechowicz 2004) and freshwater lakes and ponds (Beisner et al. 2006; Shurin et al. 2009). In a few cases, they have compared these measures for different taxa in the same ecosystem (e.g., Beisner et al. 2006; Astorga et al. 2012). In addition, by regressing community similarity against environmental and spatial distance, one can estimate the relative effects of local environmental variation (and species-sorting) versus dispersal limitation in determining the change in community species composition with distance (e.g., Tuomisto et al. 2003; Green et al. 2004). Ecologists are beginning to elucidate differences in dispersal tendencies related to species traits (e.g., body size; Beisner et al. 2006; Shurin et al. 2009). However, we are still a long way from knowing the dispersal rates of a significant fraction of the constituent species for any given metacommunity.

An insightful approach for empirically estimating the impact of species dispersal on community structure and metacommunity dynamics was employed by Luc De Meester, Karl Cottenie, and their colleagues in a study of a series of 34 interconnected ponds in Belgium. The authors estimated that an average of 3,600 zooplankton individuals per hour dispersed though the overflows and rivulets that connect each pair of ponds in this system, a rate corresponding to a population turnover time of 13 days (Michels et al. 2001). Despite this relatively high rate of species dispersal, the ponds exhibited distinct plankton communities that were predictably related to the local environment, particularly to the presence or absence of fish and macrophytes (Cottenie et al. 2003; Cottenie and De Meester 2003; Vanormelingen et al. 2008). Moreover, these strong and predictable influences of the biotic environment on zooplankton community structure were confirmed via a transplant experiment (Cottenie and De Meester 2004). Thus, this metacommunity showed the clear impact of species-sorting, despite relatively high rates of species dispersal. The effects of dispersal, however, were not insignificant. Dispersal increased the species richness of cladoceran zooplankton by an average of three species (a mass effect), and dispersal made the response of the community to the local environment more predictable (i.e., the local communities were acting as adaptive systems).

The Neutral Perspective

Ecologists have long focused on the importance of niche differences in maintaining biodiversity. Niche differences not only allow species to coexist (thus maintaining biodiversity), but their proliferation, through the processes of character displacement and adaptive radiation, is hypothesized to be the principal generator of species diversity over evolutionary time. As Schluter (2000: 2) explains, "Adaptive radiation is the evolution of ecological diversity within a rapidly multiplying lineage. It is the differentiation of a single ancestor into an array of species that inhabit a variety of environments and that differ in traits used to exploit those environments."

Thus, the idea that species differ in the ways in which they use the environment (i.e., in their niches) seems fundamental to community ecology and to the evolution and maintenance of species diversity. That is why,

(A)

(B)

(C)

Figure 13.8 (A) Barro Colorado Island, Panama, was formed by the construction of the Panama Canal, which dammed the Chagres River and created Gatun Lake. (B) Scientists have mapped and repeatedly censused all woody plants larger than 1 cm in diameter at breast height in the Forest Dynamics Plot, a 50 ha experimental plot of tropical rainforest on Barro Colorado Island. (C) Mapped distributions of small and large trees of the species *Ocotea whitei* in the Forest Dynamics Plot. The distribution of *O. whitei* is significantly associated with slope at both life stages. (Photographs © Christian Ziegler/Minden Pictures; C from Comita et al. 2007.)

when Steve Hubbell's *Unified Neutral Theory of Biodiversity and Biogeography* was published in 2001, it created quite a stir. In Hubbell's vision, niches are absent and all species are functionally equivalent. Stated more formally, species are assumed to be demographically identical on a per capita basis in terms of their vital rates of birth, death, and dispersal (Hubbell 2006).

Before concluding that Hubbell is some harebrained theoretician who has never set foot in the field or seen the variety of life firsthand, we need to realize that he has a long history of working in one of the most diverse ecosystems in the world—the tropical rainforest. Hubbell was among the first to recognize the value of mapping all the trees in a tropical forest, and he did this (along with Robin Foster and other colleagues) in a 50 ha forest plot on Barro Colorado Island in Panama (**Figure 13.8**). This plot contained about 300 species of trees and about 250,000 individuals with a stem diameter of at least 1 cm at breast height (Hubbell and Foster 1983; Condit 1998). Since that time (1980), dozens of other 50 ha forest plots from around the world have been mapped, yielding a wealth of new information (see the Center for Tropical Forest Science website, www.ctfs.si.edu).

The evolution and maintenance of high species diversity in tropical wet forests is an enduring puzzle to ecologists (see Chapter 2). It seems implausible that the 300 or so tree species on Barro Colorado Island could occupy 300 different niches. How can there be that many ways for trees to differ in habitat, resource use, etc.? Rather than searching for subtle niche differences between tropical tree species, Hubbell took exactly the opposite approach. He wondered instead whether tree diversity might be maintained if the species were functionally equivalent; that is, if their interactions were neutral to any differences between species. This shift is very similar to what happened to population genetics in the 1960s, when the application of electrophoresis revealed an unexpectedly high degree of allele polymorphism. Many population geneticists concluded that observed levels of allele polymorphism were too high to be maintained by natural selection. Therefore, Kimura (1968) and others reached the conclusion that most alleles are selectively neutral—mutations arise that are neutral in their effects, and the frequencies of such mutations in a population fluctuate at random. In the ensuing years, a large body of theory developed around the idea of neutral molecular evolution, and the idea of a "molecular clock," which proposes that the rate of molecular evolution is constant, was one of the outcomes (Zuckerkandl and Pauling 1965). The neutral theory of biodiversity (Bell 2001; Hubbell 2001) is analogous to the neutral theory of population genetics; Leigh (1999, 2007) and Nee (2005) provide insightful discussions of the parallels between the two theories.

Assumptions of the neutral theory

In their initial development of the neutral theory of biodiversity, Hubbell and Foster (1986) proposed a model for tropical forests based on distur-bance and subsequent recolonization by species with similar competitive abilities. Their model makes several assumptions:

1. The number of individuals in the community is constant. Space is a limiting resource, and all space is occupied. Therefore, gains in abun-dance by one species must be balanced by losses in the abundances of other species. Biodiversity is a "zero-sum game."

2. All individuals and species have an equal probability of colonizing open space. The colonizer of a site vacated by death (disturbance) is a random draw from the species present, with the probability of being selected equal to a species' relative abundance.

3. Death occurs at a constant and fixed rate.

Under these assumptions, the long-term (equilibrium) outcome of competi-tion in a local community is a random walk to dominance by a single species. However, if the community contains a large number of individuals that die at a relatively slow rate, this random walk to single-species dominance can take a very long time (in fact, maybe too long a time, a point we return to later). Hubbell and Foster argued that, although their model did not result in the equilibrium coexistence of species, the time over which species are maintained in the local community may be as long as the time it takes for new species to arise via speciation.

In his book, Hubbell (2001) extended the neutral model presented in Hubbell and Foster (1986) to include local communities that are connected by the immigration of species to a much larger metacommunity. This was a significant advance. In Hubbell's words (2001: 5), the metacommunity contains "all trophically similar individuals and species in a regional collection of local communities." Each local community, as in the original Hubbell and Foster model, consists of J total individuals whose offspring compete for sites that become available when an individual dies. Every individual (regardless of species) in the local community has an equal probability of colonizing an open site (the "neutrality" assumption). In addition, all open sites are assumed to be filled, and community size is constant (the "zero-sum" assumption). Further, offspring from the local community compete for sites not only with other individuals from the local community, but also with immigrants from the metacommunity. The parameter m defines the probability that an open site will be colonized by an immigrant from the metacommunity. Dispersal limitation occurs when $m < 1$.

New species arise in the metacommunity via speciation. Thus, the metacommunity provides a pool of species at a regional level; the dispersal of these species into the local community may counteract the steady loss of species through random drift. The total number of individuals in the metacommunity (J_M) is assumed to be large and constant, and the number of species in the metacommunity is in stochastic balance between the rate of extinction and the rate of speciation.

Hubbell (2001) considers two speciation mechanisms, "point mutation" and "random fission." In the "point mutation" model (to which Hubbell devotes most of his analysis), ν (Greek nu) is the per capita speciation rate (the probability of a speciation event per individual in the metacommunity). Hubbell then goes on to define θ, the rate at which new species are produced in the metacommunity, as

$$\theta = 2\rho A_M \nu$$

where ρ is the mean number of individuals per unit area in the metacommunity and A_M is the area of the metacommunity. Thus, the total number of individuals in the metacommunity (N) is ρA_M, and θ, which Hubbell refers to as the "fundamental biodiversity number," is the rate of speciation. For species with discrete generations, $\theta = 2N\nu$; for species with overlapping generations, $\theta = N\nu$ (Leigh 1999).

Testing the predictions of the neutral theory

Had Hubbell proposed his neutral theory without linking it to real-world data, it probably would have had little impact. Indeed, neutral models for communities were not new (e.g., see Caswell 1976). However, one of the strengths of Hubbell's theory is that it makes specific predictions about patterns of biodiversity that can be compared with data from natural systems. Most notably, his model predicts the expected distribution of species abundances in both the local community and the metacommunity. As we noted in Chapter 2, species abundance distributions (SADs) were one of the first biodiversity patterns examined by ecologists (Fisher et al. 1943; Preston 1948). However, until the publication of Hubbell's book, niche-based theories of species diversity had been strangely silent about the predicted forms of SADs (except for a brief flurry of activity in the 1960s; see Chapter 2). Hubbell's theory, assuming speciation by point mutation, predicts that

Figure 13.9 Hubbell's neutral model and Preston's lognormal model both provide excellent fits to the empirical data on the species abundance distribution of tropical trees in the Forest Dynamics Plot on Barro Colorado Island (see Figure 13.8). The bars are observed numbers of tree species, binned in $\log_2$ abundance classes (meaning that species abundance doubles in each class). (After Volkov et al. 2003.)

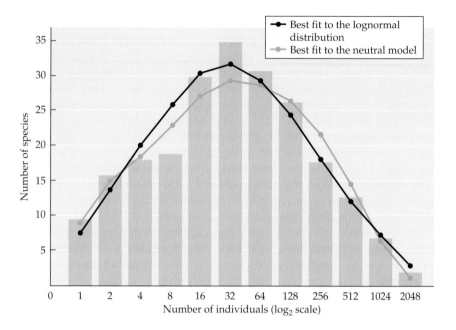

the SAD for a metacommunity will correspond to Fisher's log series distribution. In fact, θ is equivalent to the parameter α in Fisher's log series. For a local community, which receives species from the metacommunity at a migration rate of m, the SAD is similar to a lognormal distribution (see Magurran 2005 for an excellent discussion of SADs and the neutral theory).

Hubbell (2001; see also Volkov et al. 2003) examined SADs from a number of local communities, particularly those from tropical trees on Barro Colorado Island in Panama (**Figure 13.9**) and from birds in Great Britain. He found that the neutral model predicted those SADs very well. However, other researchers soon responded, showing that other models (including those that focus specifically on niche differentiation) also successfully predict observed SADs (i.e., Chave et al. 2002; McGill 2003). After much analysis, the bottom line is that a variety of models fit the data on SADs almost equally well (Chave 2004; McGill et al. 2006). Further, McGill et al. (2006: 1415) argues that "the (neutral theory) is unusually flexible in its ability to fit SADs … because each parameter is independent. J sets the scale, θ controls the shape to the right of the mode and m controls the shape to the left of the mode." Therefore, although Hubbell's initial focus on predicting SADs is laudable (SADs were too long forgotten by community ecologists), it is now clear that looking at goodness of fits to empirical data will not allow us to distinguish among competing theories of biodiversity.

One of the strengths of the neutral theory is that it makes quantitative and testable predictions about species abundances in space and time. If the abundance of each species within a community drifts randomly through time, then we can specify the expected difference in species composition between communities separated in space (beta diversity), and we can specify how long it should take species, on average, to change in abundance within a community (including the time to extinction and the time to become com-

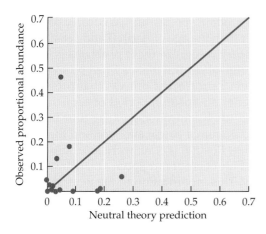

Figure 13.10   A comparison between species-specific predictions of the neutral model parameterized for sessile marine species in the rocky intertidal on Tatoosh Island, Washington State, and the actual proportional abundances of species as observed in experimental plots. The diagonal line is the expected relationship for a perfect fit of experimental data to model predictions. (After Wootton 2005.)

mon after invasion). Ecologists have examined both of these predictions in detail. For example, several studies have compared the species compositions of communities separated in space in tropical forests (Terborgh et al. 1996; Condit et al. 2002) and in temperate forests (Gilbert and Lechowicz 2004). In each case, the authors found that species turnover over short distances was greater than that predicted by neutral theory, as was the number of widespread species shared among communities. Working at smaller spatial scales, Wootton (2005) devised an experimental test of the predictions of the neutral theory and applied it to the community of sessile organisms found in the marine rocky intertidal zone on Tatoosh Island, off the coast of Washington State. He found, as have others, that the neutral theory provided a good fit to the general shape of the species abundance distribution in the community. However, when Wootton tried to use the theory to explain the dynamics of individual species over the 7–11 years' duration of the experiment, it did a poor job of predicting the observed community response (**Figure 13.10**). In another test of the neutral theory's predictions at small spatial scales, Adler (2004) compared patterns of species abundances among plots in a Kansas grassland, looking at the joint effects of space and time, and found that the neutral theory failed to explain the observed patterns.

Ricklefs (2006b) argues that the most unambiguous tests of Hubbell's models address time itself. Population geneticists long ago worked out the time required for a neutral mutation to increase or decrease to a certain level within a population of size N (Fisher 1930; Kimura and Ohta 1969). Drift is a slow process, and the larger the population, the longer it takes. Ecologists and evolutionary biologists have applied the theories of population genetics to estimate the expected life spans of the most common species in a community, the expected time to extinction, and the time to community equilibrium under the assumptions of the neutral model. In all cases, the predicted time scales are inconsistent with those in real communities—species come to dominance, and species disappear, much more quickly in nature than predicted based on a theory of neutral dynamics (Leigh 1999, 2007; Lande et al. 2003; Nee 2005; Ricklefs 2006b). For example, estimates based on neutral dynamics would place the ages of common tropical tree species back before the origin of the angiosperms (Nee 2005), and would make the time to extinction for abundant species such as oceanic plankton or tropical insects or trees greater than the age of Earth (Lande et al. 2003).

Proponents of the neutral theory recognize these problems and have proposed a modification to the original neutral theory whereby speciation in the metacommunity occurs via a protracted process that produces a recognizable new species only after a transition period of some number of generations (Rosindell et al. 2010). By introducing a lag time between the origination of incipient species from a single ancestor and the actual formation of recognizable species, Rosindell et al. were able to produce more realistic speciation rates and numbers of rare species than the original formulation of the neural theory. This modification to the neutral theory, however, doesn't completely solve the problem of unrealistically great ages for common species (Rosindell et al. 2010). In addition, Desjardins-Proulx and Gravel (2012) showed that modeling speciation as a more explicit allopatric or parapatric process within the neutral theory severely limits the number of species that can be supported. Thus, the neutral theory's incompatibility with realistic modes and rates of speciation presents a fundamental challenge to its ability to explain biodiversity.

The value of the neutral theory

You may wonder why so much of this chapter has been devoted to Hubbell's neutral theory of biodiversity, given that the weight of empirical evidence is against it. I think there are at least four reasons why the neutral theory is worth studying in detail:

1. The theory has been enormously influential; it has caused ecologists to reexamine areas of ecology that were almost forgotten (e.g., SADs). It will continue to play a major role in community ecology for years to come, as a null model if nothing else (Gotelli and McGill 2006; Leigh 2007).

2. The neutral theory is mathematically elegant and tractable. Because of this, it can be applied to many problems in ecology outside the reach of current niche theory. Thus, for these problems, we can see what the solution would look like if species were "ecologically neutral."

3. A number of ecologists have suggested that real communities represent a continuum of niche and neutral interactions; that is, that communities are composed of a mix of species, some with niche-based dynamics and some with neutral dynamics (e.g., Gravel et al. 2006; Leibold and McPeek 2006; Adler et al. 2007; Chase 2007; McPeek 2008; Ellwood et al. 2009). We will explore this idea in more detail in Chapter 14.

4. Because it brushes aside species differences, neutral theory focuses our direct attention on how dispersal and regional species pools may determine local community structure. Thus, it embodies the essence of the metacommunity perspective.

Interestingly, recent work by Zillio and Condit (2007) and Chisholm and Pacala (2010) suggests that the assumption of "neutrality" itself may not be the dominant force behind the success of the neutral theory in explaining SADs. McGill and Nekola (2010) reach a similar conclusion, suggesting that of the three most important aspects of the neutral theory of biodiversity, neutrality appears dispensable, while dispersal limitation and the input of species from the metacommunity (regional replacement) appear to be critical in determining SADs for local communities.

While the neutral theory is fundamentally flawed in assuming that all species are ecologically equivalent, it has pointed us in new directions, it has highlighted critical processes for the maintenance of biodiversity, and it has invigorated our science. As Hubbell (2006: 1388) noted, "Perhaps the [unified neutral theory's] most significant and lasting contribution will be its explicit incorporation of linkages between the ecological processes of community assembly on local scales and evolutionary and biogeographic process on large scales, such as speciation and phylogeography." Or, as Ricklefs (2006b: 1430) put it, "Except for its abandonment of ecology, Hubbell's view of the world incorporates much of the regional and historical perspective that I would advocate. The fact that many ecologists have been attracted to this idea gives hope that the discipline is ready to embrace a more comprehensive concept of the species composition of ecological systems."

Conclusion

By viewing communities as interconnected entities within a landscape, metacommunity ecology focuses our attention on the importance of dispersal, regional species pools, and demographic stochasticity (in the case of neutral theory) in determining species composition and species abundances within communities. These exciting developments will move us toward a better understanding of how biodiversity and community structure are determined at both local and regional scales. However, while the theory of spatial ecology marches forward at a heady pace, empirical studies lag far behind; Amarasekare (2003: 1109) notes that, while "theory is in advance of data in most areas of ecology, nowhere is it more apparent than in spatial ecology." This is not surprising; there are enormous challenges in quantifying the important ecological processes that occur across spatial scales (e.g., dispersal rates) and in linking theoretical concepts such as "patch" and "local community" to real-world entities that we can define and study. Despite these challenges, great progress has been made in a very short time.

Summary

1. A metacommunity is a set of local communities linked by dispersal of one or more of their constituent species.

2. Ecologists have studied metacommunities from several perspectives. Differences between these perspectives hinge on the degree of environmental heterogeneity and species niche differences found in the metacommunity.

3. The patch dynamics perspective extends metapopulation models (Chapter 12) of species dynamics in a system of homogeneous patch types to include more than two species.

4. The mass effects perspective applies a multispecies version of source–sink dynamics and rescue effects in a system of heterogeneous patch

types. A model of source–sink metacommunities devised by Mouquet and Loreau shows that the rate of dispersal between local communities may affect diversity at multiple spatial scales. At a dispersal rate of zero, each local community is dominated by the best local competitor, and gamma and beta diversities are at their maximum. As dispersal rate increases, alpha diversity increases rapidly as species are rescued from competitive exclusion by immigration. Beta diversity decreases, because local communities become more similar in their composition, whereas gamma diversity remains constant. When the dispersal rate is very high, the metacommunity functions as a single large community in which the regionally best competitor excludes all other species. Empirical studies provide general support for the predictions of Mouquet and Loreau's model.

5. The species-sorting perspective emphasizes differences in species' abilities to utilize different patch types in a heterogeneous environment. Like the mass effects perspective, species-sorting assumes a heterogeneous environment and interspecific niche differences. It also assumes dispersal occurs at a low rate, such that dispersal has no direct effect on population abundances or the outcome of species interactions within local communities (i.e., there are no mass effects), but dispersal rate is high enough that all species are able to reach every locality where they are capable of persisting.

6. A low but nonzero rate of dispersal potentially allows communities to track environmental change and to respond to changing environments as "complex adaptive systems." The metacommunity provides a source of renewal (and new species) to local communities, and its importance for sustaining ecosystem function in changing environments may be large.

7. The rate of dispersal between local communities is a key parameter in all metacommunity models but is difficult to measure in nature. For some taxa, dispersal rates can be estimated by observing their rates of colonization of new habitats. An alternative to directly observing dispersal is to measure some proxy that correlates with dispersal, such as the degree of dissimilarity in species composition between local communities.

8. The neutral perspective posits that niche differences between species are unimportant. Species are assumed to be demographically identical in terms of their per capita birth, death, and dispersal rates.

9. In Hubbell's neutral theory of biodiversity, every individual in the local community (regardless of species) has an equal probability of colonizing an open site (the "neutrality" assumption). All open sites are assumed to be filled, and community size is constant (the "zero-sum" assumption). Individuals compete for sites with other individuals from their local community, and also with immigrants from the metacommunity. New species arise in the metacommunity via speciation. Dispersal of these species into the local community may counteract the loss of species through random drift; thus, the metacommunity provides a pool of species at a regional level.

10. The neutral theory successfully predicts real-world species abundance distributions, but other models, including those based on niche differentiation, predict them equally well. Furthermore, species come to dominance, and species disappear, much more quickly in nature than predicted by neutral dynamics. Despite these issues, neutral theory focuses our attention directly on how dispersal rates and regional species pools may determine local community structure, and thus embodies the essence of the metacommunity perspective.

Species in Changing Environments

Ecology and Evolution

14 Species Coexistence in Variable Environments

Stable coexistence ... requires important ecological differences between species that we may think of as distinguishing their niches ... [along] four axes: resources, predators (and other natural enemies), time, and space.

Peter Chesson, 2000: 348

The only constant in life is change.

Heraclitus, c. 505 BCE

A freshwater lake may contain 30 or more species of phytoplankton, a coral reef a hundred or more species of fish, and a single hectare of tropical rainforest over 300 species of trees. Ecologists have long puzzled over this wealth of biodiversity, especially in situations where the number of available niches seems inadequate to account for the coexistence of so many species (e.g., Hutchinson 1961; Sale 1977; Hubbell and Foster 1986). Hutchinson (1961) was the first to express this dilemma formally and to offer a solution. In a paper entitled "The paradox of the plankton," he asked how so many species of freshwater phytoplankton, all of which require the same few limiting resources, coexist in an environment that appears to have little habitat structure (the open water of a lake). Hutchinson suggested that a solution to the paradox of the plankton might be found in recognizing that the physical and biotic environment of the lake changes over time. That is, under a given set of environmental conditions, we expect there to be winners and losers in competition and that, given enough time, competitive exclusion may occur. But what if the environment changes before we reach this outcome? Might it be possible for a large number of species to coexist in an environment that is constantly changing, such that no one species or group of species is favored at all times?

We now know that there are many more opportunities for niche partitioning among phytoplankton species than Hutchinson or others envisioned at the time; phytoplankton, for example, may require different ratios of nutrients (Tilman 1982; Sommer 1989), may use different wavelengths of light (Stomp et al. 2004), and may suffer different rates of mortality from grazers (McCauley and Briand 1979). However, Hutchinson's puzzlement over the paradox of the plankton, and his suggestion that environmental fluctuations may promote species diversity, was a groundbreaking insight. It turns out that the specific mechanism of coexistence proposed by Hutchinson—that is, that species will coexist if the time to environmental change roughly matches the time to competitive exclusion—is insufficient (Chesson and Huntly 1997). However, his insight that temporal fluctuations in the environment can modify the predictions of classic competition models and may promote species coexistence was dead on.

In this chapter, we will explore in detail how environmental variation impacts species coexistence and community structure. In doing so, we will link up many of the ideas that we have discussed earlier in different contexts, including exploitative competition, metacommunities, neutral theory, and local versus regional dynamics. The organizing framework of this chapter is one proposed by Peter Chesson (2000a), focusing on the *equalizing and stabilizing properties of species coexistence mechanisms*.

Properties of Species Coexistence Mechanisms

When we consider the number and types of species found in a community, an important question is whether their coexistence is stable or unstable. **Stable coexistence** means that species tend to recover from low densities and that species densities do not show long-term trends; in other words, species abundance may fluctuate over time, but there are mechanisms in place that prevent species from disappearing. **Unstable coexistence** means that, although species may coexist within a community for long periods as a result of slow rates of competitive exclusion, there are no mechanisms that promote the recovery of species when they become rare; thus, species will eventually disappear and community richness will decline. Hubbell's (2001) neutral theory of biodiversity assumes unstable species coexistence. In Hubbell's model (described in Chapter 13), species within a local community are lost due to slow competitive exclusion, and long-term diversity in the local community is maintained only by the immigration of species from the regional metacommunity.

Mechanisms that promote species coexistence may be equalizing or stabilizing (Chesson 2000a). **Equalizing mechanisms** reduce fitness differences between species. For example, Hubbell's (2001) neutral model assumes that there are no functional differences between individuals, regardless of species. Therefore, in the neutral model, all species in the community respond in the same way to the environment, and all species are demographically equivalent. The more similar species are in their fitness—i.e., the greater the equalizing mechanisms—the longer it takes for competition (or any other interaction) to favor one species at the expense of another. However, equalizing mechanisms by themselves cannot facilitate the recovery of species from low densities and thus cannot lead to stable coexistence. As we know from our study of Hubbell's neutral model (Chapter 13), in the absence of immigration, the number of species in a local community will undergo slow random drift to eventual dominance by a single species. Stable coexistence requires **stabilizing mechanisms** that favor the recovery of species when they become rare. However, the more similar species are in their fitness (i.e., the greater the equalizing mechanisms), the less is required of stabilizing mechanisms to promote coexistence (Chesson 2000a).

Stabilizing mechanisms include traditional mechanisms of species coexistence such as resource partitioning and frequency-dependent predation. These "niche-based" mechanisms act to make intraspecific density dependence stronger than interspecific density dependence. Recall from our studies of competition (Chapter 7) that the strength of intraspecific competition relative to interspecific competition is key to promoting stability among interacting species. Stabilizing mechanisms such as resource partitioning and frequency-dependent predation operate whether the environment fluctuates or not; these stabilizing mechanisms are **fluctuation-independent mechanisms**. Other mechanisms that stabilize the coexistence of species depend critically on fluctuations in population densities and environmental factors (Chesson 1985, 2000a). Such stabilizing factors are **fluctuation-dependent mechanisms**, and we examine them below.

Fluctuation-Dependent Mechanisms of Species Coexistence

Chesson (2000a) recognized two classes of fluctuation-dependent stabilizing mechanisms: those involving *the relative nonlinearity of competition* and those involving *the storage effect*. You are already familiar with the first of these classes. In Chapter 7, we saw that two (or more) consumer species can coexist on a single limiting resource if there are nonlinearities in the consumers' functional responses (e.g., Figure 7.5) and if there are fluctuations in resource abundance. This stabilizing mechanism operates because some species are favored when resource abundances are high and other species are favored when resource abundances are low. Variation in the physical environment (i.e., seasonal changes) can cause resource abundances to fluctuate over time. In the case of biotic resources (living species), the interaction between consumers and resources can also generate endogenous temporal variations in resource abundance—i.e., stable limit cycles.

When consumer and resource densities fluctuate in a stable limit cycle, two (or more) consumer species can coexist indefinitely on a single resource (Armstrong and McGehee 1980). The conditions allowing for this coexistence are, first, that the species differ in the nonlinearity of their functional response to the habitat's limiting factor (e.g., limiting resource) such that one species is favored by fluctuations in the factor and the other is favored by constancy, and second, that the species favored by fluctuations makes the factor more constant over time, while the species favored by constancy causes the factor to fluctuate. A trade-off between species performance at high and low resource levels can result in these conditions; for example, in Figure 7.5, species 2 is favored by resource constancy because it has the lowest R^*, whereas species 1 is favored by resource fluctuations because it has a higher growth rate at high resource levels. We discussed an example of such a trade-off in Chapter 7 when we considered two types of consumers: gleaners and opportunists (Frederickson and Stephanopoulos 1981; Grover 1990). Species interactions due to shared predation (apparent competition; see Chapter 10) may also be stabilized by fluctuations in predator densities when at least one of the consumer species has a nonlinear mortality response to predator numbers (Kuang and Chesson 2008).

The second type of fluctuation-dependent stabilizing mechanism, *the storage effect*, was discovered by Chesson in the early 1980s, when he set out to rigorously explore a hypothesis, proposed by Peter Sale, that temporal variation in recruitment could allow the coexistence of many species of fish competing for territories on a coral reef (Sale 1977). Chesson and colleagues have since shown that the storage effect is an important stabilizing mechanism that can operate in a wide variety of taxa and communities, and not just in situations where species compete for space (e.g., Chesson 2000b; Kuang and Chesson 2010). However, we will use competition for space among reef fishes to introduce the storage effect, as it provides a clear and intuitive example of this stabilizing mechanism. We will then expand our discussion to show the generality of the mechanism, although we will steer away from the mathematical details. Those wishing a rigorous mathematical treatment of the storage effect in variable environments should consult Chesson (1994, 2000a,b and references therein).

The storage effect

The life histories of many coral reef fishes include an adult phase that is sedentary and territorial, and thus tightly tied to a single reef, as well as a widely dispersed larval phase, which can colonize multiple reefs. Competition for space on a reef is intense, but an individual that succeeds in obtaining a territory on the reef is likely to retain that territory until it dies. Sale (1977) hypothesized that this system could allow for the coexistence of a large number of fish species, even if those species were very generalized in their diets and habitat use, so long as there was random temporal variation in the availability of territories (caused by the deaths of adults) and in the settlement of larval fish of different species on those territories. Sale (1977: 354) envisioned that "reef fishes are adapted to this unpredictable supply of space in ways which make interspecific competition for space a lottery in which no species can consistently win. Thus, the high diversity of reef fish communities may be maintained because the unpredictable environment prevents development of an equilibrium community."

Chesson and Warner (1981) formalized Sale's verbal argument in a more rigorous mathematical model that they initially termed the "lottery model," based on the way competition for space was determined. Since that time, the theory behind the lottery model has been expanded to include a much broader domain of study and is now more generally referred to as models of the storage effect, in reference to the mechanism of coexistence.

Figure 14.1 illustrates a simplified version of the lottery model for two fish species. Species 1 initially holds 90% of the 100 territories on the reef, and species 2 holds the remaining 10% of the territories. Each species suffers 10% adult mortality between recruitment events. The key features of Chesson and Warner's model as applied to this competition for space between reef fish are:

1. Adult fish hold territories on the reef, and an individual holds its territory until it dies. Adults are iteroparous (reproduce more then once), and each species has overlapping generations.

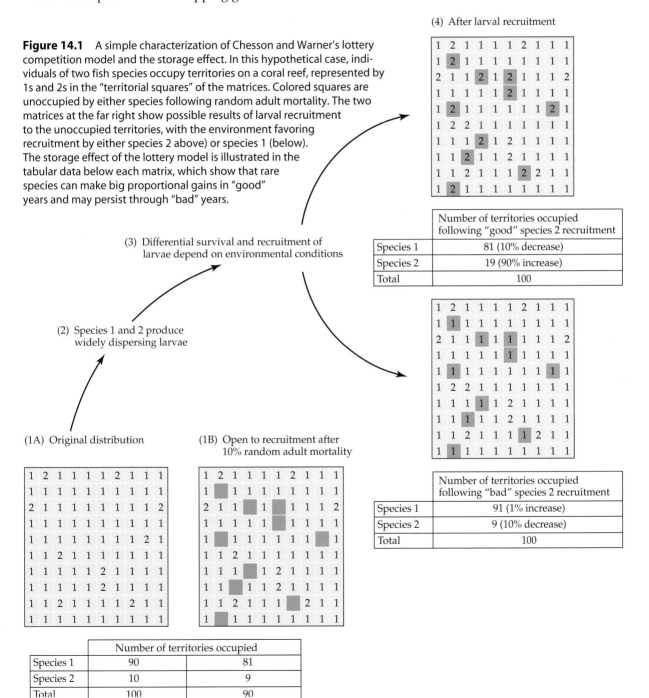

Figure 14.1 A simple characterization of Chesson and Warner's lottery competition model and the storage effect. In this hypothetical case, individuals of two fish species occupy territories on a coral reef, represented by 1s and 2s in the "territorial squares" of the matrices. Colored squares are unoccupied by either species following random adult mortality. The two matrices at the far right show possible results of larval recruitment to the unoccupied territories, with the environment favoring recruitment by either species 2 above) or species 1 (below). The storage effect of the lottery model is illustrated in the tabular data below each matrix, which show that rare species can make big proportional gains in "good" years and may persist through "bad" years.

(4) After larval recruitment

	Number of territories occupied following "good" species 2 recruitment
Species 1	81 (10% decrease)
Species 2	19 (90% increase)
Total	100

(3) Differential survival and recruitment of larvae depend on environmental conditions

(2) Species 1 and 2 produce widely dispersing larvae

	Number of territories occupied following "bad" species 2 recruitment
Species 1	91 (1% increase)
Species 2	9 (10% decrease)
Total	100

(1A) Original distribution

(1B) Open to recruitment after 10% random adult mortality

	Number of territories occupied	
Species 1	90	81
Species 2	10	9
Total	100	90

2. To survive and reproduce, a larval fish must obtain a territory. Space is limiting, and there are always more larvae attempting to establish territories than space allows. Allocation of territories is on a first-come first-served basis: the first individual to arrive at a suitable site can establish a territory there.

3. Larvae are produced in abundance and dispersed; therefore, the number of larvae seeking territories at a site is not directly related to the number of adults already present at that site.

4. Larval survival depends on environmental conditions. Environmental conditions vary over time, and species differ in their responses to those conditions, such that species have relatively "good" and "bad" years for larval survival and recruitment.

How would each species respond to a "good" recruitment year, in which one species obtains all the available territories and the other species obtains none? Differences in good and bad recruitment years among species are expected, for example, if variation in the physical environmental affects the number of larvae surviving to potentially recruit to the reef. The hypothetical data in Figure 14.1 illustrate that a good recruitment year has a disproportionate positive effect on the species that is at lower abundance. This temporal variation in recruitment, coupled with a relatively long-lived adult stage that can "store" the population contributions of good years (or, similarly, buffer against population loss in bad years), allows temporal fluctuations to act as a stabilizing mechanism—a concatenation of qualities that has come to be known as the **storage effect**.

The storage effect model is potentially relevant to any group of organisms that have over-lapping generations and a relatively long-lived life stage. Annual plants with long-lived seeds (Pake and Venable 1995; Sears and Chesson 2007; Angert et al. 2009) and zooplankton with diapausing (resting) eggs (Cáceres 1997) have been used to test its predictions. Chesson and colleagues have further shown that the storage effect is quite general and that it can act to promote species coexistence in a wide variety of situations where there is temporal or spatial variation in the environment (Chesson 2000b; Chesson and Kuang 2010).

Environmental variation (fluctuation) may come from many factors, including resource availability, rainfall, temperature, and predation levels. For the storage effect to function, this variation must change the birth, survival, or recruitment rates of species from year to year (or place to place). Critically important is that species differ in their responses to this environmental variation—that is, they must differ in their niches (Chesson 1991). Environmental variation that affects all species equally, such as density-independent mortality caused by disturbance, harsh environmental conditions, or nonselective predation, is not a stabilizing mechanism (Chesson and Huntly 1997; Chase et al. 2002). In order for environmental variation to stabilize species coexistence, species must respond differently to the environment, such that some species are favored (e.g., have high larval survivorship) at particular times or places and other species are not. This is the first component of the storage effect: *species-specific environmental responses*.

The second component of the storage effect is a long-lived life stage (e.g., diapausing eggs, dormant seeds, long-lived adults). A long-lived life stage allows species to store the positive effects of "good" years and thus buffer the negative effects of "bad" years. Otherwise, a species would go extinct after a short run of "bad" years. Chesson (2000a) calls this second component *buffered population growth*.

The third component of the storage effect is *covariance between environment and competition*. That is, the effects of competition must covary with the environment such that negative environmental effects decrease competition and positive environmental effects increase competition. It is easy to see this covariance operating in the simple reef fish example presented in Figure 14.1. Because the two species respond differently to the environment (component 1), we expect that intraspecific competitive effects will be strengthened when the environment favors a species' population growth and will be reduced when the environment is unfavorable for a species and it becomes rare.

Adler et al. (2006a) tested for the three components of the storage effect in a long-term (30-year) data set that mapped individual prairie plants in permanent plots in Kansas, focusing their analysis on the three grass species that made up more than 95% of the vegetative cover. All three species had long life spans and thus met the requirement for buffered population growth (**Figure 14.2A**). In addition, there were differential species-specific responses to environmental variation (e.g., variation in precipitation and mean annual temperature) such that pairwise correlations in species' population growth rates were absent or weak (**Figure 14.2B**). Lastly, Adler and colleagues found evidence for the third component of the storage effect: the effects of competition were more severe in more favorable years (i.e., covariation between environment and competition). Using a population growth model parameterized for each species, they showed that neighboring plants limited a species' population growth in favorable years, but had weak or even facilitative effects in unfavorable years (**Figure 14.2C**). Thus, all three requirements of the storage effect were met. Adler et al. (2006a) also showed via simulations that each species' low-density population growth rate was higher in a variable environment than in a constant environment (**Figure 14.3**), further evidence that coexistence in these prairie grasses was strongly promoted by the storage effect.

Adler et al. (2009) applied a similar analysis to a sagebrush steppe community. However, in this case, they found little covariation between environmental conditions and competition, leading them to conclude that climate variability had only a weak effect on species coexistence in this ecosystem. A handful of other studies have documented aspects of the storage effect operating in desert annuals (Pake and Venable 1995; Angert et al. 2009), freshwater zooplankton (Cáceres 1997), coral reef fishes (Chesson and Warner 1981), and tropical trees (Runkle 1989). However, more studies are needed that test all three components of the storage effect (including the empirically challenging test of whether competition limits growth more in favorable than in unfavorable years). Until then, we remain somewhat in limbo; the storage effect has been shown to be an important stabilizing mechanism in a variety of theoretical models, and it is very likely to be a

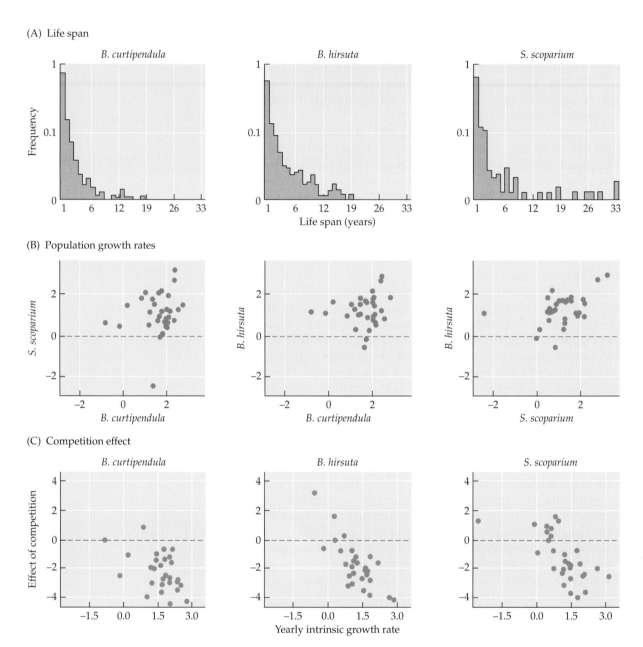

Figure 14.2 Evidence that the three components of the storage effect are operating in a community of coexisting prairie grasses dominated by *Bouteloua curtipendula*, *B. hirsuta*, and *Schizachyrium scoparium*. (A) Each of the three species has the potential for long life spans, which buffers population growth. (B) Weak correlation of population growth rates between species pairs demonstrates that the species respond differently to environmental variation. (C) For each species, competition has stronger negative effects on population growth in more favorable years (years with higher intrinsic growth rates). (After Adler et al. 2006a.)

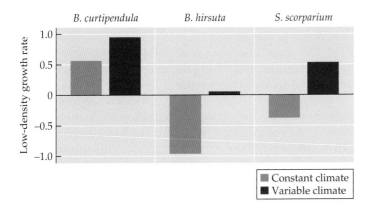

Figure 14.3 Estimated long-term low-density growth rates in constant and variable climate conditions based on model simulations for the three species of prairie grass in Figure 14.2. All three performed better in variable than in constant environments, indicating that the storage effect may be acting to stabilize species coexistence. (After Adler et al. 2006a.)

potent force in nature, but it is hard to confirm empirically (although, to be fair, explicit tests of the mechanisms of species coexistence are rare in any system; Siepielski and McPeek 2010).

The intermediate disturbance hypothesis

The **intermediate disturbance hypothesis** (**IDH**; Grime 1973a,b; Connell 1978) is perhaps the best-known and most commonly cited explanation for the maintenance of species diversity in a fluctuating environment, aside from the storage effect. In brief, the IDH proposes that intermediate levels of environmental disturbance will allow more species to coexist than low or high rates of disturbance. Initial statements of the IDH were quite general (Grime 1973a,b; Connell 1978), with the basic idea being that disturbance interrupts the process of succession, such that good colonizers are favored at high rates of disturbance, good competitors are favored at low rates of disturbance, and intermediate rates of disturbance promote the coexistence of both types of species. "Disturbance" may be any factor that causes the destruction of biomass (Grime et al. 1987) and makes space and resources available for use by new individuals (Roxburgh et al. 2004). "Intermediate" most often refers to the frequency of disturbance (its time scale), but the term has also been applied to the intensity, extent, and duration of disturbance as well. Here we will consider only the frequency of disturbance over time. Ecologists have most often applied the IDH to species coexistence in patchy environments, where disturbance resets succession within a patch by opening up space and freeing up resources, thereby allowing a number of species to coexist in a system of patches at different successional states (e.g., Sousa 1979a,b). This application of the IDH to a patchy environment, which has been termed the "successional mosaic hypothesis" (Chesson and Huntly 1997), generally interprets species coexistence as the result of a trade-off between competitive ability and dispersal ability (see the discussion of this competition/colonization trade-off in Chapters 12 and 13). The IDH has also been applied to environments without explicit spatial structure (e.g., Bartha et al. 1997; Collins and Glenn 1997). In a spatially homogeneous context, disturbance affects all organisms simultaneously, regardless of their spatial location (Roxburgh et al. 2004). Because there is no spatial structure, the competition/colonization trade-off does not apply in this case, although other niche differences between species may allow for differential responses to disturbance.

Roxburgh et al. (2004; see also Shea et al. 2004) provide an excellent discussion of the mechanisms underlying the effects of intermediate disturbance frequencies on species coexistence in spatially heterogeneous or homogeneous environments. They show that coexistence in a patchy environment depends on the competition/colonization trade-off; without this trade-off, the superior competitor wins at all disturbance frequencies at which species can maintain positive population growth. With this trade-off, coexistence is promoted at intermediate disturbance levels. Intermediate disturbance frequencies can also promote coexistence in spatially homogeneous environments, but only if niche differences exist between species that differentiate their responses to disturbance (e.g., life history differences that trade off with competitive ability). When these niche differences are present, disturbance may promote coexistence via the two general classes of mechanisms discussed earlier in this chapter: relative nonlinearity of competition and the storage effect (Chesson 2000a; Roxburgh et al. 2004). In the absence of such niche differences, disturbance can only slow the rate of competitive exclusion (as in the models of Huston 1979; Hubbell and Foster 1986); it cannot lead to long-term coexistence.

There are some excellent empirical studies demonstrating that species richness may be increased by intermediate levels of disturbance. One of the best of these studies is the classic work by Sousa (1979a,b), who showed that the number of algal species living on rocks in a intertidal boulder field near Santa Barbara, California, peaked at intermediate boulder sizes. Sousa convincingly demonstrated that boulders of intermediate size were disturbed (rolled over) by storms at intermediate frequencies, and that this intermediate level of disturbance prevented them from being dominated by the best algal competitors. In this system, there appears to be a trade-off between competitive ability and dispersal ability. However, despite the intuitive appeal of the IDH and its long tenure in ecological thought, there is surprisingly little evidence supporting its general importance in nature. In a literature review, Shea et al. (2004) found that only 18% of 250 studies supported the IDH. Similarly, in their meta-analysis of 116 species richness–disturbance relationships, Mackay and Currie (2001) found that only 16% of those relationships showed a peaked response. Bongers et al. (2009) recently examined the IDH in tropical forests, the ecosystem for which the IDH was first proposed (Connell 1978). Using a large data set of tree inventories from 2504 one-hectare plots, Bongers et al. (2009: 798) found that "while diversity indeed peaks at intermediate disturbance levels little variation is explained outside dry forests, and that disturbance is less important for species richness patterns in wet tropical rain forests than previously thought."

As we have seen, environmental fluctuations that affect species equally are not stabilizing, and they do not by themselves promote the long-term coexistence of species. This distinction, however, is not always recognized. Instead, it is common to hear ecologists express the idea that environmental disturbance or harshness reduces the intensity of competition and thereby promotes species coexistence. The intermediate disturbance hypothesis is sometimes portrayed in this light, as are the ideas expressed by Huston (1979) and Menge and Sutherland (1987). However, as Chesson (2000a: 354) emphatically notes, "no credence can be given to the idea that disturbance promotes stable coexistence by simply reducing population densities to levels where competition is weak" (see also Holt 1985; Chesson and Huntly 1997; Chase et al. 2002). Mortality by itself cannot promote species coexistence, because its positive effect of lowering competition is countered by its negative effect on population growth. For environmental fluctuations (including disturbance) to stabilize species coexistence, they must affect species differentially. This differential effect on species need not be immediate, however, but may act in a chain of events that differentially affect species' recruitment and/or mortality.

Niche-Based and Neutral Processes in Communities

As we have seen above, species coexistence is promoted by (1) mechanisms that reduce the fitness differences between species (equalizing mechanisms) and (2) by mechanisms that make intraspecific density dependence stronger than interspecific density dependence (stabilizing mechanisms). Equalizing mechanisms may slow the time to competitive exclusion, however, stabilizing mechanisms that allow species to recover when they become rare are necessary for the long-term coexistence of species. Moreover, stabilizing mechanisms can be fluctuation-dependent (e.g., the storage effect) or fluctuation-independent (e.g., resource partitioning), and they may involve density-dependent feedback loops that act through competitors or predators or both (Chesson and Kuang 2008).

Given the potential for both equalizing and stabilizing mechanisms to promote species coexistence and enhance diversity, how important is each type of mechanism in nature? This question is similar, although not identical, to recent inquiries into the relative roles of "niche-based" and "neutral" processes in structuring communities (e.g., Leibold and McPeek 2006; Holt 2006; Siepielski et al. 2010). As we saw in Chapter 13, the neutral theory of biodiversity (Hubbell 2004) assumes that all individuals have the same demographic properties regardless of species identity. This individual-level equivalence necessarily results in species-level equivalence (Chesson and Rees 2007). Thus, we can think of the neutral theory as describing a system in which all species have the same average fitness (i.e., equalizing mechanisms are present simply because we assume that all individuals are equivalent). However, the neutral model lacks stabilizing mechanisms for the same reason: individuals are not differentially affected by the presence of conspecifics or heterospecifics, so there can be no difference in the strengths of intraspecific and interspecific density dependence.

To what extent does species identity matter in the dynamics of communities? Neutral theory says that it does not matter. Niche-based theory says it does. Ecologists are just beginning to address this question. One way to discriminate between niche-based and neutral processes is to manipulate both the total and relative abundances of co-occurring species occupying

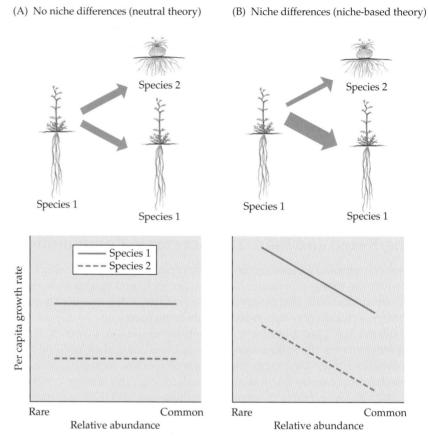

(A) No niche differences (neutral theory)   (B) Niche differences (niche-based theory)

Figure 14.4 Niche differences (such as differences in rooting depths in plants) cause species to limit the growth of conspecifics more than the growth of other species and thereby promote species coexistence. If there are no niche differences, species limit themselves and their competitors equally. (A) When niche differences are absent, a species' relative abundance has no effect on its per capita growth rate. (B) When niche differences are present, a species' per capita growth rate declines with an increase in its relative abundance in the community. Arrow widths represent the degree to which individuals limit one another. (After Levine and HilleRisLambers 2009.)

the same trophic level (i.e., potential competitors). If neutral mechanisms govern the dynamics of co-occurring species, then we would expect each species' performance to depend on total species abundance, but not on its own specific abundance. That is, all species should show similar density-dependent responses to manipulations of total abundance, but no response to manipulations of relative abundances when total abundance is held constant (**Figure 14.4A**). On the other hand, if niche-based mechanisms govern the dynamics of co-occurring species, we would expect each species to respond strongly to changes in its relative abundance (**Figure 14.4B**). Species responses to density manipulations are most appropriately measured as changes in per capita population growth rates, as in Figure 14.4, although other measures correlated with per capita population growth rate are also informative.

This simple approach allows us to assess whether niche-based or neutral processes are operating in a community, although it does not distinguish their relative importance (Adler et al. 2007; Levine and HilleRisLambers 2009; Siepielski et al. 2010).

Mark McPeek and colleagues used this experimental approach to explore niche-based and neutral mechanisms of species coexistence in damselflies, focusing on interactions at the larval stage. Larval damselflies of the genus *Enallagma* are generalist predators that live in the shallow waters of ponds and lakes. In North America, the damselfly genus *Enallagma* (38 species) differs ecologically from its sister genus *Ischnura* (14 species): larval *Enallagma* are better at avoiding predation than are larval *Ischnura*, but *Ischnura* larvae are better at converting prey into their own biomass than are *Enallagma* (McPeek 1998). Thus, there are pronounced niche differences between the genera that promote coexistence at this level. Within the genus, species are very similar: all of them spend 10–11 months of the year as aquatic larvae living in the vegetation of lakes. From 5 to 12 *Enallagma* species may co-occur in lakes throughout eastern North America (Johnson and Crowley 1980; McPeek 1998). Negative density-dependent feedbacks are well documented at the larval stage: *Enallagma* larvae experience density-dependent growth rates due to competition and density-dependent predation by fish and invertebrate predators (McPeek 1998). Do these density-dependent feedbacks act to regulate individual species in a stabilizing fashion? This is the question that McPeek and colleagues set out to address.

McPeek and his colleagues (Siepielski et al. 2010) experimentally manipulated the absolute and relative abundances of two *Enallagma* species in the field. They found that the species responded strongly to manipulations of total abundance, but showed no responses to changes in their relative abundances when total abundance was held constant (the responses measured were larval per capita mortality and growth rates over a 3-month period). Furthermore, in a survey of 20 lakes located close to one another, they found that per capita mortality and growth rates of multiple *Enallagma* species were not correlated with the species' relative abundances in a lake, and that the pattern of co-occurrence of species among lakes did not differ from what would be expected by chance. Thus, Siepielski et al. (2010) concluded that within the genus *Enallagma*, the species appear to be ecologically equivalent.

A problem with explaining diversity via equalizing mechanisms alone, however, is the question of how ecologically similar species arise and become established if they have no competitive advantage when rare. The evidence suggests that *Enallagma* diversification may not have involved the invasion of intact communities by rare species. Much of the diversification in the genus *Enallagma* occurred during periods of repeated glaciation during the Quaternary, when the landscape was fragmented into isolated lake communities and species may have arisen and expanded their ranges into relatively unoccupied lakes (Turgeon et al. 2005). Moreover, it appears that speciation in this damselfly group was driven primarily by sexual selection for differentiation in reproductive structures, with little ecological differentiation among species (McPeek et al. 2008). Thus, average fitness equality and very long times to competitive exclusion (helped by large population sizes and low levels of dispersal between lakes) may explain much of the diversity in *Enallagma* in the absence of strong niche differences and little evidence for stabilizing mechanisms (McPeek 2008).

In contrast to the damselflies, two recent studies of plant communities showed that the relative abundances of species with differences in average fitness were maintained by stabilizing mechanisms. Levine and HilleRis-Lambers (2009) constructed an experimental community of annual plant species that occur naturally on serpentine soils in California. They varied the relative abundances of ten focal species in field plots while keeping the total abundance of all plants in the plots constant. If niche properties stabilize coexistence, then we would expect each species' population growth rate to decrease with an increase in its relative abundance. However, if species identity does not matter and the system dynamics are neutral, then a species' growth rate should not change with a change in its relative abundance as long as total abundance remains constant. The most abundant species in the annual plant community showed strong declines in population growth with increasing relative abundance, demonstrating strong stabilizing mechanisms at work. Other species showed weaker declines in growth rates with increased relative abundance, whereas the growth rates of the three rarest species actually increased with an increase in relative abundance. The authors concluded that niche differences stabilized species diversity, especially for the dominant species in the community.

Levine and HilleRisLambers (2009) and Adler et al. (2010) also tested for niche-based and neutral mechanisms in natural plant communities by fitting demographic models to species' observed population dynamics and then using these models to look at the predicted response of the community when stabilizing mechanisms were removed. Both studies found that removing the effects of stabilizing mechanisms caused diversity within the communities to decline rapidly. In the serpentine plant community studied by Levine and HilleRisLambers (2009), species diversity (measured as a function of species richness and species evenness, or Shannon's index of diversity) decreased after only two simulated generations when niche-based stabilizing mechanisms were removed (**Figure 14.5**). Stabilizing mechanisms were also prevalent in the sagebrush steppe community studied by Adler et al. (2010). In this study, the authors analyzed spatially explicit models fit to 22 years of demographic data to (1) quantify the stabilizing effect of niche differences, (2) compare the relative strengths of stabilizing mechanisms and fitness differences, and (3) partition the stabilizing mechanisms into fluctuation-dependent and fluctuation-independent processes. They found that average fitness differences between species were small and that stabilizing mechanisms were much larger than required to overcome the fitness differences. They also showed that fluctuation-dependent mechanisms were less important than fluctuation-independent mechanisms in this community. Thus, Adler et al. (2010) concluded that niche differences are crucial to maintaining species diversity in this community.

The demographic modeling approach used by Levine and HilleRis-Lambers (2009) and Adler et al. (2010) is a powerful tool for quantifying the relative importance of niche-based and neutral mechanisms in maintaining

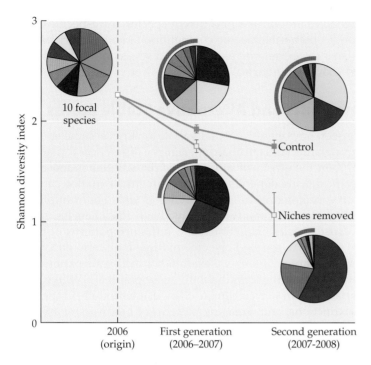

Figure 14.5   Two simulated generations (2006–2007, 2007–2008) of change in the diversity and composition of a serpentine plant community stabilized by niche differences ("control") versus one in which the demographic influence of niche differences was removed. Pie charts show the average proportion of total community seed mass for each of 10 focal species in each treatment and generation. Arcs at the edges mark the collective abundances of the seven rarest species. Species diversity (data points) was measured as a function of species richness and species evenness, or Shannon's diversity index (mean ±1 SE). Diversity decreased after only two simulated generations when niche-based stabilizing mechanisms were removed. (After Levine and HilleRis-Lambers 2009.)

biodiversity in communities. However, it does not identify the niche differences between species that promote coexistence (e.g., resource partitioning, frequency-dependent predation, the storage effect). Instead, it evaluates the collective importance of multiple coexistence mechanisms acting in concert and points to whether further investigation into those mechanisms would be worthwhile. Clark (2010), reasoning along somewhat different lines, suggests that niche differences acting along many axes ("high niche dimensionality") are in fact crucial for the coexistence of diverse communities such as forest trees (see also Clark et al. 2007). Using 6–18 years of data on tree dynamics in 11 forests in the southeastern United States, Clark modeled the responses of individual trees and tree species to environmental variation. He concluded that individuals responded in different ways to environmental fluctuations, depending on fine-scale variation in resources (e.g., moisture, nutrients, and light) and their genotypic differences. But, because individuals of the same species responded more similarly to one another than to individuals of different species, intraspecific competition was stronger than interspecific competition—the hallmark of stabilizing mechanisms of coexistence (see further discussion in Chesson and Rees 2007 and Clark and Agarwal 2007).

Regime Shifts and Alternative Stable States

We have seen how environmental variation may promote species coexistence through mechanisms that act via niche differences (broadly defined) to strengthen the effects of intraspecific density dependence relative to interspecific density dependence. Environmental variation can have other important consequences for communities as well. For example, over geologic time scales, climatic fluctuations drive glacial cycles, shift species ranges, cause species extinctions, and influence patterns of biodiversity at global scales. We touched on some of these long-term effects of climate variation in Chapter 2 when we examined broad-scale patterns in species diversity, including the latitudinal diversity gradient. We live today in an era of rapid, human-driven climate change, where shifts in species ranges, species extinctions, and changing patterns of biodiversity are observable even in human lifetimes. The scope of global climate change and its impacts on ecological systems is beyond what we can cover in a book focused on patterns and processes in community ecology. However, the study of species interactions and the mechanisms that structure communities leads naturally to the questions of whether communities might occur in alternative states, depending on the outcomes of species interactions, and whether environmental variation can cause communities to shift from one state to another.

The idea that communities or ecosystems might occur in alternative stable states was first proposed theoretically by Lewontin (1969), explored mathematically by May (1977b), and linked empirically to ecology by Holling (1973) and Sutherland (1974). In theory, even small environmental changes can lead to dramatic shifts in community state. Ecologists are rightfully concerned about these shifts, as they have important implications for the preservation of biodiversity, the management of biotic resources, and the restoration of ecosystems.

Before beginning our discussion of this important topic, we need to define some terms. **Alternative stable states** describes a situation where communities may exist in different configurations (e.g., of species richness, species composition, food web structure, size structure) and these different community configurations represent different equilibrium states (May 1977b; Beisner et al. 2003). Alternative stable states may be defined mathematically as different basins of attraction (May 1977b; Scheffer et al. 2001). The concept of alternative stable states is often represented graphically as a landscape of hills and valleys (**Figure 14.6**). The community is represented by a "ball" on the landscape, and different valleys (or basins) represent different potential community states. A community tends to remain in a stable state (i.e., in a valley) until environmental disturbance or change causes it to shift to a new state (Figure 14.6).

A change from one community state to another is termed a **regime shift** or **phase shift**, and the critical threshold at which a system undergoes a

Figure 14.6 "Hill and valley" illustration of alternative stable states. A community or ecosystem (represented by the ball) will remain in the same state (valley or basin) unless (1) it is perturbed by some environmental disturbance to shift to another state, or (2) the landscape changes shape due to changes in the external environment. The steepness of a valley corresponds to the size of the perturbation required to shift the community into a new state and to the speed at which a community recovers from a perturbation.

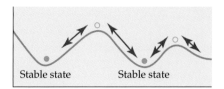

Stable state   Stable state

regime shift is referred to as a **tipping point**. Pimm (1984) defined the strength of perturbation needed to shift a community from one state to another as **resistance** and the speed at which a community recovers from a perturbation as **resilience**. The concept of resilience, however, has been used in multiple ways since it was first employed by Holling (1973). It is now common to see resilience defined either as (1) **ecological resilience**—the magnitude of a perturbation that a system can withstand before undergoing a regime shift, or as (2) **engineering resilience**—the time it takes for a system to recover from a perturbation (see reviews in Gunderson 2000; Nyström et al. 2008).

There are many examples of communities or ecosystems showing dramatic regime shifts in response to environmental change. These include shifts from coral-dominated reefs to reefs dominated by fleshy macroalgae (McCook 1999; Nyström et al. 2008; Hughes et al. 2010), shifts from clear to turbid shallow lakes (Scheffer 1998; Scheffer and van Nes 2007), shifts from grass-dominated to tree-dominated savannas (Wilson and Agnew 1992; Bond 2010), shifts in arid ecosystems from vegetated communities to deserts (Rietkerk et al. 2004; Kéfi et al. 2007a), abrupt shifts in fish communities and sudden collapses in harvested fish stocks (Hutchings and Reynolds 2004; Daskalov et al. 2007; Persson et al. 2007), and even shifts in the microbial community of the human gut following surgery or treatment with antibiotics (Hartman et al. 2009; Dethlefsen and Relman 2011). Many of these examples of regime shifts are the result of anthropogenic environmental changes that cause ecosystems to shift to an undesirable state. For example, increased grazing may reduce vegetation cover in arid ecosystems, reducing the facilitation of local water availability by plants and shifting the system toward desertification (Schlesinger et al. 1990; Rietkerk et al. 2004). Thus, an important question in both basic and applied ecology is how predictable regime shifts are and whether they are reversible (Suding et al. 2004; Scheffer 2009). Simple graphic models can help us understand why regime shifts occur and why they may be difficult to reverse. The presentation that follows borrows from the work of Scheffer and colleagues (e.g., Scheffer et al. 2001; Scheffer 2009) and Kéfi (2008).

We expect that a slow and steady change in environmental conditions will result in a smooth transition from one community state to the next—for example, the loss or gain of species in response to a change in temperature or nutrients. Such a community response to environmental change might be very gradual and linear, as shown in **Figure 14.7A**. Or, a community might be relatively insensitive to environmental changes across a range of conditions, but respond strongly around some threshold condition (**Figure 14.7B**). In this case, we might observe a change in community state that is

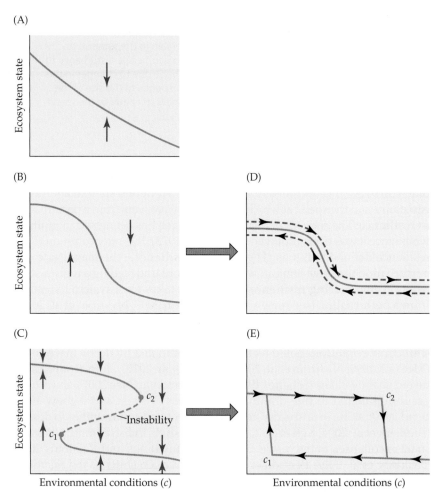

Figure 14.7 Communities and ecosystems may respond smoothly or abruptly to changing environmental conditions. Abrupt responses may lead to regime shifts and alternative stable states. (A–C) Responses in the equilibrium state of a community (blue curve) to a change in environmental conditions can be classified into three types: smooth (nearly linear) responses (A); abrupt (highly nonlinear) responses (B); and catastrophic responses (C), resulting in alternative stable states. Red arrows indicate the system's movement toward its equilibrium state. In (A) and (B), only one equilibrium community state exists for each environmental condition. In (C), however, the equilibrium curve is folded back on itself, so that multiple equilibria exist for each environmental condition. Equilibria on the dashed line are unstable. If the system is on the upper branch of the equilibrium curve near point c_2, a slight increase in environmental conditions (a move to the right on the x axis) will cause the system state to jump catastrophically to the lower branch (an alternative stable state). (D–E) A change in environmental conditions over time can reveal alternative stable states. (D) As environmental conditions decrease and then increase (black arrowheads), the state of the system responds smoothly along the solid line (although there may be regions of rapid change). (E) A gradual increase in environmental conditions causes the community to shift state at a tipping point (c_2). If the change in the environmental condition is subsequently reversed, the system shows hysteresis, and a shift back occurs only when environmental conditions return to the tipping point (c_1). The existence of such hysteresis loops is strong evidence for the existence of alternative stable states. (After Scheffer et al. 2001, 2009.)

sufficiently large and abrupt to be considered a regime shift. A more extreme kind of change at a threshold occurs when a system has alternative stable states (**Figure 14.7C**). In this case, the curve that describes the community's response to environmental change folds back on itself. When this occurs, the system has two alternative stable states, separated by an unstable equilibrium that marks the border between basins of attraction.

A variety of mechanisms can theoretically lead to alternative stable states in communities, including positive feedbacks, nonlinear species interactions, size-based refuges from predation, and priority effects (e.g., Lewontin 1969; Chase 2003; van Nes and Scheffer 2004; Kéfi et al. 2007b). We won't go into the theory underlying the formation of alternative stable states, but it is important that we understand how environmental variation can affect communities that have alternative stable states. There is a crucial distinction between the abrupt response exhibited by the system in Figure 14.7B and the alternative stable states shown in Figure 14.7C. We can best appreciate this distinction by following the response of a community as environmental conditions change.

In the situation depicted by Figure 14.7B, a change in environmental conditions (along the x axis) will cause the equilibrium state of the community to track smoothly along the path shown by the arrows in **Figure 14.7D**. There may be regions of more or less rapid community change, but a change in environmental conditions in one direction, followed by a reversal of that change, will cause the state of the community to transition smoothly along a single path. One the other hand, if a community exists in alternative stable states, there is a tipping point (a point of critical transition) between community states (Figure 14.7C). When the community is in the state represented by the upper branch of the folded curve in Figure 14.7C, it cannot pass to the lower branch smoothly. Instead, a change in environmental conditions past the tipping point c_2 will cause the community state to shift abruptly to the lower branch. In the theory of dynamic systems, this instantaneous shift from one system state to another is known as a *bifurcation point*. If the environmental conditions return to their previous state (i.e., they move back along the x axis in Figure 14.7C), there is a lag in the community response. Only after the environmental conditions reach the tipping point c_1 does the community shift back to its previous state. This delayed response is illustrated in **Figure 14.7E**, where we see that the state of the community tracks along two different paths in response to an environmental change in one direction followed by a reversal of that change. This delayed response to a forward and backward change in environmental conditions is termed **hysteresis**, and it is a distinguishing feature of systems that exist in alternative stable states.

Another way to visualize the transitions between alternative stable states is to use the familiar hill and valley diagram. **Figure 14.8** uses this approach to show how a vegetated community in an arid environment might shift to a desert state due to overgrazing. The community is represented by a ball in a landscape of environmental conditions (hills and valleys). Starting with environmental conditions near c_1, we see that the landscape contains only one stable state, which has a deep basin of attraction (panel A at the bottom of Figure 14.8). At this point, the community is stable and will respond

Figure 14.8 Two ways in which a community may shift from one stable state to another. The upper diagram is a bifurcation plot similar to that in Figure 14.7C. The solid portions of the equilibrium curve correspond to stable states and the dashed portion to instability. The x axis represents environmental conditions, c. Tipping points c_1 and c_2 bound the parameter space where two stable states co-occur (bistability). Panels A–F show how the stability landscape changes with changing environmental conditions; the ball represents the community state. For small values of c, the landscape has only one valley. With increasing c, a second valley appears (at c_1), and the community can exist in either of two alternative states. At tipping point c_2, the first valley disappears and the landscape is again reduced to a single valley. Once the system has made the shift, recovery to the initial state may occur if environmental conditions return to a point below c_1 (hysteresis). (After Kéfi 2008.)

to a perturbation by quickly returning to its original state. However, as environmental conditions change (moving from c_1 to c_2), the shape of the landscape changes. At the point along the x axis where the environmental condition is $c = 2.3$, there are two alternative stable states (panel C). A sufficiently strong perturbation will push the community from one alternative state to the other (represented by the "dotted" ball in panel C). However, in the face of a small perturbation, the community will remain in its current state. As environmental conditions move farther along the x axis toward condition $c = 2.5$ (approaching the tipping point c_2 in the top diagram), the basin of attraction in which the community rests becomes very small and shallow (panel D). At this point, the system will be slow to recover from a perturbation, and even a small perturbation may send it into a new state. At tipping point c_2 ($c = 2.7$, panel E), the community shifts abruptly to the alternative stable state (the lower arm of the equilibrium curve in the top diagram). If we were to now reverse the environmental conditions and move from c_2 toward c_1, the community would remain in its new stable state until environmental conditions again reached c_1; in other words, the system displays hysteresis.

You might wonder why we should be so concerned about whether communities respond to environmental change as predicted by the theory of alternative stable states. After all, the difference between a nonlinear community response (Figure 14.7B) and a shift between alternative stable states (Figure 14.7C) might seem to be only a matter of degree. In a sense, that is true, making it difficult in practice to distinguish between a strong nonlinear response to environmental change and a true shift to an alterna-

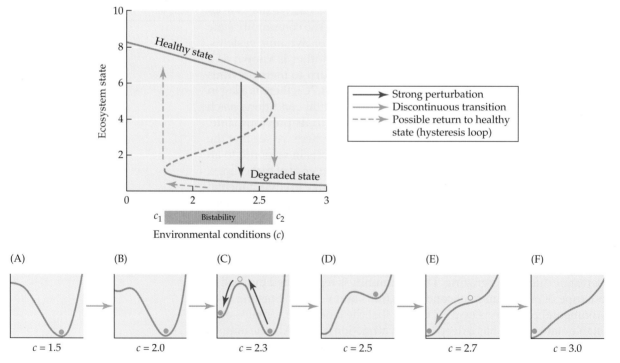

tive stable state (Scheffer and Carpenter 2003). For this and other reasons, some ecologists have long questioned the evidence for alternative stable states in nature (Connell and Sousa 1983). However, there are good reasons why ecologists and conservation biologists should be concerned about the existence of alternative stable states. If communities or ecosystems have tipping points at which they shift abruptly to different states in response to environmental change, this has obvious consequences for predicting the impacts of climate change and environmental degradation. More importantly, if communities or ecosystems occur in alternative stable states, reversing the impacts of environmental change becomes very difficult; that is, returning the environment to its previous condition will not cause the community or ecosystem to return to its previous state if there is hysteresis. In Hawaii, for example, the invasion of woodlands by non-native grasses promotes fire and alters nitrogen cycling, which positively affects the invasive grasses at the expense of native shrubs, creating an internally reinforced state that is very difficult to reverse (Mack and D'Antonio 1998; Mack et al. 2001). Suding et al. (2004), Kéfi (2008), Martin and Kirkman (2009), and Firn et al. (2010) provide additional examples of degraded or changed ecosystems that are resilient to traditional restoration efforts in ways that suggest alternative stable states and necessitate new management practices to disrupt feedbacks and address constraints on ecosystem recovery.

How prevalent are alternative stable states in nature? As mentioned above, showing definitively that alternative stable states exist in any real community or ecosystem is a challenge (Connell and Sousa 1983; Scheffer and Carpenter 2003). Documenting hysteresis in response to a slow forward and backward change in environmental conditions is perhaps the strongest test (Scheffer and Carpenter 2003). This approach has been used in a few experimental studies (e.g., Handa et al. 2002; Schmitz et al. 2004). However, Schröder et al. (2005), in a literature review, found no field experiments that had manipulated a natural system over sufficient time to demonstrate full hysteresis. Experimental manipulations are difficult to apply at large spatial scales and over long periods. Instead, ecologists have followed long-term community responses to environmental changes induced by human activities, "natural experiments," or a combination of events. For example, Scheffer and Carpenter (2003) noted that vegetation cover in a shallow lake in the Netherlands demonstrated hysteresis in its response to an increase and subsequent decrease in nutrient concentrations (**Figure 14.9A**). Daskalov et al. (2007) found hysteresis loops in the responses of some marine organisms to long-term changes in fishing pressure in the Black Sea (**Figure 14.9B, C**), although some organisms they studied did not respond in this way.

In our own work, my colleagues and I examined changes in the zooplankton (cladoceran) community of a Michigan lake in response to a long-term (16-year) decrease and then increase in planktivorous fish density (Mittelbach et al. 2006). Planktivore densities in Wintergreen Lake, Michigan, have varied over two orders of magnitude following the elimination of two fish species (largemouth bass and bluegill) due to a low-oxygen winterkill in 1978 (a natural event) and the sequential reintroduction of those species in the following years (bass in 1986 and bluegill in 1997). The loss of bass

Figure 14.9 Examples of ecosystems showing hysteresis loops and potential alternative stable states. (A) Vegetation cover in a shallow lake changed with nutrient (phosphorus) concentration; arrows show the direction of change over time. Note how the abrupt response in vegetation cover resembles that predicted for systems displaying alternative stable states (see Figure 14.7E). (B, C) Time series of changes in resource biomass in response to changes in consumer abundance in the Black Sea. (B) Change in phytoplankton biomass as a function of zooplankton biomass. (C) Change in planktivorous fish biomass as a function of fishing mortality. Numbers on the plots are years. In each case, the trajectory of resource biomass over time in response to changing consumer abundance suggests a hysteresis loop. (A after Scheffer and Carpenter 2003; B, C after Daskalov et al. 2007.)

(A)

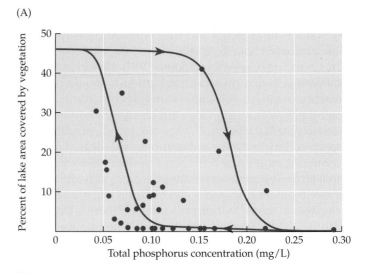

(B)

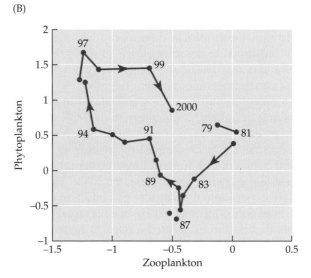

(C)

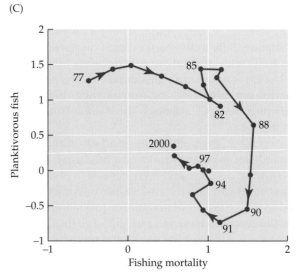

(the top predator in the lake) led to a dramatic increase in the abundance of planktivorous golden shiners, resulting in a "high-planktivory" state. When bass were reintroduced, they consumed all the golden shiners (see Figure 10.8), resulting in a "low-planktivory" state. The reintroduction of bluegill returned the lake to a "high-planktivory" state.

The change in planktivorous fish densities caused dramatic shifts in the species composition of the zooplankton community. The "low-planktivory" zooplankton community was dominated by two species of large-bodied *Daphnia*, whereas the "high-planktivory" zooplankton community was composed of a suite of small-bodied species (**Figure 14.10**). Moreover, these two community states (high and low planktivory) shared no zooplankton species in common (i.e., there was complete species turnover). Plotting the

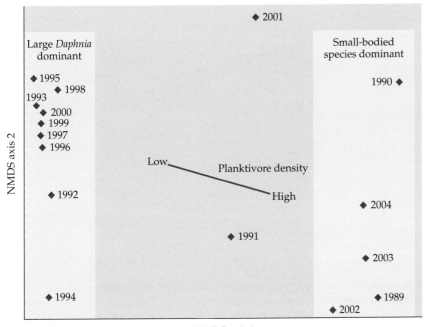

Figure 14.10 The zooplankton (cladoceran) community of Wintergreen Lake, Michigan, shifts between two states, depending on the density of planktivorous fish. Changes in zooplankton species abundances over the years 1989–2004 were visualized using non-metric multidimensional scaling (NMDS), an ordination technique which reveals that the zooplankton community (represented by the red diamonds) occurs in two distinct states (gold shading). In years when the lake contained few planktivorous fish, the zooplankton community was dominated by large-bodied *Daphnia* (left column). In years of abundant planktivores, the zooplankton community was made up of small-bodied zooplankton (e.g., *Ceriodaphnia* and *Bosmina*), and large *Daphnia* were completely absent (right column). The years 1991 and 2001 had intermediate planktivore density and were transitional between the two zooplankton community states. The angle and length of the joint plot line of planktivore density (center) indicates the direction and strength of the relationship of planktivore density with the NMDS scores (Pearson's $r = 0.74$), and shows that the separation in zooplankton community types was strongly correlated with the density of planktivorous fish. (After Mittelbach et al. 2006.)

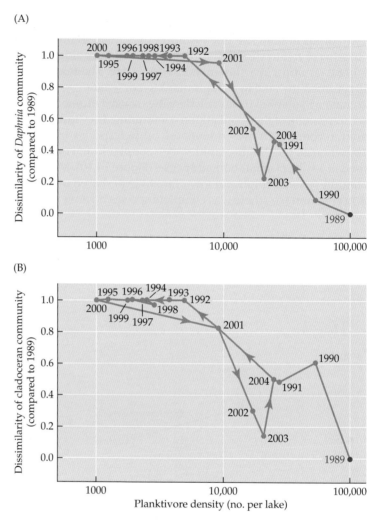

Figure 14.11   Zooplankton community composition changes smoothly with changes in planktivore density in Wintergreen Lake, Michigan, 1989–2004. Change in community state refers to the degree of dissimilarity in species composition relative to the initial community state in 1989 (red), measured as Bray–Curtis distances, where 0 indicates identical species composition and 1.0 indicates no species in common (maximum dissimilarity). Time trajectories are indicated by arrows and years placed next to data points. (A) Change over time for the *Daphnia* community. (B) Change over time for the entire cladoceran assemblage (eight species). (After Mittelbach et al. 2006.)

dissimilarity of the zooplankton community from the initial community state in 1989 for each year as a function of planktivore density allows us to see the pattern of community disassembly and reassembly over time and to look for hysteresis in the community response (**Figure 14.11**). This long-term data set shows that the pronounced regime shift in community composition occurred smoothly and predictably; the replacement of species followed a very similar trajectory through both a decrease and an increase in planktivore abundance, with little suggestion of hysteresis.

Although there are a few good examples of alternative stable states in nature (e.g., Scheffer et al. 2001; Schmitz et al. 2006; Persson et al. 2007; Schooler et al. 2011), the number of well-documented cases remains limited. Moreover, the range of responses to environmental perturbations is much broader than the simple question of whether a system returns to its pre-perturbation state or not (e.g., Schröder et al. 2012). At this point, I am inclined to side with those who question the widespread existence of true alternative stable states. It seems more likely that alternative stable states represent the end of a continuum of responses to environmental change (e.g., Fung et al. 2011); for example, I refer back to Figure 14.7. If the "fold" in the equilibrium curve in Figure 14.7C is relaxed and made smaller the response of the system becomes smoother and more like the response shown in Figure 14.7B. The shift between community states in Figure 14.7B is still large and rapid around the tipping point—there is a regime shift—but there are no alternative stable states. Thus, whether a community responds to a changing environment with a steep threshold response or with a bifurcation is important theoretically, but this distinction may be difficult to demonstrate empirically, and in the end it may make little practical difference to applied management (Kinzig et al. 2006; Suding and Hobbs 2009). Moreover, environmental responses that predict impending regime shifts may apply equally well to systems characterized by continuous and by discontinuous transitions (e.g., Kéfi et al. 2007b). This does not make the search for such "leading indicators" of regime shifts any less important (Scheffer et al. 2009).

Conclusion

Hutchinson's (1961) early idea that competitive exclusion may be prevented in an environment that shifts to favor different species at different times has been replaced by more rigorous mathematical models showing how variable environments may promote species coexistence via relative nonlinearity of competition and the storage effect. In both cases, niche differences among species play important stabilizing roles. For environmental fluctuations (including disturbance) to stabilize species coexistence, they must differentially affect species (Chesson 1991). On the other hand, environmental variation that affects all species equally has been shown to prolong the time to competitive exclusion at best, without changing competitive outcomes or leading to long-term coexistence. However, when population sizes are large and species differences are small, competitive exclusion can take a very long time (Hubbell 2001). Looking broadly at the factors the promote species coexistence in communities, we see two types of mechanisms at work: (1) equalizing mechanisms that reduce the fitness differences between species, and (2) stabilizing mechanisms that make intraspecific density dependence stronger than interspecific density dependence. Ecologists have just begun to quantify the relative importance of stabilizing and equalizing mechanisms in real communities, as well as the importance of niche-based and neutral processes.

We live today in an era of rapid environmental change, and understanding how communities and ecosystems respond to changing environments is an area of intense research. Theory predicts that communities may respond to gradual environmental change via smooth transitions, or via regime shifts—abrupt and startling shifts between states. Ecologists are exploring the causes and consequences of regime shifts, looking for early warning signals of critical transitions (Scheffer et al. 2009; Carpenter et al. 2011; Dakos et al. 2011), and examining the potential impacts of regime shifts and tipping points on restoration and conservation (Suding and Hobbs 2009).

Summary

1. Species within a community may co-occur for long periods as a result of slow rates of competitive exclusion, but if there are no mechanisms promoting the recovery of species that become rare, species coexistence will be unstable, species will eventually disappear, and community richness will decline. Stable coexistence requires that species tend to recover from low densities and that species densities do not show long-term trends.

2. Equalizing mechanisms of species coexistence reduce fitness differences between species. However, equalizing mechanisms by themselves cannot facilitate the recovery of species from low densities and thus cannot lead to stable coexistence.

3. Stabilizing mechanisms act to make intraspecific density dependence stronger than interspecific density dependence. The greater the equalizing mechanisms, the less is required of stabilizing mechanisms to promote coexistence.

4. Fluctuation-independent stabilizing mechanisms operate whether the environment fluctuates or not.

   Fluctuation-dependent stabilizing mechanisms depend on fluctuations in population densities and environmental factors to stabilize coexistence. Two classes of fluctuation-dependent stabilizing mechanisms can be recognized: those involving the relative nonlinearity of competition, and those involving the storage effect.

5. The three components of the storage effect are: (1) differential, species-specific responses to environmental variation; (2) buffered population growth, in which a long-lived life stage allows species to "store" the positive effects of "good" recruitment years and buffer the negative effects of "bad" years; and (3) covariance between environment and competition such that intraspecific competitive effects are strengthened when the environment favors a species' population growth and reduced when the environment is unfavorable and the species becomes rare.

6. The intermediate disturbance hypothesis proposes that disturbance interrupts the process of succession, so that good colonizers are favored at high rates of disturbance, good competitors are favored at low rates of disturbance, and intermediate rates of disturbance promote the coexistence of both types of species. There is surprisingly little evidence supporting the general importance of this hypothesis in nature.

7. One way to discriminate between niche-based and neutral mechanisms of coexistence is experimental manipulation of the total and relative abundances of co-occurring species occupying the same trophic level. If neutral mechanisms govern the dynamics of co-occurring species, then each species' performance should depend on total abundance, but not on its own abundance. If niche differences stabilize coexistence, then each species' population growth rate should decrease with an increase in its relative abundance. This simple approach allows us to assess whether niche-based or neutral processes are operating in a community, although it does not distinguish their relative importance.

8. A community's response to slow and steady environmental change might be gradual and linear. Or, a community might be relatively insensitive to environmental changes across a range of conditions, but respond strongly around some tipping point. Such a large, abrupt change is referred to as a regime shift. The strength of perturbation needed to shift a community from one state to another is called resistance, and the speed at which a community recovers from a perturbation may be referred to as resilience, although resilience has other definitions as well.

9. Communities may exist in alternative stable states—different configurations that represent different equilibrium states. Alternative stable states may be defined mathematically as different basins of attraction and represented graphically as a landscape of hills and valleys. A community tends to remain in a stable state (in a valley or basin) until a perturbation shifts it to another state, or until the landscape changes shape due to a change in the external environment. The community's delayed response to a forward and backward change in environmental conditions is termed hysteresis. There are a few good examples of alternative stable states in nature, although the number of well-documented cases remains limited.

15 Evolutionary Community Ecology

Nothing in biology makes sense except in the light of evolution.
Theodosius Dobzhansky, 1964: 449

While it has long been recognized that adaptive evolution occurs in an ecological context, it is now clear that evolutionary change feeds back directly and indirectly to demographic and community processes.
Scott Carroll et al., 2007: 390

Despite the potential for a directional influence of evolution on ecology ... we still don't know if the evolution–ecology pathway is frequent and strong enough in nature to be broadly important.
Thomas W. Schoener, 2011: 426

Ecology and evolution go hand in hand, as noted by Hutchinson (1965) in his famous essay entitled The Ecological Theater and the Evolutionary Play. Hutchinson's metaphor describes a traditional way of viewing the interplay between ecological and evolutionary processes: ecology provides the selective environment in which evolution acts. Because evolution occurs over relatively long time scales, ecologists had thought it unlikely that contemporary evolutionary events could affect the dynamics of populations or the outcome of species interactions. This view is changing, however. In this chapter, we will see that in some communities, the time scales of selection, adaptation, and ecological interactions are very similar, resulting in eco-evolutionary feedbacks that may affect population dynamics, species diversity, community structure, and even ecosystem functioning. Such cases of rapid evolution are increasingly well documented in a variety of ecosystems (Schoener 2011).

In addition to considering the consequences of contemporary evolution, ecologists are using the information contained in phylogenies to infer how the evolutionary relationships between species can influence the structure of communities. The application of phylogenetic analysis to community ecology, also known as **community phylogenetics**, is an active field of study, driven by rapid advances in the availability of phylogenetic data and new analytical tools (Webb et al. 2002; Cavender-Bares et al. 2009; Pausas and Verdú 2010). In the second half of this chapter, we will explore some of the ways in which phylogenetic analyses are being used to understand the role of evolution in determining the structure and functioning of present-day communities, keeping in mind the limitations inherent in phylogenetic analyses (Webb et al. 2002; Losos 2011). Finally, near the end of this chapter, we will revisit the process of adaptive radiation (discussed in Chapter 8) as an example of how species interactions and natural selection lead to niche filling and community development.

Rapid Evolution and Eco-Evolutionary Dynamics

There are countless ways in which ecology and evolution may feed back on each other, and biologists have long appreciated how the interplay between ecology and evolution can lead to the generation of biodiversity over relatively long time scales (e.g., Darwin 1859; Lewontin 2000). Adaptive radiation is a classic example of how ecological interactions can drive the process of speciation (Schluter 2000; Grant and Grant 2008; Losos 2010), resulting in an increase in the diversity of form and function within a community, which in turn has important consequences for community and ecosystem processes. Only recently, however, have ecologists come to recognize that evolutionary changes in species traits driven by selection over a relatively few generations can result in eco-evolutionary feedbacks in contemporary time. Post and Palkovacs (2009: 1629) define **eco-evolutionary feedbacks** as "the cyclical interaction between ecology and evolution such that changes in ecological interactions drive evolutionary change in organismal traits that, in turn, alter the form of ecological interactions." Post and Palkovacs (2009) focus on the ecological consequences of eco-evolutionary feedbacks in species that have especially strong effects on communities and ecosystems (e.g., keystone species, foundation species, and ecosystem engineers). Other studies of eco-evolutionary feedbacks have focused on how rapid evolutionary changes in species traits affect population dynamics and the outcome of species interactions over ecological time (e.g., Hairston et al. 2005; Strauss et al. 2008; terHorst et al. 2010; Carlson et al. 2011).

Rapid evolution and its consequences for population dynamics and species interactions

Theory suggests that evolutionary changes in prey vulnerability can markedly affect the stability of predator–prey interactions, particularly the tendency for populations to cycle (Ives and Dobson 1987; Abrams and Matsuda 1997; see the review in Abrams 2000). For example, if traits that confer reduced vulnerability to predators come at the cost of lower population growth in the prey (e.g., there is a trade-off between being well defended and being a good competitor), and if predators have a saturating (type II) functional response, then the evolution of reduced vulnerability will generally increase the propensity for prey and predator populations to cycle (Abrams 2000). Might such effects of eco-evolutionary feedbacks on predator–prey dynamics occur in the real world? Hairston, Ellner, and colleagues (Yoshida et al. 2003; Becks et al. 2010) developed an elegant experimental system using two species of freshwater plankton—the green alga *Chlamydomonas reinhardtii* and the grazing rotifer *Brachionus calyciflorus*—to answer this question.

Phytoplankton (algae) possess a number of traits that can reduce their vulnerability to grazing by zooplankton (such as rotifers), including spines, toxic chemicals, gelatinous coatings, and especially increased size. One common way in which algae can increase in size is to form colonies. *Chlamydomonas reinhardtii* normally grow as individual biflagellate cells. When grazing zooplankton are present at high densities, however, *Chlamydomonas* form palmelloid colonies (clumps of dozens of unflagellated cells) that are too large for rotifers to eat. There is a cost to this defense, however, as the clumped form of *Chlamydomonas* is a poorer competitor for resources and has a lower population growth rate than the single-celled form (see Yoshida et al. 2003; Becks et al. 2010 and references therein). Becks et al. (2010) showed that the formation of colonies in *Chlamydomonas* is a heritable trait subject to natural selection. Thus, this rotifer–algal system provides an excellent model system in which to test for eco-evolutionary dynamics.

Jones and Ellner (2007) and Jones et al. (2009) showed theoretically that when prey exhibit a trade-off between defensive ability and competitive ability, qualitatively different population dynamics will result depending on whether the initial prey population contains a wide or a narrow range of genotypes along the trade-off curve. If the range of variation among the prey genotypes present is large, then oscillations in predator and prey densities will create temporal changes in the direction and rate of evolution that result in the maintenance of variation in prey defensive traits. However, if the initial range of prey genotypes present is small (alternative types are close together on the trade-off curve), then the population will evolve toward a single genotype, with the better-defended prey trait going to fixation (Becks et al. 2010). Thus, Becks et al. were able to make and test clear predictions about the role of eco-evolutionary dynamics in maintaining adaptive variation in a prey population and in generating alternative predator–prey dynamics (e.g., cycling versus stable coexistence).

Becks et al. (2010) initiated some of their laboratory microcosms with a population of *Chlamydomonas* that had not been exposed to grazing by rotifers and thus had relatively little heritable variation in their defensive ability (measured here by colony size). They found that after one full predator–prey oscillation, in which prey were driven to very low densities, the prey population evolved toward a single, moderately well-defended genotype, and predator and prey populations settled into an equilibrium (**Figure 15.1A**). This is exactly what theory predicted in the presence of eco-evolutionary feedbacks. On the other hand, when microcosms were started with a *Chlamydomonas* population that had been exposed to grazing—and thus had a large amount of genotypic variation in defensive ability—the predator and prey populations cycled, with the dominant algal genotype alternating over time between well-defended large colonies and vulnerable small colonies or single cells (**Figure 15.1B**). Again, this is exactly what the theory predicted: high predator densities should lead to strong selection for well-defended genotypes, which results in a decrease in predator abundance, which leads to the replacement of well-defended prey genotypes with less well-defended but better competitor genotypes, which leads to an increase

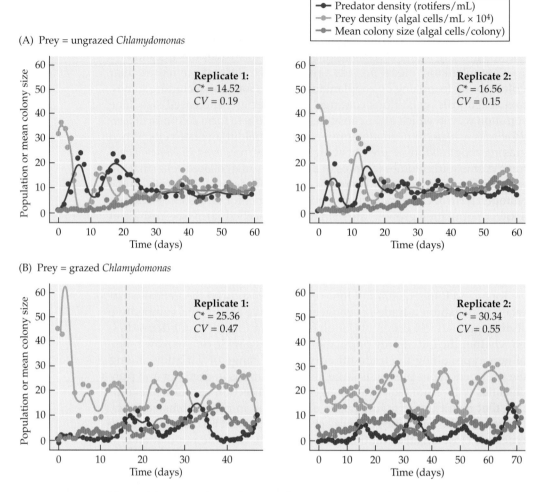

Figure 15.1 Population dynamics for predator (the rotifer *Brachionus*) and prey (the alga *Chlamydomonas)* and mean algal colony size (a measure of algal defensive ability) in two replicate chemostat runs with "ungrazed" and "grazed" *Chlamydomonas* as prey. (A) "Ungrazed" *Chlamydomonas* were obtained from laboratory cultures grown for thousands of generations in the absence of any predators. (B) "Grazed" *Chlamydomonas* came from laboratory cultures continuously exposed to predation by *Brachionus* for 6 months. $C*$ is a measure of algal defensive ability; higher values indicate greater resistance to predation. CV is the average coefficient of variation for the predator and total prey abundances (higher values indicate stronger predator–prey cycling). $C*$ and CV were computed using the dates after the vertical dashed line in each panel. (After Becks et al. 2010.)

in predator densities and the initiation of another cycle. Moreover, Becks et al. showed that average prey defensive ability (mean colony size) was greater in the cycling populations than in the populations at equilibrium (Figure 15.1), as predicted by theory.

Eco-evolutionary feedbacks in nature

Evolutionary biologists and ecologists traditionally viewed evolution by natural selection as a slow process that was unlikely to affect ecological dynamics in nature except over very long time scales (e.g., Slobodkin 1961). However, the experimental studies of Hairston, Ellner, and colleagues described above are a striking example of how rapid evolution and eco-evolutionary feedbacks may affect population dynamics, the stability of species interactions, and the maintenance of genetic variability within populations. Few studies in any area of community ecology exhibit such close and elegant ties between theoretical predictions and experimental tests in a model system. However, as we discussed in Chapter 8, ecologists want to know whether species in nature (which are exposed to a variety of uncontrolled abiotic and biotic effects) also respond as theory and model systems predict. In the case of eco-evolutionary feedbacks, Thompson (1998: 329) stated that the critical question "is whether … the persistence of interactions and the stability of communities truly rely upon ongoing rapid evolution … or whether such rapid evolution is ecologically trivial." To date, relatively few studies have examined the ecological consequences of rapid evolution in the field, although the number is increasing.

There are many examples of evolutionary change in natural populations over a few hundred generations or less (see Hendry and Kinnison 1999; Reznick and Ghalambor 2005 for reviews). Human activities (e.g., species introductions, selective harvesting, and environmental change) have provided the grist for most of these examples of rapid evolution. Collectively, these studies "open the door to the possibility that evolutionary dynamics can affect ecological dynamics—in principle" (Schoener 2011: 427). Fussmann et al. (2007), in an initial review of studies of eco-evolutionary community dynamics, cited eight studies (two field studies and six laboratory studies) that met many of their criteria for eco-evolutionary feedbacks. More recently, Schoener (2011) summarized four field studies of eco-evolutionary dynamics that have appeared since Fussmann and colleagues' (2007) review. These four studies, summarized below, showed that recent adaptive changes in phenotype had significant effects on community dynamics or ecosystem functioning, or both.

Bassar et al. (2010) found that guppies (*Poecilia reticulata*) from Trinidadian streams with few or abundant predators evolved different morphological phenotypes and feeding behaviors. Guppies from high-predation streams exhibited greater food selectivity, consuming more invertebrates and less algae and detritus, than guppies from low-predation streams. These heritable differences in guppy phenotypic traits were correlated with ecological differences in guppy population density and population size structure between high- and low-predation streams. When Bassar et al. experimentally varied guppy phenotypes and abundances independently in field mesocosms (holding population size structure constant), they found that phenotypic differences between populations had marked effects on several ecosystem-level properties; for example, mesocosms containing guppies from high-predation streams had greater algal standing stocks, lower invertebrate biomasses, lower leaf decomposition rates, and higher NH_4 excretion rates than mesocosms containing guppies from low-predation streams. Thus, the evolutionary responses of guppies to high and low predator densities led to marked ecosystem changes in experimental mesocosms in the field.

The interactive effects of contemporary evolution and ecology on ecosystem functioning exhibited in the guppy study are mirrored in the three other field studies cited by Schoener (2011):

1. Recently evolved species of three-spined sticklebacks (*Gasterosteus aculeatus*; see description of their adaptive radiation in Chapter 8) have differential effects on community and ecosystem properties (primary production, dissolved organic matter, prey species diversity) in experimental mesocosms (Harmon et al. 2009).

2. In New England, the construction of dams has resulted in the formation of migratory and landlocked forms of the alewife (*Alosa pseudoharengus*). These fish, which can act as keystone predators in lakes, have evolved morphological differences in response to being resident or migratory as the result of eco-evolutionary interactions with the lake zooplankton community (Post et al. 2008; Palkovacs and Post 2009).

3. Experimentally generated differences in primrose (*Oenothera biennis*) genotypes have been shown to affect the abundance and diversity of insects living on these plants (Johnson et al. 2009; see also studies cited by Hughes et al. 2008).

Schoener (2011) notes that these field experimental studies, as well as observational field studies analyzing long-term data (e.g., Grant and Grant 2008; Ezard et al. 2009), provide valuable insights into the potential consequences of eco-evolutionary feedbacks in nature. However, no experimental field study to date has demonstrated the multigenerational, ongoing process of eco-evolutionary feedback to the same convincing degree that laboratory experiments have (e.g., Fussmann et al. 2005; Becks et al. 2010; terHorst et al. 2010).

Quantifying the ecological consequences of rapid evolution

Recently, Ellner et al. (2011) developed analytical methods to quantify the relative effects of evolutionary trait change, nonheritable trait change (i.e., phenotypic plasticity), and the environment on the outcome of species interactions and the structure and functioning of communities and ecosystems. Their approach to studying the ecological consequences of rapid evolution is an advance over earlier analytical methods developed by these and other collaborators (i.e., Hairston et al. 2005), which assumed that all trait change was the result of evolution (i.e., no phenotypic plasticity). Ellner et al. (2011) applied their analytical methods to a number of empirical studies of fish, birds, and zooplankton. They found that the contribution of rapid evolution to the outcomes of ecological interactions can be substantial, but that this contribution can vary widely among study systems. Moreover, they found that rapid evolution of traits often acts to oppose or reduce the impact of environmental change, suggesting that rapid evolution may be most important when it is least evident. For example, in analyzing the results of the guppy experiments by Bassar et al. (2010) described above, Ellner et al. (2011) showed that half of the ecosystem response variables measured were affected more by guppy evolutionary responses to predation than by predator-driven changes in guppy densities, and that in a number of cases the strong contribution of evolutionary change was opposite in direction from the ecological contribution (**Figure 15.2**).

Many questions remain about the importance of rapid evolution to the outcome of ecological interactions, population dynamics, and the structure and functioning of ecological systems. Addressing these questions in natural systems won't be easy (Ellner et al. 2011; Schoener 2011). Nevertheless, the study of rapid evolution is making impressive progress. One question that has not been addressed is how species diversity within a community may affect the potential for rapid evolution within species or its ecological consequences. For example, consider the rotifer–alga predator–prey system studied by Hairston, Ellner, and colleagues. In this system, we saw that when prey exhibit a trade-off between defensive ability and competitive ability, qualitatively different dynamics will result depending on whether the prey population contains a wide or a narrow range of genotypes along the trade-off curve. But what if the prey community contains multiple species arrayed along the trade-off curve of defensive and competitive ability, rather than the single prey species studied by Becks et al. (2010)? We might expect the range of trait space occupied by multiple species to be far greater than the range of trait space occupied by multiple genotypes of a single species. How would this affect the potential for rapid evolution and its impact on population dynamics and species interactions? Would

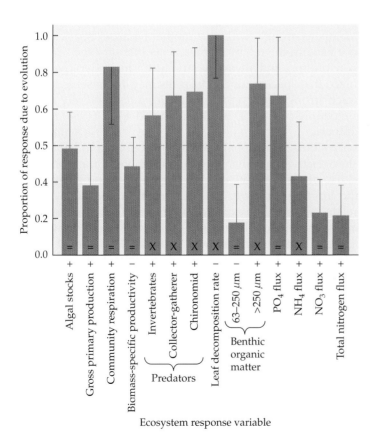

Figure 15.2 Elner and colleagues analyzed the experiments of Bassar et al. (2010) to quantify the ecosystem effects of the evolutionary and ecological responses of guppy populations to high and low predator densities. Exposure to predators caused an evolutionary change in guppy phenotypes and an ecological change (roughly a twofold decrease) in guppy population density. Histogram bar heights measure the absolute value of the "evolution" contribution to an ecosystem response divided by the sum of the absolute values of the "evolution" and "ecology" contributions (i.e., the proportion of the total response that is due to "evolution"). The range of possible values is 0–1; a value of 0.5 (shown by the horizontal dashed line) indicates an equal effect of guppy evolution and guppy density on the ecosystem response variable. Symbols below the bars indicate the "ecology" contribution to the ecosystem response variable was positive or negative; symbols within the bars indicate whether the "ecology" and "evolution" contributions are in the same (=) or opposite (X) directions. (After Ellner et al. 2011.)

we see eco-evolutionary dynamics between predator and prey, or would we see species sorting and dominance by different species under different environmental conditions, or some combination of the above? These and other questions in eco-evolutionary dynamics are just beginning to be explored (de Mazancourt et al. 2008).

Community Phylogenetics

No ecological study can fail to benefit in some way from an understanding of the phylogenetic relationships of its taxa.

—C. O. Webb et al. 2002: 497

A **phylogeny** describes the hypothesized pattern of evolutionary relationships among a set of organisms (generally represented as a branching tree with living taxa at the tips of the branches). A simple phylogenetic tree with the major features labeled is presented in **Box 15.1**. Although evolutionary relationships and phylogenies have been studied for a very long time (Darwin produced one of the first evolutionary trees in his book *On The Origin of Species*), the application of phylogenetic analysis to the study of comparative biology is surprisingly recent, beginning in earnest in the 1980s (Brooks 1985; Felsenstein 1985; Losos 2011). The application of phylogenetics to the study of communities is even more recent. A comment by Darwin (1859: 76), however, suggested one of the first applications of evolutionary analysis to community ecology. Darwin noted that closely related species might compete strongly with one another because "species of the same genus have usually, though by no means invariably, some similarity in habits and constitution, and always in structure." In modern parlance, Darwin recognized that species' functional traits may be conserved across phylogenies. Such strong phenotypic similarities among closely related species are an example of what Derrickson and Ricklefs (1988) termed **phylogenetic effects.**

If species with similar functional traits use resources and habitats in similar ways, close relatives may therefore compete more strongly with one another than with more distant relatives. Thus, we would expect interspecific competition and limiting similarity to lead to communities that contain fewer closely related species than expected by chance. This hypothesis has been termed the "competitive-relatedness hypothesis" (Cahill et al. 2008) or the "phylogenetic limiting similarity hypothesis" (Violle et al. 2011). On the other hand, if closely related species have similar environmental requirements and ecological needs, we might expect the selective effects of environmental factors, referred to as **habitat filtering**, to lead to communities that contain more closely related species than expected by chance. Webb et al. (2002), in one of the first papers to link phylogenetic methods to community ecology, suggested that phylogenetic analysis could help resolve this question.

If strong interspecific competition determines which species coexist in a community, then we would expect community composition to show **phylogenetic overdispersion**; that is, species should be more evenly dispersed in phylogenetic space than would occur by chance (i.e., in a community

BOX 15.1

A **phylogenetic tree**, also known as a phylogeny, illustrates the inferred evolutionary relationships among a group of organisms (taxa) that are descended from a common ancestor. The tips of the tree represent descendant taxa (A–F in the example below), these are often species, but could be higher taxonomic units such as genera or families. A **node** represents a taxonomic unit (terminal nodes are existing taxa and internal nodes represent shared ancestors). Speciation events are inferred to occur at the internal nodes. The branches of the tree define the relationships among the taxa; the pattern of branching is referred to as the tree's **topology**. Branch lengths may or may not have meaning. If a phylogeny is constructed based on molecular data, such as DNA nucleotide substitutions, then the branch lengths may represent the number of nucleotide substitutions. If the substitution rate is calibrated to time (a molecular clock), then the branch lengths may represent evolutionary time in years. In this example, a scale bar denotes the branch length corresponding to 0.1 units of time. Most phylogenies include an **outgroup**, or distantly related taxon , which is useful for rooting the tree (i.e., determining the group's common ancestor, or **root**) and defining the relationships among the taxa of interest (species F is the outgroup in this example). A **monophyletic lineage** within a phylogeny—one that includes an ancestor and all its descendants—is called a **clade** (the dashed line encircles one of the clades in this example). By definition, a clade can be separated from the root of the tree by cutting a single branch. Two descendants that split from the same node are called **sister taxa**. In this example, species A and B are sister species (they are each other's closest relatives). Likewise, the clade including species A and B and the clade including species C, D, and E, are sister clades (the two clades share a node). For more information on phylogenies and phylogenetic methods, see Baum 2008 and the following online sources: University of California Museum of Paleontology, *Understanding Evolution*, http://www.evolution.berkeley.edu/; National Center for Biology Information, *A Science Primer*, http://www.ncbi.nlm.nih.gov/About/primer/phylo.html.

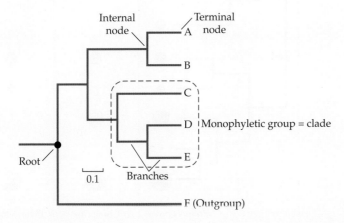

Figure 15.3 An illustration of how processes at the regional and local levels may interact to determine the phylogenetic structure of a community. The red dots represent a quantitative trait (e.g., seed size, specific leaf area, mouth size) and the size of the dot represents the value of that trait (e.g., species with similar-sized dots have similar trait values). In the regional species pool (left panels), the trait may be (A) conserved or (B) convergent across a phylogeny. The results of two different community assembly processes, habitat filtering and limiting similarity (interspecific competition) are illustrated in the right-hand portion of the figure. The shaded boxes represent four hypothetical communities (numbered 1–4), each containing five species selected from the regional pool of ten species. Habitat filtering favors species with similar trait values (communities 1 and 3), thus generating phenotypic clustering within communities; the phylogenetic structure within a community depends on whether traits are conserved or convergent. Limiting similarity prevents species with similar trait values from co-occurring, producing phenotypic overdispersion in communities 2 and 4. The phylogenetic structure is overdispersed in community 2, but can be clustered, random, or overdispersed when traits are convergent (community 4). (After Pausas and Verdú 2010.)

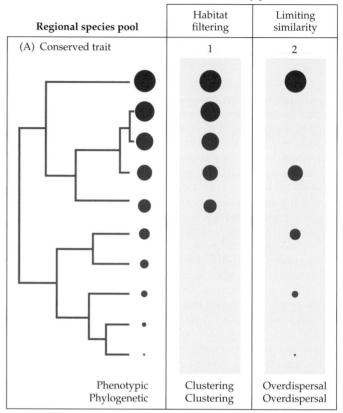

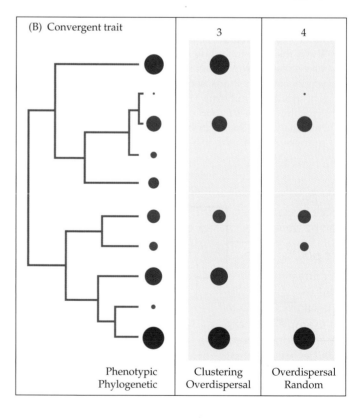

assembled by a random draw from the species pool) (Webb et al. 2002; Kraft et al. 2007; Cavender-Bares et al. 2009; **Figure 15.3A**). The concept of phylogenetic overdispersion is similar to the concept of niche overdispersion discussed in Chapter 8 (Hutchinsonian ratios, shown in Figure 1.3, are an example of niche overdispersion). On the other hand, if closely related species share similar environmental requirements, then we would expect communities to contain more closely related species than would occur by chance, or **phylogenetic clustering**. Thus, phylogenetic analysis (along with phenotypic analysis; see below) has the potential to help resolve this question and reveal the relative importance of competition versus habitat filtering in determining the species composition of communities.

Are closely related species stronger competitors?

Tests of the competitive-relatedness hypothesis to date have yielded mixed results (Mayfield and Levine 2010). There have been few direct tests of the assumption that closely related species are stronger competitors than distantly related species. In a comprehensive analysis, Cahill et al. (2008) found that the relationship between the intensity of interspecific competition and the evolutionary relatedness of the competitors was weak in 142 plant species surveyed, especially compared with the strong and consistent relationships observed between competitive ability and plant functional traits such as size. Cahill and colleagues note that one explanation for this result is that plant traits related to competitive ability may not be strongly conserved across phylogenies, at least not among the species they examined. On the other hand, Burns and Strauss (2011) found that more closely related plant species were more similar in their ecologies (e.g., rates of germination and early survival) and (under some conditions) were stronger competitors. Violle et al. (2011) conducted a multigenerational laboratory experiment with 10 freshwater protist species (ciliate bacterivores), which demonstrated a strong negative relationship between phylogenetic distance and the frequency of competitive exclusion. More closely related protist species were less likely to coexist in pairwise competition (**Figure 15.4A**). In addition, the time to competitive exclusion (**Figure 15.4C**) and the impact of competition on population density (in cases of species coexistence) were both correlated with phylogenetic distance. Thus, Violle et al.'s 2011 study provides strong support for the idea that more closely related species are stronger competitors, at least in the laboratory. Interestingly, the probability of competitive exclusion was better predicted by phylogenetic distance than by differences in a key functional trait (mouth size; **Figure 15.4B**).

The phylogenetic structure of communities

In contrast to the few studies that have tested the assumptions of the competitive-relatedness hypothesis, many studies have examined the community patterns predicted by this hypothesis (summarized in Figure 15.3A). For example, Graham et al. (2009) compared the phylogenetic structures of 189 hummingbird communities in tropical Ecuador. Hummingbird communities are excellent systems in which to test the competitive-relatedness hypothesis; these birds compete strongly for nectar, and there is abundant evidence linking hummingbird traits such as bill length, bill curvature, and

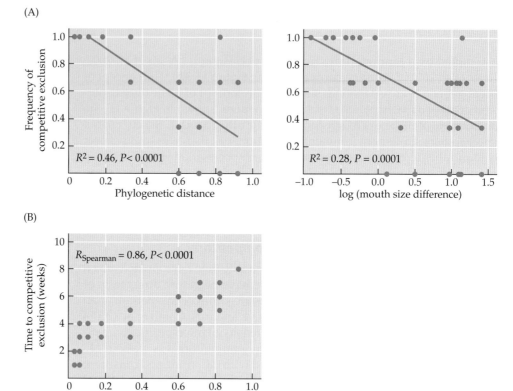

Figure 15.4 Competitive exclusion was measured in 8-week laboratory experiments conducted for all two-species combinations of ten freshwater protist species. (A) The frequency of competitive exclusion declined as the phylogenetic distance (left) and niche difference (mouth size; right) between species increased. Lines and statistics are based on linear regressions. (B) The time to competitive exclusion increased as a function of phylogenetic distance between the competing species. Results are listed for Spearman's rank correlation. (After Violle et al. 2011.)

wing length/body size ratio to a bird's ability to hover and harvest nectar in different environments. There is also some evidence that these functional traits are evolutionarily conserved. In Ecuador, hummingbirds are found in a variety of habitats, ranging from moist tropical lowlands to seasonally arid lowlands to high mountains (up to 5000 m), and more than 25 species of hummingbirds may co-occur at a site. Graham et al. found that 37 of 189 hummingbird communities (28%) exhibited significant phylogenetic clustering, but only one community (2%) showed significant phylogenetic overdispersion.

The rarity of significant phylogenetic overdispersion is somewhat surprising, given the strong evidence for interspecific competition in this taxonomic group. However, when Graham et al. (2009) analyzed the spatial distribution of the communities that fell in the top tenth percentile of the most phylogenetically dispersed communities and the top tenth percentile of the most phylogenetically clustered communities, they found an interesting spatial pattern: the most phylogenetically dispersed communities were found in the moist lowlands, whereas the most phylogenetically clustered communities were found in the harsh environments of the high Andes and in the seasonally dry lowlands west of the Andes (**Figure 15.5**). Other

(A) Benign habitats

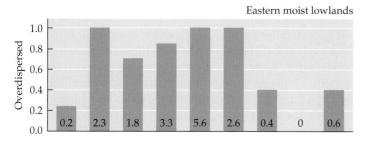

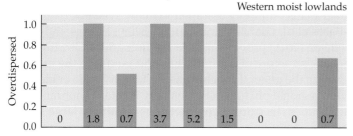

(B) Challenging habitats

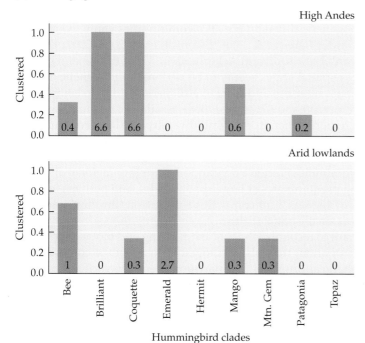

Figure 15.5 Hummingbird communities in Ecuador show the greatest phylogenetic dispersion in the environmentally benign moist lowlands; they show the greatest phylogenetic clustering in the challenging environments of the high Andes and the dry western lowlands. (A) Histograms show hummingbird clades represented in the top tenth percentile of phylogenetically overdispersed communities that exist in the eastern and western moist lowlands. (B) Histograms show the clades in the top tenth percentile of phylogenetically clustered communities existing in the high Andes and the dry regions west of the Andes. Hummingbird clades are listed along the x axis. The y axis shows the proportion of communities where a given clade is represented; the numbers in the bars are the mean numbers of species per community. Overdispersed and clustered communities tend to be represented by different clades in a manner consistent with their ecologies. (After Graham et al. 2009.)

studies also have found more phylogenetic clustering (as well as phenotypic clustering) in harsh environments than in benign environments (e.g., Swenson and Enquist 2007; Cornwell and Ackerly 2009), supporting the hypothesis that harsh environments provide a stronger filter on the traits of species than do more benign environments (Weiher and Keddy 1995). The overall level of support for this hypothesis is still unknown, however, as there have been relatively few empirical tests.

The application of phylogenetic analysis to understanding the relative contributions of habitat filtering and interspecific competition (or other biotic interactions such as facilitation) to the structuring of communities is an active area of research. Note that contrasting predictions can be made depending on whether traits are evolutionarily conserved or evolutionarily convergent (see Figure 15.3B). Such contrasting predictions highlight the importance of testing for a phylogenetic signal in the traits of interest in any phylogenetic analysis of community assembly. Losos (2011) provides a particularly thoughtful discussion of this topic and suggests that phylogenetic effects are weaker than commonly assumed in many systems.

Cavender-Bares et al. (2009), Vamosi et al. (2009), Pausas and Verdú (2010), and Swenson (2011) provide insightful summaries of the literature on the phylogenetic structure of communities and guides to the methods employed. Although the general conclusions are tentative, the evidence to date suggests that regional phylogenetic diversity is filtered nonrandomly into local communities. In studies conducted at relatively large spatial scales, phylogenetic clustering within a community is generally greater than predicted based on a random draw from the regional species pool (i.e., communities show positive habitat filtering). However, studies that have looked within very localized communities have found greater than expected phylogenetic diversity (i.e., the communities show overdispersion; e.g., Cavender-Bares et al. 2006; Swenson et al. 2006, 2007; Helmus 2007; Vamosi et al. 2009; Kraft and Ackerly 2010). Thus, the general pattern observed across spatial scales seems to be regional phylogenetic clustering (due to habitat filtering) and local overdispersion (due to competitive interactions), Emerson and Gillespie (2008). Similar results may apply to the filtering of regional phenotypic (trait) diversity into local communities, although the outcome depends on the types of traits examined (Swenson 2011). Thus, the current bottom line is that phylogenetic and phenotypic diversity in communities is often a nonrandom sample of regional diversity: habitat filtering predominates at large spatial scales, and perhaps in harsh environments or regions, and competitive exclusion/limiting similarity predominates at small spatial scales.

It is important to recognize that habitat filtering and biotic interactions (including competition) may also act together to assemble communities; thus, phylogenetic approaches that partition the relative effects of these different factors may prove most useful in the future (Pausas and Verdú 2010). For example, when Helmus et al. (2007) examined the distribution of fish species in the family Centrachidae (sunfishes) across 890 lakes in Wisconsin, they found no phylogenetic pattern. However, after accounting for the species' shared responses to environmental variables (habitat filtering), the authors found that closely related species were less likely to co-occur in lakes sharing the same environmental characteristics, a result most likely driven by interspecific competition. Finally, Mayfield and Levine (2010) have suggested that we need to refine our expectations for how phylogenetic relatedness influences species coexistence and the phylogenetic structure of communities based on a more nuanced view of interspecific competition that acknowledges the roles of both equalizing and stabilizing mechanisms in determining species coexistence (see Chapter 14).

Phylogenetic niche conservatism

The idea that closely related species are ecologically similar underlies another current topic in evolutionary ecology: the concept of phylogenetic niche conservatism, or simply niche conservatism (Wiens and Graham 2005; Losos 2008a; Warren et al. 2008; Wiens et al. 2010). **Phylogenetic niche conservatism** can be defined as "the tendency of species to retain ancestral ecological characteristics" (Wiens and Graham 2005: 519). In a sense, niche conservatism looks at the ecological and evolutionary consequences of a slow rate of change in species' functional traits. Wiens and Graham (2005) list four potential causes of niche conservatism: (1) stabilizing selection for functional traits, (2) limited genetic variation in the traits of interest, (3) gene flow from the center of a population's range to the edge, which may limit opportunities to adapt to new environmental conditions, and (4) pleiotropy and trade-offs among traits.

There has been considerable controversy over the concept of niche conservatism: whether it is a pattern or a process, how best to test for its existence, and at what spatial and temporal scales it applies (Losos 2008a,b, 2011; Warren et al. 2008; Wiens 2008; Wiens et al. 2010). As Wiens and Graham (2005) pointed out, we expect functional traits to be conserved to some extent among related species, but rarely will they be identical. Therefore, proponents of niche conservatism argue that the most important question is not whether niches are more or less conserved than some particular process (e.g., the rate of divergence predicted by a Brownian motion model of evolution) would lead us to expect, but rather, what the consequences of niche conservatism are for ecological and evolutionary patterns and processes. For example, those who study niche conservatism are interested in how the conservation of traits affects the formation of regional biotas, the distribution of species across landscapes, the generation of diversity gradients (e.g., the latitudinal diversity gradient), and the maintenance of biodiversity (Wiens et al. 2010). Interpreting the role of niche conservatism in generating biodiversity patterns is challenging, however, because the consequences of niche conservatism depend on the spatial and temporal scale of the questions explored. The difficulty is best illustrated by an example.

Figure 15.6A shows a hypothetical phylogeny for a group of eight species. In this case, six of the eight extant species are found in tropical habitats and two species are found in temperate habitats. Evolutionary relationships among the species (inferred, say, from some combination of molecular data, fossils, and species traits) suggest that the earliest ancestor of the group was found in tropical habitats and that the two species currently found in temperate habitats are the result of a recent niche shift from tropical to temperate habitats. If the six tropical species are more similar in their functional traits than are the two temperate species, we might conclude the following: First, the conservation of ancestral traits (niche conservatism) is responsible for the present-day pattern of species distribution across habitats: six of eight extant species are found in the habitat of origin (the tropics), and they share similar traits. Second, niche conservatism, combined with increased time for speciation, has resulted in the greater current species richness in tropical than in temperate habitats.

Figure 15.6 A hypothetical phylogeny illustrating the difficulties in interpreting patterns of niche conservatism. (A) Six of the eight extant species in this clade are found in tropical habitats and two species are found in temperate habitats. The evolutionary relationships among the species suggest that the earliest ancestor of the clade was found in tropical habitats, and that the two species currently found in temperate habitats are the result of a recent niche shift from tropical to temperate habitats. Assuming that the six tropical species are more similar in their functional traits than are the two temperate species, we might conclude that phylogenetic niche conservation contributes to the present-day pattern of species distribution across habitats. (B) An expanded phylogeny for the same clade that includes two distantly related species with a temperate-zone distribution. This expanded phylogeny suggests that the earliest common ancestor of the clade occurred in temperate habitats and that the diverse tropical component resulted from a very early niche shift from the temperate zone to the tropics, followed by a high rate of diversification in the tropics compared with the temperate zone and a subsequent recent niche shift back to the temperate zone by two species. These two phylogenies illustrate the difficulties in interpreting the underlying cause of the difference in tropical and temperate species richness in this clade: niche conservatism and time for speciation, as suggested by (A), or differences in rates of diversification between habitats, as suggested by (B).

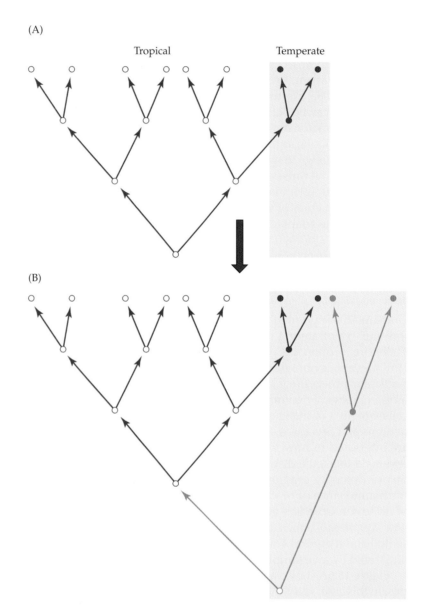

Including a more distantly related group of species in the phylogeny, however, reveals a potentially different story (**Figure 15.6B**). Now we see that the earliest common ancestor of the clade occurred in temperate habitats and that the diverse tropical component of the clade was the result of a very early niche shift, followed by a higher rate of diversification in the tropics than in temperate habitats. If the phylogeny depicted in Figure 15.6B traces the true evolutionary history of the group, then it appears that a shift from temperate to tropical habitat very early in the clade's history was crucial in producing the current pattern of species diversity across habitats. Moreover, the current high tropical diversity is the result of early colonization of the tropics, a higher rate of tropical diversification, and the tendency of the tropical clade to remain in this habitat after the initial habitat shift.

The example in Figure 15.6 is hypothetical. However, Losos (2008b) and Wiens (2008) discuss a real-world example of lizard species distributions in the Caribbean and in the continental United States that illustrates some of the same challenges in interpreting the role of niche conservatism in structuring species assemblages at different spatial and temporal scales. Condamine et al. (2012) provide an impressive example of how fossil-calibrated molecular phylogenetic data, combined with ancestral state reconstruction (based on host plant associations) and ancestral area reconstruction (based on paleogeographic data and species ranges) can yield a synthetic picture of the underlying evolutionary and ecological drivers of broad-scale diversity patterns in swallowtail butterflies (Papilionidae).

The extent to which species traits are conserved evolutionarily and the degree to which niche conservatism determines contemporary distributions of species and clades are active areas of research that have yielded significant insights into patterns of biodiversity, including the latitudinal diversity gradient (e.g., Wiens et al. 2009) and the composition and richness of local communities (e.g., Harrison and Grace 2007). How species and communities will respond to global climate change is also intimately tied to the question of niche conservatism and the degree to which species can adapt to changing environments (Pearman et al. 2008). Rapid evolution of adaptive traits may ameliorate, in part, the negative impacts of global climate change by allowing species to respond to changing environments (Thomas et al. 2001; Jump and Peñuelas 2005; Lavergne et al. 2010). On the other hand, evolutionary conservation of climatic tolerances could require species to shift their ranges or go extinct (Thuiller et al. 2005; Parmesan 2006). As Sexton et al. (2009: 430) note, "Rapid climate change will teach us volumes about the stability or mobility of range limits." Global climate change is also likely to "teach us volumes" about the ability of species to adapt to new environments via rapid evolution or to be constrained by niche conservatism to environments rooted in their evolutionary past.

Adaptive Radiation, Niche Filling, and Community Assembly

Adaptive radiation—the diversification by adaptation of a single phylogenetic lineage into a variety of forms occupying different ecological niches (Simpson 1953; Schluter 2000)—is a prominent feature of evolution. In Chapter 8 we examined adaptive radiations driven by interspecific competition through the mechanisms of niche divergence and character displacement. Classic examples of adaptive radiations include Darwin's finches on the Galápagos archipelago (see Figure 8.5; Grant and Grant 2008); cichlid fishes in the African Great Lakes (Fryer and Iles 1972; Seehausen 2006); *Anolis* lizards on islands in the Caribbean (see Figure 8.6; Losos 2009); and silverswords (plants; **Figure 15.7**), honeycreepers (birds), and *Drosophila* (fruit flies) in the Hawaiian Islands (Carlquist et al. 2003; Pratt 2005; Losos and Ricklefs 2009). It is no accident that these classic examples of adaptive

(A) *Madia sativa* (ancestral)

(B) *Argyroxiphium sandwicense*

(C) *Wilkesia gymnoxiphium*

(D) *Dubautia arborea*

Figure 15.7 The Hawaiian silversword alliance is a spectacular example of an adaptive radiation of plants. All silverswords are descended from a single ancestral tarweed (probably similar to the extant tarweed *Madia sativa*, shown in panel A) that colonized the Hawaiian Islands from western North America some 5 million years ago (Baldwin and Sanderson 1998). The three genera and 30-odd silversword species endemic to Hawaii show an extravagant diversity of life forms, including monocarpic rosettes (B, C), and woody vines and trees (D). They occur in habitats from rainforests to deserts. (A © Steffen Hauser/botanikfoto/Alamy; B © Science Photo Library/Alamy; C © Dani Carlo/AGE Fotostock; D courtesy of Forest and Kim Starr.)

radiation come from isolated islands and lakes. Adaptive radiations are strongly associated with ecological opportunity and often occur when an ancestral species colonizes a relatively unoccupied environment where resources are underutilized (Losos 2010). Other ways in which a lineage may "stumble upon an array of unexploited niches" (Schluter 2000: 163) are to survive an extinction event or to acquire a novel evolutionary character—a "key innovation"—that allows the lineage to exploit resources in a new way (Simpson 1953; Roelants et al. 2011).

The hallmarks of adaptive radiation are the speed of diversification and the wide variety of forms and species produced from a single ancestor (e.g., the silverswords in Figure 15.7). However, the same processes that drive adaptive radiations, most notably strong ecological interactions, are probably involved in the evolutionary diversification of most lineages (Schluter 2000; Sobel et al. 2009; Losos 2010). Thus, studying adaptive radiations lends critical insight into the general processes that produce biodiversity. Adaptive radiations typically show a burst of speciation and morphological diversification early in their history and a slowing of diversification thereafter (Gavrilets and Losos 2009; Mahler et al. 2010). Evidence from evolutionary radiations in general shows a similar tempo of diversification. Many fossil groups show a peak in morphological diversity soon after their appearance in the fossil record (Foote 1997) or differentiate widely and rapidly following a mass extinction (Erwin 2001). Phylogenetic analyses from a wide variety of taxa also show a general pattern of diversification "slowdowns" over time in the evolution of a lineage (e.g., Phillimore and Price 2008; Rabosky and Lovette 2008), although there are exceptions to this pattern (e.g., Condamine et al. 2012).

Bursts of taxonomic and morphological diversification early in a clade's history, followed by slowdowns suggest strong ecological controls on speciation, with diversification rates declining as niches are filled (Rabosky 2009; Rabosky and Alfaro 2010; but see counterpoint in Wiens 2011). In an extensive analysis of 245 species-level molecular phylogenies for both animals and plants, McPeek (2008) found that most clades showed a pattern of diversification slowdown. However, McPeek attributed this slowing in diversification not to a reduction in speciation rate as niches filled, but rather to increased rates of competitive exclusion and species extinction. By combining a model of competitive interactions and species coexistence with a model of clade diversification, McPeek showed that clades that diversify by strong niche differentiation exhibit decelerating lineage accumulation over time, whereas clades that diversify primarily by modes of speciation that generate little or no ecological diversification (e.g., sexual selection) show accelerating diversification. The predictions of the model were broadly consistent with the patterns of diversification observed in molecular phylogenies (McPeek 2008) and in the fossil record (McPeek 2007). If speciation and extinction rates vary strongly over time due to ecological interactions, this makes the estimation of diversification rates from molecular phylogenies problematic, potentially compromising the use of phylogenies in addressing how biodiversity develops over time and between regions (see Chapter 2; Losos 2011).

Are adaptive radiations predictable in terms of the types of ecological niches filled by species in similar environments? This question can be addressed in part by comparing independent radiations of species on separate islands or island groups with similar environmental conditions (see Losos and Ricklefs 2009 for an excellent summary). In the best-documented comparison of multiple adaptive radiations, *Anolis* lizards on the four islands of the Greater Antilles (Cuba, Hispaniola, Jamaica, and Puerto Rico) have diversified independently to produce the same set of habitat specialists (ecomorphs) on each island (see Figure 8.6; Losos 2009). Similarly, land snails in the genus *Mandarina* have diversified into the same four niches (arboreal, semi-arboreal, sheltered ground, and exposed ground) on different islands in the Bonin Islands near Japan (Chiba 2004; reviewed in Losos 2009). In postglacial lakes in the Northern Hemisphere (lakes can be viewed as "islands" in a terrestrial landscape), fish species have repeatedly diversified into pelagic (open water) and benthic (bottom habitat) specialists (Schluter 2000; see also Chapter 8). Although these case studies show similarities in community development and niche filling across independent adaptive radiations, Losos (2010) argues from a review of the literature that replicated adaptive radiations, in the sense of strict species-for-species matching (Schluter 1990), are rare in nature and that those that occur involve closely related species. Interestingly, as Losos and Ricklefs (2009) note, islands provide excellent examples of both determinism and contingency in evolution, as the chance colonization of islands by different taxa has resulted in the evolution of markedly different communities. Some general trends in niche filling are evident, however; for example, woodpecker "equivalents" have repeatedly evolved on islands in the absence of true woodpeckers, which are poor dispersers.

Conclusion

The fields of eco-evolutionary dynamics and community phylogenetics have exploded in recent years. The number of studies demonstrating contemporary evolution and its ecological consequences is expanding rapidly, although experimental field studies are still rare. Likewise, the range of application of phylogenetic analysis to ecological questions seems unbounded. Work to date shows that the phylogenetic structure of many (possibly most) communities is nonrandom and that habitat filtering appears to play the more important role in structuring communities over large spatial scales, and that competitive exclusion/limiting similarity is more important at local scales. Similar results may apply to the filtering of regional trait diversity into local communities, although the evidence here is more limited and the results depend on the types of traits measured (Swenson 2011). Other recent applications of phylogenetics to community ecology have included (1) combining phylogenetic information with species abundance data to obtain better estimates of the elusive parameters of dispersal rates and speciation rates needed to test Hubbell's neutral theory (Jabot and Chave 2009), (2) applying phylogenetic information to the study of food webs and other ecological networks (e.g., Cattin et al. 2004; Rezende et al. 2007), (3) using phylogenetic analyses to understand patterns of species invasions (e.g., Strauss et al. 2006a,b; Gerhold et al. 2011), (4) applying phylogenetics to metacommunities (e.g., Leibold et al. 2010), and (5) understanding the importance of phylogenetic cascades (co-diversification across multiple trophic levels) in the development of high tropical diversity (Forister and Feldman 2011). Each of these areas illustrates the potential for using the rapidly expanding database of phylogenetic relationships and rapidly improving analytical techniques to better understand the role of evolution in determining the structure and functioning of communities.

It is important to remember, however, that phylogenies are hypothesized evolutionary relationships; they are not literal descriptions of the evolution of any taxon. In an elegant and thought-provoking essay, Losos (2011) discusses the limitations of the application of phylogenies to comparative biology (including ecology). He concludes that while phylogenetic analysis is a powerful tool for understanding ecology in an evolutionary context, it is not omnipotent. Therefore, phylogenetics should be combined with other investigative tools (fossils, trait analysis, studies of ongoing evolutionary processes, ecological experiments) to best understand the evolution of communities. Losos (2011: 724) concludes with the admonition that "the study of evolutionary phenomena [is like] a detective story in which there is no smoking gun and in which disparate clues have to be drawn together to provide the most plausible hypothesis of what happened." Similarly, it is important to think about the broader context of current studies of rapid evolution and its ecological consequences.

We live in an era of rapid environmental change, presenting evolutionary biologists with unique opportunities to study the processes of selection and adaptation in nature. In this chapter, we have examined examples of rapid evolution and eco-evolutionary feedbacks, documenting their effects on population dynamics in the laboratory and illustrating their potential consequences for community structure and ecosystem functioning in nature. The study of eco-evolutionary dynamics is fertile ground, both for increasing our understanding of the processes that structure communities and for predicting the consequences of climate change for species distributions and biodiversity.

An additional long-standing question in evolutionary biology is how the microevolutionary processes of selection and adaptation are related to the macroevolutionary processes of speciation, adaptive radiation, and the filling of ecological niches (Reznick and Ricklefs 2009). These macroevolutionary processes are of particular interest to ecologists because they focus directly on the production of biodiversity. Adaptive radiations highlight how species interactions, particularly interspecific competition, may lead to adaptive niche divergence and how ecological opportunity may drive the rate of diversification within lineages. However, much remains to be done to link Darwin's "struggle for existence" and adaptive divergence to a truly mechanistic understanding of how the microevolutionary processes of selection and adaptation translate into the macroevolutionary dynamics of speciation and the diversification of forms (Carroll et al. 2007; Jablonski 2008; Reznick and Ricklefs 2009). Adaptive radiations provide some of the best opportunities to study these links, and advances in molecular biology and the ability to study the genetic basis of adaptation promise to make the next few decades "a golden age in the study of adaptive radiation" (Losos 2011: 635).

Summary

1. In some communities, the time scales of selection, adaptation, and ecological interactions are very similar, resulting in eco-evolutionary feedbacks that may affect population dynamics, species diversity, community structure, and ecosystem functioning.

2. Theory suggests that evolutionary changes in prey vulnerability can markedly affect the stability of predator–prey interactions, particularly the tendency for populations to cycle, and that eco-evolutionary dynamics can maintain adaptive variation in prey populations. Laboratory experiments have borne out these predictions, but no experimental field study has yet demonstrated eco-evolutionary dynamics to the same degree.

3. New analytical methods applied to empirical studies have shown that the contribution of rapid evolution to the outcomes of ecological inter-

actions can be substantial, but that this contribution can vary widely among study systems. Moreover, rapid evolution of traits may act to oppose or reduce the impact of environmental change, suggesting that rapid evolution may be most important when it is least evident.

4. If species' functional traits are conserved across phylogenies, and if species with similar functional traits use resources and habitats in similar ways, then close relatives may compete more strongly with one another than with more distant relatives. The "competitive-relatedness hypothesis" predicts that if that is the case, then interspecific competition should lead to communities that show phylogenetic overdispersion. On the other hand, if closely related species share similar environmental requirements, then communities should show phylogenetic clustering.

5. There have been few direct tests of the assumption that closely related species are stronger competitors than distantly related species, and the evidence on this point is mixed.

6. Studies have found more phylogenetic clustering (and phenotypic clustering) in harsh environments than in benign environments, supporting the hypothesis that harsh environments provide a stronger filter on the traits of species than do more benign environments.

7. Phylogenetic and phenotypic diversity in communities is often a nonrandom sample of regional diversity. A general pattern across spatial scales seems to be phylogenetic clustering at regional scales (driven by habitat filtering) and phylogenetic overdispersion at local scales (driven by competitive interactions).

8. Habitat filtering and biotic interactions (including competition) may act together in the assembly of communities. Phylogenetic approaches that partition the relative effects of these different factors may prove useful in the future.

9. The extent to which species traits are conserved evolutionarily and the degree to which niche conservatism determines contemporary distributions of species and clades are active areas of research. Such studies are challenging, however, because the consequences of niche conservatism depend on the spatial and temporal scale of the questions explored.

10. The study of adaptive radiation lends critical insight into the processes that produce biodiversity. Adaptive radiations are marked by a burst of speciation and morphological diversification early in a clade's history and a slowing of diversification thereafter. Evidence from evolutionary radiations in general shows a similar tempo when diversification is driven by niche differentiation.

11. Adaptive radiations are strongly associated with ecological opportunity and often occur where resources are underutilized, as in a newly colonized, relatively unoccupied environment, after a mass extinction event, or when a lineage acquires a "key innovation" that allows it to exploit resources in a new way.

16 Some Concluding Remarks and a Look Ahead

Illuminating our understanding of nature, that's what it's about.

Peter A. Lawrence, 2011: 31

I am not embarrassed to be a puzzle solver.

Michael L. Rosenzweig, 1995: 373

I can imagine a student reaching the end of this book and saying, "That's all very interesting, really. I see how species diversity varies across space and time—fascinating stuff. And I understand how biodiversity can affect the functioning of communities and ecosystems and why the loss of biodiversity is such a critical concern. And I have learned a great deal about the nitty-gritty of species interactions and how these interactions impact species diversity. And I understand that species diversity in a community is a function of local interactions plus the movement of species between communities. But what I *don't* get is how everything fits together. That's what we are supposed understand, isn't it, if we want to use our ecological knowledge to help preserve the diversity of life in our world? So, just tell me—which of all the things that we studied is most important in determining the number and kinds of species I see around me?"

There are days when I share this imaginary student's frustration. In the short history of our discipline, community ecologists have identified many (perhaps most) of the processes that regulate and maintain species diversity at local and regional scales. We understand (in principle) how these processes act together to form areas of high and low biodiversity, and we can predict (in a very general way) how biodiversity will change with a changing environment. But if you ask me why the small Michigan lake outside my backyard is home to about 20 species of fish rather than 2 species or 200 species, I can't tell you, even though I have spent most of my adult life studying freshwater communities (and fish in particular). But maybe that's asking too much. A colleague once said to me that we ecologists are always lamenting our inability to predict species diversity, when in fact we should celebrate our successes. For example, ecologists understand in a deep and fundamental way the factors that regulate populations, and we use this knowledge to manage our natural resources effectively (or at least we try to—politics often gets in the way). We understand the processes that govern species interactions and that determine the outcomes of these interactions under different environmental conditions. And we understand the mechanisms that promote species coexistence, regulate biodiversity, and affect biogeochemical processes, even if we cannot yet predict precisely the number of species we will find in any one community. These are our successes. A few years back, a well-known ecologist, John Lawton (1999: 178), pronounced that "community ecology is a mess, with so much contingency that useful generalizations are hard to find." Contingency indeed abounds in ecological interactions; however, I hope that this book has shown how far community ecologists have come in lifting us out of "the mess" that Lawton perceived in the 1990s—which perhaps is what Lawton was provoking us to do all along.

I have spent almost five years writing this text, and as I look back on it now, I am still uneasy about its content. Yes, it covers a lot of ground. But there is much that is missing or barely touched upon. Whole areas of study have received only a passing glance—infective agents and disease, stoichiometry, metabolic theory, succession, invasive species, species distribution modeling, and many more. Some of these are exciting new areas in community ecology and some are old saws. I am tempted to say that there just wasn't room to include them all, but that wouldn't be honest. In reality, as Robert May's quote from the Preface notes, deciding what to emphasize in a book about community ecology is "necessarily quirky." This book represents my own quirkiness. Most importantly, it reflects my belief that community ecology is fundamentally about the study of species interactions. Therefore, it is crucial that students have a clear understanding of how consumers and resources interact within communities, and that they appreciate how these interactions may play out in the context of predators and prey, or competitors, or mutually beneficial species.

In addition, students should recognize that (except in rare cases) communities don't have distinct boundaries, nor are they closed to the immigration or emigration of species. This means that, although we might study a group of species found in a particular forest or field or lake of our choosing, we need to appreciate how that community is connected to the broader landscape in both space and time. Recognizing the connections between local and regional processes is what sets the current discipline of community ecology apart from its earlier days, when the focus was almost entirely on local processes and local diversity. Ecologists may wrestle at the moment with our concepts of "metacommunities," how to define and delimit them, how to measure the rate of species dispersal between communities, and how regional diversity arises and affects local diversity. But we will never go back to studying communities in isolation.

How, then, might we go forward? That's a big question. I wish I could stop now without sticking my neck out to discuss "future directions." But I would be remiss to end this book without at least giving it a try. Listed below are some areas that I think will occupy the attention of community ecologists for the next decade or so and that may yield big dividends in terms of understanding the processes that structure communities and govern their functioning.

Looking Ahead: Issues to Ponder

Metacommunities and the integration of local and regional processes

The study of metacommunity dynamics, broadly defined, will continue to play a large role in the development of community ecology. Hanski and colleagues once estimated that more than 50% of Finnish butterfly species may function as metapopulations (Hanski and Kuussaari 1995). Can we make such an estimate of the percentage of communities of different types that function as metacommunities? More importantly, can we determine how much of the structure and functioning of local communities is governed by the dispersal of species between communities and the size of regional species pools as compared with the impact of species interactions and species sorting within communities? At some point, the answer to these questions comes down to measuring the rates of fundamental processes. As depicted in Figure 1.5, four processes (selection, drift, speciation, and dispersal) feed into "the black box of community ecology" to determine the patterning of species diversity across space and time (Vellend 2010). Selection (species sorting in this context) and drift (random external effects on population abundances) are local processes, speciation occurs at a regional scale, and dispersal is the mechanism that ties local and regional dynamics together. That's probably as good as we can do (for now) in developing an overarching framework for community ecology. But, as I said in Chapter 1, what we really want to do is look *inside* the black box and shed some light on the workings within. To do this, we need to know more about the actual rates of species movement between communities, of species selection (sorting) and drift within communities, and of species accumulation (or loss) within regions.

Drivers of regional biodiversity

I believe that the latitudinal diversity gradient, as well as other broad-scale gradients in biodiversity, will continue to serve as useful frameworks for understanding the processes that generate regional differences in biodiversity. Ecologists are able to map patterns of biodiversity (e.g., species richness and species traits) onto climatic variables, landforms, and historical and evolutionary processes (e.g., paleoclimates, phylogenetic patterns) with a level of precision unimaginable 20 years ago (e.g., Belmaker et al. 2012; Condamine et al. 2012). This capability has revitalized the study of biodiversity at broad spatial scales and has helped fuel the integration of biogeography, ecology, and evolution (Ricklefs and Jenkins 2011). The next decade or so could reveal answers to a question that has hounded ecologists for 200 years: Why are the tropics so species-rich? But determining causal relationships between regional biodiversity and a host of potential drivers that tend to be spatially correlated is difficult, especially at spatial scales where many manipulative experiments are difficult to impossible. Further insights into the evolutionary and ecological factors driving broad-scale biodiversity patterns will come from creative efforts to apply a combination of tools (e.g., phylogenetics, historical biogeography and paleontology, distributional mapping, trait analyses; e.g., Belmaker et al. 2012; Condamine et al. 2012) to understand why biodiversity is distributed heterogeneously across Earth.

Although experimental studies over broad spatial scales are challenging, they can reveal much about the drivers of biodiversity. I have highlighted a few examples of replicated experiments conducted at multiple sites in previous chapters (e.g., see Figure 3.5, Hector et al. 2002; Figure 9.8, Callaway et al. 2002). Other recent examples include the manipulation of predator presence and absence in intertidal communities at four sites, from Connecticut to Panama, to compare the strength of biotic interactions across latitudes (Freestone et al. 2011a,b); and nutrient and herbivore manipulations in small grassland plots worldwide to examine top-down and bottom-up controls of plant diversity (Adler et al. 2011). Interestingly, the latter study is an example of a grassroots collaboration of 50 or more scientists who recognized the value of conducting the same set of experiments at multiple sites across the globe and did so with very little external funding (Stokstad 2011). I hope we see many more of these collaborative, question-driven, cross-site experiments in the next decade—hopefully not all done on a shoestring budget, but rather with strong support from the National Science Foundation and other agencies.

Niche-based and neutral dynamics

The neutral theory (Hubbell 2001) provides a useful null model for examining how patterns of species diversity across space and time should vary as a function of dispersal rates and speciation rates in the absence of any niche differences between species. Empirically sorting out the contributions of niche-based and neutral processes in determining the structure of communities (Holt 2006; Leibold and McPeek 2006) and the diversity of taxa within evolutionary clades (McPeek 2008) is an important area of ongoing and future research. Chesson's (2000) characterizing of the processes affecting species coexistence as stabilizing and equalizing mechanisms provides a useful way to probe neutral and niche-based dynamics. Likewise, McGill's (2011) provocative suggestion that many observed biodiversity patterns are statistical consequences of the clustering of individuals within a species, the unequal distribution of abundance among species, and sampling phenomena challenges our understanding of the drivers of macroecological patterns.

Pathogens, parasites, and natural enemies

A colleague asked me if there would be a chapter on "disease ecology" in my book, and was clearly disappointed when I said no. However, I do believe we are about to see a boom in research on the effects of pathogens and natural enemies in communities. In particular, there is a lot of current excitement over the idea that species-specific pathogens may promote species coexistence and diversity within communities (e.g., Ricklefs 2011), even though the foundation for this idea was set long ago. In the early 1970s,

Janzen (1970) and Connell (1971) suggested that the extraordinarily high diversity of tropical forest trees was due (in part) to low seedling recruitment near a parent tree. They hypothesized that seeds and seedlings found at high densities near a parent tree were especially vulnerable to attack from specialized natural enemies such as pathogens (fungi, bacteria) and insect herbivores, and that this phenomenon would lead to strong negative density dependence. Density-dependent mortality due to species-specific pathogens and other natural enemies is now commonly referred to as a "Janzen–Connell effect." By limiting the recruitment of offspring near a parent plant, Janzen–Connell effects may promote species coexistence and increase species diversity within communities.

Evidence for Janzen–Connell effects is accumulating in a variety of ecosystems, including tropical forests (e.g., Bagchi et al. 2010; Comita et al. 2010; Mangan et al. 2010; Shixiao et al. 2012) and temperate grasslands (e.g., Bever et al. 2010; de Kroon et al. 2012), and the diversity-promoting effects of species-specific natural enemies have been suggested as a factor in animal communities as well (e.g., Ricklefs 2011). Recent studies also suggest an important role for soil pathogens in the relationship between plant diversity and ecosystem functioning (see Chapter 3): reductions in primary productivity at low plant species richness may be driven in part by the negative effects of species-specific soil pathogens on plants grown in monoculture or at low species richness (e.g., Maron et al. 2011; Schnitzer et al. 2011; de Kroon et al. 2012). However, as Ricklefs (2011: 2045) notes, pathogens are difficult to study in nature, and "the impacts of pathogens on the coexistence, distribution and abundance of competing species in regional communities are poorly known." I expect that we will know a lot more about the role of pathogens, diseases, and other natural enemies in structuring ecological communities in the coming years, and that theory will advance to place competition and predation on more equal footing in determining species coexistence (Chesson and Kuang 2008; Chesson 2011).

Biodiversity and ecosystem functioning

The study of how biodiversity affects ecosystem functioning has been one of the most active fields in community ecology for almost 20 years. It's hard to know whether this momentum and rate of discovery will be maintained for another decade or longer. Yet there continue to be important advances. As noted above, soil pathogens have recently been identified as playing an important role in driving the diversity–productivity relationship in plant communities. Another recent discovery, revealed in a study that synthesized the results of multiple plant biodiversity experiments conducted across multiple years suggests that the commonly observed "plateauing" of ecosystem functioning at low to moderate plant species richness may be misleading. Isbell et al. (2011) found that 84% of the 147 grassland plant species studied in 17 biodiversity experiments promoted ecosystem functioning at least once, and that different species promoted ecosystem functioning in different years and in different environmental contexts. Thus, estimates of the number of species needed to maintain ecosystem functioning and services may have been greatly underestimated by studies looking at only a few ecosystem functions across relatively short time scales and in only a few environments or climate change scenarios. A number of unanswered questions about biodiversity and ecosystem functioning are outlined at the end of Chapter 3, and I won't repeat them here. Suffice it to say that the field shows no evidence of slowing, and I suspect that many important connections between biodiversity, ecosystem functioning, and ecosystem services remain to be discovered.

Eco-evolutionary feedbacks and community phylogenetics

Few ecologists a decade ago would have appreciated the potential for rapid evolutionary change to influence population dynamics and the outcome of species interactions in ecological time. Yet, as Chapter 15 showed, eco-evolutionary dynamics are common in many systems, even if we don't yet know whether "the evolution–ecology pathway is frequent and strong enough in nature to be broadly important" (Schoener 2011: 426). Schoener's question about the importance of eco-evolutionary feedbacks in nature is crucial, and I suspect that efforts to answer it will occupy the attention of many evolutionary community ecologists in the coming years. The application of phylogenetic methods to understanding what evolutionary patterns exist within communities and how phylogenetic relatedness influences species coexistence in communities and metacommunities is likely to be another active area of research (e.g., Leibold et al. 2010).

One important question that is not often raised with regard to rapid evolution and eco-evolutionary dynamics is how much of the evolutionary change observed over short (i.e., ecological) time scales is meaningful in the long run. Evolutionary biologists have found evidence for significant, ongoing directional selection in many (perhaps most) populations in nature (e.g., Kingsolver et al. 2001; Kingsolver and Diamond 2011), yet stasis in species traits is the general rule—that is, species tend to look the same year after year. Futuyma (2010: 1875) and others argue that such "sluggish evolution" and organism-wide stasis are the result of processes that limit adaptation, and that "without reproductive isolation (speciation), interbreeding will cause the adaptations of local populations to be ephemeral." I raise this issue as a cautionary flag for the study of rapid evolution and its consequences for species interactions and the structuring of communities. If the direction of selection on species traits varies over time and/or across environments, or if there is significant gene flow between populations, then rapid evolution may have minimal impact on the outcome of species interactions in the long run. Again, it is a question of rates.

Climate change and its effects on species distributions and species interactions

How climate change will affect the distribution and abundance of species, their interactions, and biodiversity and ecosystem functioning is *the* question facing ecologists in the next decade and beyond. Here again, most of the important questions and their answers focus on rates of change. How rapidly will species adapt to changing environmental conditions? Will the movements of species (and perhaps communities) in response to climate change be rapid enough to prevent extinctions? Are movement corridors and assisted dispersal useful tools for the conservation of species faced with fragmented landscapes and changing climates? Or are we playing Russian roulette with species introductions? How much will climate change impact the functioning of communities by changing relationships among species? For example, Harley (2011) showed that the upper distributional limits of predator and prey did not shift in unison with temperature change in the marine intertidal zone of the Pacific Northwest, leading to a thermally forced reduction in enemy-free space for mussels along rocky shores and a 50% decrease in the vertical extent of mussel beds.

In Closing, I'd Like to Say…

… it's an exciting time to be a community ecologist. We have the power to illuminate our understanding of the natural world as never before, to help conserve Earth's biodiversity, and to help provide the knowledge needed to develop a more sustainable world. I look forward to seeing what ecologists can accomplish in the coming decade, and I hope this book might contribute in some small way toward facilitating those accomplishments, perhaps by inspiring more young scientists to take up the challenge.

Literature Cited

Aarssen, L. W. 1997. High productivity in grassland ecosystems: Affected by species diversity or productive species? Oikos 80: 183–184.

Abrams, P. A. 1975. Limiting similarity and the form of the competition coefficient. Theoretical Population Biology 8: 356–375.

Abrams, P. A. 1977. Density-independent mortality and interspecific competition: A test of Pianka's niche overlap hypothesis. American Naturalist 111: 539–552.

Abrams, P. A. 1982. Functional responses of optimal foragers. American Naturalist 120: 382–390.

Abrams, P. A. 1983a. Arguments in favor of higher order interactions. American Naturalist 121: 887–891.

Abrams, P. A. 1983b. The theory of limiting similarity. Annual Review of Ecology and Systematics 14: 359–376.

Abrams, P. A. 1984. Variability in resource consumption rates and the coexistence of competing species. Theoretical Population Biology 25: 106–124.

Abrams, P. A. 1986. Character displacement and niche shift analyzed using consumer–resource models of competition. Theoretical Population Biology 29: 107–160.

Abrams, P. A. 1987. Indirect interactions between species that share a predator: Varieties of indirect effects. Pages 38–54 in W. C. Kerfoot and A. Sih, editors. *Predation: Direct and Indirect Impacts on Aquatic Communities.* University Press of New England, Hanover and London.

Abrams, P. A. 1988. Resource productivity–consumer species diversity: Simple models of competition in spatially heterogeneous environments. Ecology 69: 1418–1433.

Abrams, P. A. 1989. Decreasing functional responses as a result of adaptive consumer behavior. Evolutionary Ecology 3: 95–114.

Abrams, P. A. 1990. The effects of adaptive behavior on the type-2 functional response. Ecology 71: 877–885.

Abrams, P. A. 1992. Why don't predators have positive effects on prey populations? Evolutionary Ecology 6: 449–457.

Abrams, P. A. 1993. Effects of increased productivity on abundances of trophic levels. American Naturalist 141: 351–371.

Abrams, P. A. 1995. Implications of dynamically variable traits for identifying, classifying, and measuring direct and indirect effects in ecological communities. American Naturalist 146: 112–134.

Abrams, P. A. 1996. Evolution and the consequences of species introductions and deletions. Ecology 77: 1321–1328.

Abrams, P. A. 2000. The evolution of predator–prey interactions: Theory and evidence. Annual Review of Ecology and Systematics 31: 79–105.

Abrams, P. A. 2001a. Describing and quantifying interspecific interactions: A commentary on recent approaches. Oikos 94: 209–218

Abrams, P. A. 2001b. The effect of density-independent mortality on the coexistence of exploitative competitors for renewing resources. American Naturalist 158: 459–470.

Abrams, P. A. 2004a. Trait-initiated indirect effects due to changes in consumption rates in simple food webs. Ecology 85: 1029–1038.

Abrams, P. A. 2004b. When does periodic variation in resource growth allow robust coexistence of competing consumer species? Ecology 85: 372–382.

Abrams, P. A. 2007. Defining and measuring the impact of dynamic traits on interspecific interactions. Ecology 88: 2555–2562.

Abrams, P. A. 2008. Measuring the impact of dynamic antipredator traits on predator–prey–resource interactions. Ecology 89: 1640–1649.

Abrams, P. A. 2009a. Determining the functional form of density dependence: Deductive approaches for consumer-resource systems having a single resource. American Naturalist 174: 321–330.

Abrams, P. A. 2009b. The implications of using multiple resources for consumer density dependence. Evolutionary Ecology Research 11: 517–540.

Abrams, P. A. 2010. Implications of flexible foraging for interspecific interactions: Lessons from simple models. Functional Ecology 24: 7–17.

Abrams, P. A. and L. R. Ginzburg. 2000. The nature of predation: Prey dependent, ratio dependent, or neither? Trends in Ecology and Evolution 15: 337–341.

Abrams, P. A. and H. Matsuda. 1997. Prey adaptation as a cause of predator–prey cycles. Evolution 51: 1742–1750.

Abrams, P. A. and H. Matsuda. 2005. The effect of adaptive change in the prey on the dynamics of an exploited predator population. Canadian Journal of Fisheries and Aquatic Sciences 62: 758–766.

Abrams, P. A. and M. Nakajima. 2007. Does competition between resources change the competition between their consumers to mutualism? Variation on two themes by Vandermeer. American Naturalist 170: 744–757.

Abrams, P. A. and J. D. Roth. 1994. The effects of enrichment of three-species food chains with nonlinear functional responses. Ecology 75: 1118–1130.

Abrams, P. A. and L. Rowe. 1996. The effects of predation on the age and size of maturity of prey. Evolution 50: 1052–1061.

Abrams, P. A., B. A. Menge, G. G. Mittelbach, D. Spiller and P. Yodzis. 1996. The role of indirect effects in food webs. Pages 371–395 in G. Polis and K. O. Winemiller, editors. *Food Webs: Integration of Patterns and Dynamics.* Chapman and Hall, New York.

Abrams, P. A., C. E. Brassil and R. D. Holt. 2003. Dynamics and responses to mortality rates of competing predators undergoing predator-prey cycles. Theoretical Population Biology 64: 163–176.

Abrams, P. A., C. Rueffler and R. Dinnage. 2008. Competition-similarity relationships and the nonlinearity of competitive effects in consumer-resource systems. American Naturalist 172: 463–474.

Abramsky, Z., E. Stauss, A. Subach, B. P. Kotler and A. Riechman. 1996. The effect of barn owls (*Tyto alba*) on the activity and microhabitat selection of *Gerbillus allenbyi* and *G. pyramidum.* Oecologia 105: 313–319.

Adler, P. B. 2004. Neutral models fail to reproduce observed species–area and species–time relationships in Kansas grasslands. Ecology 85: 1265–1272.

Adler, P. B., J. HilleRisLambers, P. C. Kyriakidis, Q. F. Guan and J. M. Levine. 2006a. Climate variability has a stabilizing effect on the coexistence of prairie grasses. Proceedings of the National Academy of Sciences 103: 12793–12798.

Adler, P. B., J. HilleRisLambers and J. M. Levine. 2006b. Weak effect of climate variability on coexistence in a sagebrush steppe community. Ecology 90: 3303–3312.

Adler, P. B., J. HilleRisLambers and J. M. Levine. 2007. A niche for neutrality. Ecology Letters 10: 95–104.

Adler, P. B., S. P. Ellner and J. M. Levine. 2010. Coexistence of perennial plants: An embarrassment of niches. Ecology Letters 13: 1019–1029.

Adler, P. B. and 57 others. 2011. Productivity is a poor predictor of plant species richness. Science 333: 1750–1753.

Akçakaya, R., R. Arditi and L. R. Ginzburg. 1995. Ratio-dependent predation: An abstraction that works. Ecology 76: 995–1004.

Akcay, E. and J. Roughgarden. 2007. Negotiation of mutualism: Rhizobia and legumes. Proceedings of the Royal Society of London B 274: 25–32.

Allee, W. C., O. Emerson, T. Park and K. Schmidt. 1949. *Principles of Animal Ecology.* W. B. Saunders, Philadelphia, PA.

Allen, A. P. and J. F. Gillooly. 2006. Assessing latitudinal gradient in speciation rates and biodiversity at the global scale. Ecology Letters 9: 947–954.

Allen, A. P. and E. P. White. 2003. Effects of range size on species-area relationships. Evolutionary Ecology Research 5: 493–499.

Allen, A. P., J. H. Brown and J. F. Gillooly. 2002. Global biodiversity, biochemical kinetics, and the energetic-equivalence rule. Science 297: 1545–1548.

Allen, A., J. Gillooly, V. Savage and J. Brown. 2006. Kinetic effects of temperature on rates of genetic divergence and speciation. Proceedings of the National Academy of Sciences 103: 9130–9135.

Allesina, S. 2011. Predicting trophic relations in ecological networks: A test of the Allometric Diet Breadth Model. Journal of Theoretical Biology 279: 161–168.

Allesina, S. and M. Pascual. 2008. Network structure, predator–prey modules, and stability in large food webs. Theoretical Ecology 1: 55–64.

Alonso, D., R. S. Etienne and A. J. McKane. 2006. The merits of neutral theory. Trends in Ecology and Evolution 21: 451–457.

Amarasekare, P. 2003. Competitive coexistence in spatially structured environments: A synthesis. Ecology Letters 6: 1109–1122.

Amarasekare, P., M. F. Hoopes, N. Mouquet and M. Holyoak. 2004. Mechanisms of coexistence in competitive metacommunities. American Naturalist 164: 310–326.

Ambrose, S. H. 1998. Late Pleistocene human population bottlenecks, volcanic winter, and differentiation of modern humans. Journal of Human Evolution 34: 623–651.

Anderson, M. J. and 13 others. 2011. Navigating the multiple meanings of β diversity: A roadmap for the practicing ecologist. Ecology Letters 14: 19–28.

Anderson, R. M. and R. M. May. 1991. *Infectious Diseases of Humans: Dynamics and Control.* Oxford University Press, New York.

Andrewartha, H. G. and L. C. Birch. 1954. *The Distribution and Abundance of Animals.* University of Chicago Press, Chicago, IL.

Angert, A. L., T. E. Huxman, P. L. Chesson and D. L. Venable. 2009. Functional tradeoffs determine species coexistence via the storage effect. Proceedings of the National Academy of Sciences 106: 11641–11645.

Antonelli, A. and I. Sanmartin. 2011. Why are there so many plant species in the Neotropics? Taxon 60: 403–414.

Arditi, R. and L. R. Ginzburg. 1989. Coupling in predator–prey dynamics: Ratio dependence. Journal of Theoretical Biology 139: 311–326.

Armstrong, R. A. 1976. Fugitive species: Experiments with fungi and some theoretical considerations. Ecology 57: 953–963.

Armstrong, R. A. and R. McGehee. 1976. Coexistence of two competitors on one resource. Journal of Theoretical Biology 56: 499–502.

Armstrong, R. A. and R. McGehee. 1980. Competitive exclusion. American Naturalist 115: 151–170.

Arrhenius, O. 1921. Species and area. Journal of Ecology 9: 95–99.

Astorga, A., J. Oksanen, M. Luoto, J. Soininen, R. Virtanen and T. Muotka. 2012. Distance decay of similarity in freshwater communities: Do macro- and microorganisms follow the same rules? Global Ecology and Biogeography 21: 365–375.

Aunapuu, M. and 9 others. 2008. Spatial patterns and dynamic responses of arctic food webs corroborate the exploitation ecosystems hypothesis (EEH). American Naturalist 171: 249–262.

Axelrod, R. and W. D. Hamilton. 1981. The evolution of cooperation. Science 211: 1390–1396.

Bagchi, R., T. Swinfield, R. E. Gallery, O. T. Lewis, S. Gripenberg, L. Narayan and R. P. Freckleton. 2010. Testing the Janzen-Connell mechanism: Pathogens cause overcompensating density dependence in a tropical tree. Ecology Letters 13: 1262–1269.

Baguette, M. 2004. The classical metapopulation theory and the real, natural world: A critical appraisal. Basic and Applied Ecology 5: 213–224.

Baird, D. and H. Milne. 1981. Energy flow and the Ythan Estuary, Aberdeenshire, Scotland. Estuarine and Coastal Shelf Science 13: 455–472.

Baird, D. and R. E. Ulanowicz. 1989. The seasonal dynamics of the Chesapeake Bay ecosystem. Ecological Monographs 59: 329–364.

Baldwin, B. G. 2006. Contrasting patterns and processes of evolutionary change in the tarweed–silversword lineage: Revisiting Clausen, Keck, and Hiesey's findings. Annals of the Missouri Botanical Garden 93: 64–93.

Baldwin, B. G. and M. J. Sanderson. 1998. Age and rate of diversification of the Hawaiian silversword alliance (Compositae). Proceedings of the National Academy of Sciences 95: 9402–9406.

Balvanera, P., A. B. Pfisterer, N. Buchmann, J.-S. He, T. Nakashizuka, D. Raffaelli and B. Schmid. 2006. Quantifying the evidence for biodiversity effects on ecosystem functioning and services. Ecology Letters 9: 1146–1156.

Barabasi, A.-L. 2009. Scale-free networks: A decade and beyond. Science: 412–413.

Bartha, S., T. Czaran and I. Scheuring. 1997. Spatiotemporal scales of non-equilibrium community dynamics: A methodological challenge. New Zealand Journal of Ecology 21: 199–206.

Bascompte, J. 2009. Disentangling the web of life. Science 325: 416–419.

Bascompte, J. and P. Jordano. 2007. Plant-animal mutualistic networks: The architecture of biodiversity. Annual Review of Ecology, Evolution, and Systematics 38: 567–593.

Bascompte, J. and F. Rodríguez-Trelles. 1998. Eradication thresholds in epidemiology, conservation biology and genetics. Journal of Theoretical Biology 192: 415–418.

Bascompte, J., P. Jordano, C. J. Melián and J. M. Olesen. 2003. The nested assembly of plant–animal mutualistic networks. Proceedings of the National Academy of Sciences 100: 9383–9387.

Bascompte, J., C. J. Melián and E. Sala. 2005. Interaction strength combinations and the overfishing of a marine food web. Proceedings of the National Academy of Sciences 102: 5443–5447.

Bassar, R. D. and 10 others. 2010. Local adaptation in Trinidadian guppies alters ecosystem processes. Proceedings of the National Academy of Sciences 107: 3616–3621.

Bastolla, U., M. A. Fortuna, A. Pascual-Garcia, A. Ferrera, B. Luque and J. Bascompte. 2009. The architecture of mutualistic networks minimizes competition and increases biodiversity. Nature 458: 1018–1020.

Baum, D. 2008. Reading a phylogenetic tree: The meaning of monophyletic groups. Nature Education 1(1).

Bawa, K. S. 1990. Plant-pollinator interactions in tropical rain forests. Annual Review of Ecology and Systematics 21: 399–422.

Beckerman, A. P., O. L. Petchey and P. H. Warren. 2006. Foraging biology predicts food web complexity. Proceedings of the National Academy of Sciences 103: 13745–13749.

Beckerman, A. P., O. L. Petchey and P. J. Morin. 2010. Adaptive foragers and community ecology: Linking individuals to communities and ecosystems. Functional Ecology 24: 1–6.

Becks, L., S. P. Ellner, L. E. Jones and N. G. Hairston, Jr. 2010. Reduction of adaptive genetic diversity radically alters eco-evolutionary community dynamics. Ecology Letters 13: 989–997.

Begon, M., C. R. Townsend and J. L. Harper. 2006. Ecology: From Individuals to Ecosystems. Blackwell Publishing, Oxford.

Beisner, B. E., D. T. Haydon and K. Cuddington. 2003. Alternative stable states in ecology. Frontiers in Ecology and the Environment 1: 376–382.

Beisner, B. E., P. R. Peres-Neto, E. S. Lindstrom, A. Barnett and M. Lorena Longhi. 2006. The role of environmental and spatial processes in structuring lake communities from bacteria to fish. Ecology 87: 2985–2991.

Bell, G. 2001. Neutral macroecology. Science 293: 2413–2218.

Belmaker, J., C. H. Sekercioglu and W. Jetz. 2012. Global patterns of specialization and coexistence in bird assemblages. Journal of Biogeography 39: 193–203.

Benard, M. F. 2004. Predator-induced phenotypic plasticity in organisms with complex life histories. Annual Review of Ecology, Evolution, and Systematics 35: 651–673.

Bender, E. A., T. J. Case and M. E. Gilpin. 1984. Perturbation experiments in community ecology: Theory and practice. Ecology 65: 1–13.

Benke, A. C., J. B. Wallace, J. W. Harrison and J. W. Koebel. 2001. Food web quantification using secondary production analysis: Predaceous invertebrates of the snag habitat of a subtropical river. Freshwater Biology 46: 329–346.

Berlow, E. L. 1999. Strong effects of weak interactions in ecological communities. Nature 398: 330–334.

Berlow, E. L., S. A. Navarrete, C. J. Briggs, M. E. Power and B. A. Menge. 1999. Quantifying variation in the strengths of species interactions. Ecology 80: 2206–2224.

Berlow, E. L. and 13 others. 2004. Interaction strengths in food webs: Issues and opportunities. Journal of Animal Ecology 73: 585–598.

Berlow, E. L., U. Brose and N. D. Martinez. 2008. The "Goldilocks factor" in food webs. Proceedings of the National Academy of Sciences **105**: 4079–4080.

Berryman, A. A. 1987. Equilibrium or nonequilibrium: Is that the question? Bulletin of the Ecological Society of America **68**: 500–502.

Bertness, M. D. and R. M. Callaway. 1994. Positive interactions in communities. Trends in Ecology and Evolution **9**: 191–193.

Bertness, M. D. and S. D. Hacker. 1994. Physical stress and positive associations among marsh plants. American Naturalist **144**: 363–372.

Bever, J. D. 1999. Dynamics within mutualism and the maintenance of diversity: Inference from a model of interguild frequency dependence. Ecology Letters **2**: 52–61.

Bever, J. D. 2002. Negative feedback within a mutualism: Host-specific growth of mycorrhizal fungi reduces plant benefit. Proceedings of the Royal Society of London B **269**: 2595–2601.

Bever, J. D. and 9 others. 2010. Rooting theories of plant community ecology in microbial interactions. Trends in Ecology and Evolution **25**: 468–478.

Bezemer, T. M. and W. H. van der Putten. 2007. Diversity and stability in plant communities. Nature **446**: E6–E7.

Bezerra, E. L. S., I. C. Machado and M. A. R. Mello. 2009. Pollination networks of oil-flowers: A tiny world within the smallest of all worlds. Journal of Animal Ecology **78**: 1096–1101.

Bohannan, B. J. M. and R. E. Lenski. 1997. Effect of resource enrichment on a chemostat community of bacteria and bacteriophage. Ecology **78**: 2303–2315.

Bohannan, B. J. M. and R. E. Lenski. 1999. Effect of prey heterogeneity on the response of a model food chain to resource enrichment. American Naturalist **153**: 73–82.

Bohannan, B. J. M. and R. E. Lenski. 2000. The relative importance of competition and predation varies with productivity in a model community. American Naturalist **156**: 329–340.

Bolker, B., M. Holyoak, V. Kivan, L. Rowe and O. Schmitz. 2003. Connecting theoretical and empirical studies of trait-mediated interactions. Ecology **84**: 1101–1114.

Bolnick, D. I. and E. I. Preisser. 2005. Resource competition modifies the strength of trait-mediated predator-prey interactions: A meta-analysis. Ecology **86**: 2771–2779.

Bond, W. J. 2010. Consumer control by megafauna and fire. Pages 275–285 in J. Terborgh and J. A. Estes, editors. *Trophic Cascades: Predators, Prey, and the Changing Dynamics of Nature.* Island Press, Washington, DC.

Bonenfant, C. and 11 others. 2009. Empirical evidence of density-dependence in populations of large herbivores. Advances in Ecological Research **41**: 313–357.

Bongers, F., L. Poorter, W. D. Hawthorne and D. Sheil. 2009. The intermediate disturbance hypothesis applies to tropical forests, but disturbance contributes little to tree diversity. Ecology Letters **12**: 798–805.

Booth, M. G. and J. D. Hoeksema. 2010. Mycorrhizal networks counteract competitive effects of canopy trees on seedling survival. Ecology **91**: 2294–2312.

Boughman, J. W. 2001. Divergent sexual selection enhances reproductive isolation in sticklebacks. Nature **411**: 944–947.

Boughman, J. W., H. D. Rundle and D. Schluter. 2005. Parallel evolution of sexual isolation in sticklebacks. Evolution **59**: 361–373.

Bowers, M. A. and J. H. Brown. 1982. Body size and coexistence in desert rodents: Chance or community structure? Ecology **63**: 391–400.

Brehm, G., R. K. Colwell and J. Kluge. 2007. The role of environmental and mid-domain effect on moth species richness along a tropical elevational gradient. Global Ecology and Biogeography **16**: 205–219.

Brönmark, C. and L.-A. Hansson. 2005. *The Biology of Lakes and Ponds.* Oxford University Press, Oxford.

Brönmark, C. and J. G. Miner. 1992. Predator-induced phenotypical change in body morphology in crucian carp. Science **258**: 1348–1350.

Bronstein, J. L. 2009. The evolution of facilitation and mutualism. Journal of Ecology **97**: 1160–1170.

Brook, B. W. and C. J. A. Bradshaw. 2006. Strength of evidence for density dependence in abundance time series of 1198 species. Ecology **87**: 1445–1451.

Brooker, R. W. and T. Callaghan. 1998. The balance between positive and negative plant interactions and its relationship to environmental gradients: A model. Oikos **81**: 196–207.

Brooker, R. W. and 23 others. 2008. Facilitation in plant communities: The past, the present, and the future. Journal of Ecology **96**: 18–34.

Brooks, D. R. 1985. Historical ecology: A new approach to studying evolution of ecological associations. Annals of the Missouri Botanical Garden **72**: 660–680.

Brose, U., R. B. Ehnes, B. C. Rall, O. Vucic-Pestic, E. L. Berlow and S. Scheu. 2008. Foraging theory predicts predator–prey energy fluxes. Journal of Animal Ecology **77**: 1072–1078.

Brown, J. H. 1971. Mountaintop mammals: Nonequilibrium insular biogeography. American Naturalist **105**: 467–478.

Brown, J. H. 1975. Geographical ecology of desert rodents. Pages 315–341 in M. L. Cody and J. M. Diamond, editors. *Ecology and Evolution of Communities.* The Belknap Press of Harvard University Press, Cambridge, MA.

Brown, J. H. 1981. Two decades of homage to Santa Rosalia: Toward a general theory of diversity. American Zoologist **21**: 877–888.

Brown, J. H. and A. Kodric-Brown. 1977. Turnover rates in insular biogeography: Effect of immigration on extinction. Ecology **58**: 445–449.

Brown, J. H., J. F. Gillooly, A. P. Allen, V. M. Savage and G. B. West. 2004. Toward a metabolic theory of ecology. Ecology **85**: 1771–1789.

Brown, W. L., Jr. and E. O. Wilson. 1956. Character displacement. Systematic Zoology **5**: 49–64.

Bruno, J. F., J. J. Stachowicz and M. D. Bertness. 2003. Inclusion of facilitation into ecology theory. Trends in Ecology and Evolution **18**: 119–125.

Bullock, J. M., R. J. Edwards, R. D. Carey and R. J. Rose. 2000. Geographic separation of two *Ulex* species at three spatial scales: Does competition limit species' ranges? Ecography **23**: 257–271.

Burns, J. H. and S. Y. Strauss. 2011. More closely related species are more ecologically similar in an experimental test. Proceedings of the National Academy of Sciences **108**: 5302–5307.

Butler, G. J. and G. S. K. Wolkowicz. 1987. Exploitative competition in a chemostat for two complementary, and possibly inhibitory, resources. Mathematical Biosciences **83**: 1–48.

Butler, J. L., N. J. Gotelli and A. M. Ellison. 2008. Linking the brown and green: Nutrient transformation and fate in the Sarracenia microecosystem. Ecology **89**: 898–904.

Butterfield, B. J. 2009. Effects of facilitation on community stability and dynamics: Synthesis and future directions. Journal of Ecology **97**: 1192–1201.

Butterfield, B. J., J. L. Betancourt, R. M. Turner and J. M. Briggs. 2010. Facilitation drives 65 years of vegetation change in the Sonoran Desert. Ecology **91**: 1132–1139.

Buzas, M. A., L. S. Collins and S. J. Culver. 2002. Latitudinal difference in biodiversity caused by higher tropical rate of increase. Proceedings of the National Academy of Sciences **99**: 7841–7843.

Byrnes, J., J. J. Stachowicz, K. M. Hultgren, A. R. Hughes, S. V. Olyarnik and C. S. Thornbert. 2006. Predator diversity strengthens trophic cascades in kelp forests by modifying herbivore behaviour. Ecology Letters **9**: 61–71.

Cabana, G. and J. B. Rasmussen. 1994. Modeling food chain structure and contaminant bioaccumulation using stable nitrogen isotopes. Nature **372**: 255–257.

Cáceres, C. E. 1997. Temporal variation, dormancy, and coexistence: A field test of the storage effect. Proceedings of the National Academy of Sciences **94**: 9171–9175.

Cadotte, M. W., A. M. Fortner and T. Fukami. 2006a. The effects of resource enrichment, dispersal, and predation on local and metacommunity structure. Oecologia **149**: 150–157.

Cadotte, M. W., D. V. Mai, S. Jantz, M. D. Collins, M. Keele and J. A. Drake. 2006b. On testing the competition-colonization trade-off in a multispecies assemblage. American Naturalist **168**: 704–709.

Cadotte, M. W., B. J. Cardinale and T. H. Oakley. 2008. Evolutionary history and the effect of biodiversity on plant productivity. Proceedings of the National Academy of Sciences **105**: 17012–17017.

Cahill, J. F., S. W. Kembel, E. G. Lamb and P. A. Keddy. 2008. Does phylogenetic relatedness influence the strength of competition among vascular plants? Perspectives in Plant Ecology, Evolution and Systematics **10**: 41–50.

Callaway, R. M. 1998. Are positive interactions species-specific? Oikos **82**: 202–207.

Callaway, R. M. 2007. Positive Interactions and Interdependence in Plant Communities. Springer, Dordrecht, Germany.

Callaway, R. M. and L. R. Walker. 1997. Competition and facilitation: A synthetic approach to interactions in plant communities. Ecology **78**: 1958–1965.

Callaway, R. M., R. W. Brooker, P. Choler, Z. Kikvidze, C. J. Lortie and R. Michalet. 2002. Positive interactions among alpine plants increase with stress. Nature **417**: 844–888.

Camerano, L. 1880. On the equilibrium of living beings by means of reciprocal destruction. Translated by J. E. Cohen. Pages 360–380 in S. A. Levin, editor. 1994. Frontiers in Mathematical Biology. Springer-Verlag, New York.

Campbell, D. R. and A. F. Motten. 1985. The mechanism of competition for pollination between two forest herbs. Ecology **66**: 554–563.

Cardillo, M. 1999. Latitude and rates of diversification in birds and butterflies. Proceedings of the Royal Society of London B **266**: 1221–1225.

Cardillo, M. and 7 others. 2005a. Multiple causes of high extinction risk in large mammal species. Science **309**: 1239–1241.

Cardillo, M., C. D. L. Orme and I. P. F. Owens. 2005b. Testing for latitudinal bias in diversification rates: An example using New World birds. Ecology **86**: 2278–2287.

Cardinale, B. J. 2011. Biodiversity improves water quality through niche partitioning. Nature **472**: 86–89.

Cardinale, B. J., D. S. Srivastava, J. E. Duffy, J. P. Wright, A. L. Downing, M. Sankaran and C. Jouseau. 2006. Effects of biodiversity on the functioning of trophic groups and ecosystems. Nature **443**: 989–992.

Cardinale, B. J., J. J. Weis, A. E. Forbes, K. J. Tilmon and A. R. Ives. 2006. Biodiversity as both a cause and consequence of resource availability: A study of reciprocal causality in a predator–prey system. Journal of Animal Ecology **75**: 497–505.

Cardinale, B. J. and 7 others. 2007. Impacts of plant diversity on biomass production increase through time because of species complementarity. Proceedings of the National Academy of Sciences **104**: 18123–18128.

Cardinale, B. J. and 8 others. 2011. The functional role of producer diversity in ecosystems. American Journal of Botany **98**: 572–592.

Carlquist, S., B. G. Baldwin and G. Carr. 2003. Tarweeds and Silverswords: Evolution of the Madinae. Missouri Botanical Garden Press, St. Louis.

Carlson, S. M., T. P. Quinn and A. P. Hendry. 2011. Eco-evolutionary dynamics in Pacific salmon. Heredity **106**: 438–447.

Carpenter, S. R. and J. F. Kitchell. 1993. The Trophic Cascade in Lakes. Cambridge University Press, Cambridge.

Carpenter, S. R., J. F. Kitchell and J. R. Hodgson. 1985. Cascading trophic interactions and lake productivity. BioScience **35**: 634–639.

Carpenter, S. R. and 12 others. 2011. Early warning of regime shifts: A whole-ecosystem experiment. Science **332**: 1079–1082.

Carroll, S. P., A. P. Hendry, D. N. Reznick and C. W. Fox. 2007. Evolution on ecological time-scales. Functional Ecology **21**: 387–393.

Caruso, C. M. 2002. Influence of plant abundance on pollination and selection on floral traits of Ipomopsis aggregata. Ecology **83**: 241–254.

Carvalheiro, L. G., Y. M. Buckley, R. Ventim, S. V. Fowler and J. Memmott. 2008. Apparent competition can compromise the safety of highly specific biocontrol agents. Ecology Letters **11**: 690–700.

Case, T. J. 1990. Invasion resistance arises in strongly interacting species-rich model competition communities. Proceedings of the National Academy of Sciences **87**: 9610–9614.

Case, T. J. 2000. *An Illustrated Guide to Theoretical Ecology*. Oxford University Press, New York.

Case, T. J., R. D. Holt, M. A. McPeek and T. H. Keitt. 2005. The community context of species' borders: Ecological and evolutionary perspectives. Oikos **108**: 28–46.

Caswell, H. 1976. Community structure: A neutral model analysis. Ecological Monographs **46**: 327–354.

Cattin, M.-F., L.-F. Bersier, C. Banašek-Richter, R. Baltensperger and J.-P. Gabriel. 2004. Phylogenetic constraints and adaptation explain food-web structure. Nature **427**: 835–839.

Cavender-Bares, J., A. Keen and B. Miles. 2006. Phylogenetic structure of Floridian plant communities depends on taxonomic and spatial scale. Ecology **87**: S109–S122.

Cavender-Bares, J., K. H. Kozak, P. V. A. Fine and S. W. Kembel. 2009. The merging of community ecology and phylogenetic biology. Ecology Letters **12**: 693–715.

Cayford, J. T. and J. D. Goss-Custard. 1990. Seasonal changes in the size-selection of mussels, *Mytilus edulis*, by oystercatchers, *Haemotopus ostralegus*: An optimality approach. Animal Behaviour **40**: 609–624.

Chaneton, E. J. and M. B. Bonsall. 2000. Enemy-mediated apparent competition: Empirical patterns and the evidence. Oikos **88**: 380–394.

Chapin, F. S., L. R. Walker, C. L. Fastie and L. C. Sharman. 1994. Mechanisms of primary succession following deglaciation at Glacier Bay, Alaska. Ecological Monographs **64**: 149–175.

Charnov, E. L. 1976. Optimal foraging: Attack strategy of a mantid. American Naturalist **110**: 141–151.

Chase, J. M. 2003. Experimental evidence for alternative stable equilibria in a benthic pond food web. Ecology Letters **6**: 733–741.

Chase, J. M. 2007. Drought mediates the importance of stochastic community assembly. Proceedings of the National Academy of Sciences **104**: 17430–17434.

Chase, J. M. and M. A. Leibold. 2003a. *Ecological Niches: Linking Classical and Contemporary Approaches*. University of Chicago Press, Chicago, IL.

Chase, J. M. and M. A. Leibold. 2003b. Spatial scale dictates the productivity–biodiversity relationship. Nature **416**: 427–430.

Chase, J. M., M. A. Leibold, A. L. Downing and J. B. Shurin. 2000. The effects of productivity, herbivory, and plant species turnover in grassland food webs. Ecology **81**: 2485–2497.

Chase, J. M. and 8 others. 2002. The interaction between predation and competition: A review and synthesis. Ecology Letters **5**: 302–315.

Chase, J. M. and 11 others. 2005. Competing theories for competitive metacommunities. Pages 336–354 in M. Holyoak, M. A. Leibold and R. D. Holt, editors. *Metacommunities: Spatial Dynamics and Ecological Communities*. University of Chicago Press, Chicago, IL.

Chave, J. 2004. Neutral theory and community ecology. Ecology Letters **7**: 241–253.

Chave, J., H. C. Muller-Landau and S. A. Levin. 2002. Comparing classical community models: Theoretical consequences for patterns of diversity. American Naturalist **159**: 1–23.

Chen, I.-C., J. K. Hill, R. Ohlemüller, D. B. Roy and C. D. Thomas. 2011. Rapid range shifts of species associated with high levels of climate warming. Science **333**: 1024–1026.

Cherif, M. and M. Loreau. 2007. Stoichiometric constraints on resource use, competitive interactions, and elemental cycling in microbial decomposers. American Naturalist **169**: 709–724.

Chesson, J. 1978. Measuring preference in selective predation. Ecology **59**: 211–215.

Chesson, J. 1983. The estimation and analysis of preference and its relationship to foraging models. Ecology **64**: 1297–1304.

Chesson, P. L. 1985. Coexistence of competitors in spatially and temporally varying environments: A look at the combined effects of different sorts of variability. Theoretical Population Biology **28**: 263–287.

Chesson, P. L. 1991. A need for niches? Trends in Ecology and Evolution **6**: 26–28.

Chesson, P. L. 1994. Multispecies competition in variable environments. Theoretical Population Biology **45**: 227–276.

Chesson, P. L. 2000a. Mechanisms of maintenance of species diversity. Annual Review of Ecology and Systematics **31**: 343–366.

Chesson, P. L. 2000b. General theory of competitive coexistence in spatially-varying environments. Theoretical Population Biology **58**: 211–237.

Chesson, P. L. 2011. Ecological niches and diversity maintenance. Pages 43–60 in I. Y. Pavlinov, editor. *Research in Biodiversity—Models and Applications*. Intech, Rijeka, Croatia.

Chesson, P. L. and N. Huntly. 1997. The role of harsh and fluctuating conditions in the dynamics of ecological systems. American Naturalist **150**: 519–553.

Chesson, P. L. and J. J. Kuang. 2008. The interaction between predation and competition. Nature **456**: 235–238.

Chesson, P. L. and J. J. Kuang. 2010. The storage effect due to frequency-dependent predation in multispecies plant communities. Theoretical Population Biology **78**: 148–164.

Chesson, P. L. and M. Rees. 2007. Commentary on Clark et al. (2007): Resolving the biodiversity paradox. Ecology Letters **10**: 659–660.

Chesson, P. L. and R. R. Warner. 1981. Environmental variability promotes coexistence in lottery competitive systems. American Naturalist **117**: 923–943.

Chiba, S. 2004. Ecological and morphological patterns in communities of land snails of the genus *Mandarina* from the Bonin Islands. Journal of Evolutionary Biology **17**: 131–143.

Chisholm, R. A. and S. W. Pacala. 2010. Niche and neutral models predict asymptotically equivalent species abundance distributions in high-diversity ecological communities. Proceedings of the National Academy of Sciences **107**: 15821–15825.

Chown, S. L. and K. J. Gaston. 2000. Areas, cradles and museums: The latitudinal gradient in species richness. Trends in Ecology and Evolution **15**: 311–315.

Clark, C. W. 1990. Mathematical Bioeconomics: The Optimal Management of Renewable Resources. John Wiley and Sons.

Clark, J. S. 2010. Individuals and the variation needed for high species diversity in forest trees. Science **327**: 1129–1132.

Clark, J. S. and P. K. Agarwal. 2007. Rejoinder to Clark et al. (2007): Response to Chesson and Rees. Ecology Letters **10**: 661–662.

Clark, J. S., M. Dietze, S. Chakraborty, P. K. Agarwal, I. Ibanez, S. LaDeau and M. Wolosin. 2007. Resolving the biodiversity paradox. Ecology Letters **10**: 647–662.

Clements, F. E. 1916. Plant succession: Analysis of the development of vegetation. Carnegie Institute of Washington, Publication no. 242, Washington, DC.

Cohen, J. E., F. Briand and C. M. Newman. 1990. Community Food Webs: Data and Theory. Springer-Verlag, New York.

Cohen, J. E., T. Jonsson and S. R. Carpenter. 2003. Ecological community description using the food web, species abundance, and body size. Proceedings of the National Academy of Sciences **100**: 1781–1786.

Collins, S. L. and S. M. Glenn. 1997. Intermediate disturbance and its relationship to within- and between-patch dynamics. New Zealand Journal of Ecology **21**: 103–110.

Colwell, R. K. 1985. The evolution of ecology. American Zoologist **25**: 771–777.

Colwell, R. K. and G. C. Hurtt. 1994. Nonbiological gradients in species richness and a spurious Rapoport effect. American Naturalist **144**: 570–595.

Colwell, R. K., C. Rahbek and N. J. Gotelli. 2004. The mid-domain effect and species richness patterns: What have we learned so far? American Naturalist **163**: E1–E23.

Colwell, R. K., C. Rahbek and N. J. Gotelli. 2005. The mid-domain effect: There's a baby in the bathwater. American Naturalist **166**: E149–E154.

Comita, L. S., R. Condit and S. P. Hubbell. 2007. Developmental changes in habitat associations of tropical trees. Journal of Ecology **95**: 482–492.

Comita, L. S., H. C. Muller-Landau, S. Aguilar and S. P. Hubbell. 2010. Asymmetric density dependence shapes species abundances in a tropical tree community. Science **329**: 330–332.

Condamine, F. L., F. A. H. Sperling, N. Wahlberg, J-Y. Rasplus and G. J. Kergoat. 2012. What causes latitudinal gradients in species diversity? Evolutionary processes and ecological constraints on swallowtail biodiversity. Ecology Letters **15**: 267–277.

Condit, R. 1998. Tropical Forest Census Plots. Springer-Verlag, Berlin.

Condit, R. and 12 others. 2002. Beta-diversity in tropical forest trees. Science **295**: 666–669.

Connell, J. H. 1961. The influence of interspecific competition and other factors on the distribution of the barnacle Chthamalus stellatus. Ecology **42**: 710–723.

Connell, J. H. 1971. On the role of natural enemies in preventing competitive exclusion in some marine animals and in rain forest trees. Pages 298–312 in P. J. den Boer and G. R. Gradwell, editors. Dynamics of Populations. Center for Agricultural Publications and Documentation, Wageningen, the Netherlands.

Connell, J. H. 1978. Diversity in tropical rain forests and coral reefs. Science **199**: 1302–1310.

Connell, J. H. 1983. On the prevalence and relative importance of interspecific competition: Evidence from field experiments. American Naturalist **122**: 661–696.

Connell, J. H. and R. O. Slatyer. 1977. Mechanisms of succession in natural communities and their role in community stability and organization. American Naturalist **111**: 1119–1144.

Connell, J. H. and W. P. Sousa. 1983. On the evidence needed to judge ecological stability or persistence. American Naturalist **121**: 789–824.

Connolly, J. 1986. On difficulties with replacement-series methodology in mixture experiments. Journal of Applied Ecology **23**: 125–137.

Connor, E. F. and E. D. McCoy. 1979. The statistics and biology of the species–area relationship. American Naturalist **113**: 791–833.

Coomes, D. A. and P. J. Grubb. 2003. Colonization, tolerance, competition and seed-size variation within functional groups. Trends in Ecology and Evolution **18**: 283–291.

Cornell, H. V. and J. H. Lawton. 1992. Species interactions, local and regional processes, and limits to the richness of ecological communities: A theoretical perspective. Journal of Animal Ecology **61**: 1–12.

Cornell, H. V., R. H. Karlson and T. P. Hughes. 2008. Local-regional species richness relationships are linear at very small to large scales in west-central Pacific corals. Coral Reefs **27**: 145–151.

Cornwell, W. K. and D. D. Ackerly. 2009. Community assembly and shifts in plant trait distributions across an environmental gradient in coastal California. Ecological Monographs **79**: 109–126.

Cottenie, K. and L. De Meester. 2003. Connectivity and cladoceran species richness in a metacommunity of shallow lakes. Freshwater Biology **48**: 823–832.

Cottenie, K. and L. De Meester. 2004. Metacommunity structure: Synergy of biotic interactions as selective agents and dispersal as fuel. Ecology **85**: 114–119.

Cottenie, K., N. Michels, N. Nuytten and L. De Meester. 2003. Zooplankton metacommunity structure: Regional vs. local processes in highly interconnected ponds. Ecology **84**: 991–1000.

Cottingham, K. L., B. L. Brown and J. T. Lennon. 2001. Biodiversity may regulate the temporal variability of ecological systems. Ecology Letters **3**: 340–348.

Crame, J. A. 2001. Taxonomic diversity gradients through geological time. Diversity and Distributions **7**: 175–189.

Crame, J. A. 2009. Time's stamp on modern biogeography. Science **323**: 720–721.

Cramer, N. F. and R. M. May. 1971. Interspecific competition, predation and species diversity: A comment. Journal of Theoretical Biology **34**: 289–293.

Crane, P. R. and S. Lidgard. 1989. Angiosperm diversification and paleolatitudinal gradients in Cretaceous floristic diversity. Science **246**: 675–678.

Creel, S. and D. Christianson. 2008. Relationships between direct predation and risk effects. Trends in Ecology and Evolution **23**: 194–201.

Cunningham, H. R., L. J. Rissler and J. J. Apodaca. 2009. Competition at the range boundary in the slimy salamander: Using reciprocal transplants for studies on the role of biotic interactions in spatial distributions. Journal of Animal Ecology **78**: 52–62.

Currie, D. J. 1991. Energy and large-scale patterns of animal-species and plant-species richness. American Naturalist **137**: 27–49.

Currie, D. J. 1993. What shape is the relationship between body size and population-density? Oikos **66**: 353–358.

Currie, D. J. and J. T. Fritz. 1993. Global patterns of animal abundance and species energy use. Oikos **67**: 56–68.

Currie, D. J. and V. Paquin. 1987. Large-scale biogeographical patterns of species richness of trees. Nature **329**: 326–327.

Currie, D. J. and 10 others. 2004. Predictions and tests of climate-based hypotheses of broad-scale variation in taxonomic richness. Ecology Letters **7**: 1121–1134.

Dakos, V., S. Kéfi, M. Rietkerk, E. H. von Nes and M. Scheffer. 2011. Slowing down in spatially patterned ecosystems at the brink of collapse. American Naturalist **177**: E153–E166.

Damschen, E. I., N. M. Haddad, J. L. Orrock, J. J. Tewksbury and D. J. Levey. 2006. Corridors increase plant species richness at large scales. Science **313**: 1284–1286.

Darcy-Hall, T. L. 2006. Patterns of nutrient and predator limitation of benthic algae in lakes along a productivity gradient. Oecologia **148**: 660–671.

Darcy-Hall, T. L. and S. R. Hall. 2008. Linking limitation to species composition: Importance of inter- and intra-specific variation in grazing resistance. Oecologia **155**: 797–808.

Darlington, P. J. 1957. *Zoogeography: The Geographical Distribution of Animals.* John Wiley and Sons, New York.

Darwin, C. R. 1859. On The Origin of Species by Means of Natural Selection or the Preservation of Favoured Races in the Struggle for Life. John Murray, London.

Daskalov, G. M., A. N. Grishin, S. Rodionov and V. Mihneva. 2007. Trophic cascades triggered by overfishing reveal possible mechanisms of ecosystem regime shifts. Proceedings of the National Academy of Sciences **104**: 10518–10523.

Davies, K. F., M. Holyoak, K. A. Preston, V. A. Offeman and Q. Lum. 2009. Factors controlling community structure in heterogeneous metacommunities. Journal of Animal Ecology **78**: 937–944.

Davies, T. J., V. Savolainen, M. W. Chase, J. Moat and T. G. Barraclough. 2004. Environmental energy and evolutionary rates in flowering plants. Proceedings of the Royal Society of London B **271**: 2195–2200.

Davis, J. M., A. D. Rosemond, S. L. Eggert, W. F. Cross and J. B. Wallace. 2010. Long-term nutrient enrichment decouples predator and prey production. Proceedings of the National Academy of Sciences **107**: 121–126.

Dayan, T. and D. Simberloff. 2005. Ecological and community-wide character displacement: The next generation. Ecology Letters **8**: 875–894.

Dayan, T., D. Simberloff, E. Tchernov and Y. Yom-Tov. 1990. Feline canines: Community-wide character displacement among the small cats of Israel. American Naturalist **136**: 39–60.

Dayton, P. K. 1971. Competition, disturbance, and community organization: The provision and subsequent utilization of space in a rocky intertidal community. Ecological Monographs **41**: 351–389.

Dayton, P. K. 1975. Experimental evaluation of ecological dominance in a rocky intertidal algal community. Ecological Monographs **45**: 137–159.

deKroon, H. and 7 others. 2012. Root responses to nutrients and soil biota: Drivers of species coexistence and ecosystem productivity. Journal of Ecology **100**: 6–15.

DeMaster, D. P., A. W. Trites, P. Clapham, S. Mizroch, P. Wade, R. J. Small and J. ver Hoef. 2006. The sequential megafaunal collapse: Testing with existing data. Progress in Oceanography **68**: 329–342.

de Mazancourt, C. and M. W. Schwartz. 2010. A resource ratio theory of cooperation. Ecology Letters **13**: 349–359.

de Mazancourt, C., M. Loreau and U. Dieckmann. 2005. Understanding mutualism when there is adaptation to the partner. Journal of Ecology **93**: 305–314.

de Mazancourt, C., E. Johnson and T. G. Barraclough. 2008. Biodiversity inhibits species' evolutionary responses to changing environments. Ecology Letters **11**: 380–388.

De Meester, L., W. V. Doorslaer, A. Geerts, L. Orsini and R. Stokes. 2011. Thermal genetic adaptation in the water flea *Daphnia* and its impact: An evolving metacommunity approach. Integrative and Comparative Biology **51**: 703–718.

Dempster, J. P. 1983. The natural control of populations of butterflies and moths. Biological Review **58**: 461–481.

Denison, R. F. 2000. Legume sanctions and the evolution of symbiotic cooperation by rhizobia. American Naturalist **156**: 567–576.

Dennis, B. and B. Taper. 1994. Density dependence in time series observations of natural populations: Estimation and testing. Ecological Monographs **64**: 205–224.

Denno, R. F., M. S. McClure and J. R. Ott. 1995. Interspecific interactions in phytophagous insects—competition reexamined and resurrected. Annual Review of Entomology **40**: 297–331.

Denslow, J. 1980. Gap partitioning among tropical rainforest trees. Biotropica (supplement) **12**: 47–55.

De Roos, A. M., L. Persson and E. Mc-Cauley. 2003. The influence of size-dependent life-history traits on the structure and dynamics of populations and communities. Ecology Letters **6**: 473–487.

Derrickson, E. M. and R. E. Ricklefs. 1988. Taxon-dependent diversification of life-history traits and the perception of phylogenetic constraints. Functional Ecology **2**: 417–423.

Desjardins-Proulx, P. and D. Gravel. 2012. How likely is speciation in neutral ecology? American Naturalist **179**: 137–144.

Dethlefsen, L. and D. A. Relman. 2011. Incomplete recovery and individualized responses of the human distal gut microbiota to repeated antibiotic perturbation. Proceedings of the National Academy of Sciences **108**: 4554–4561.

Diamond, J. 1973. Distributional ecology of New Guinea birds. Science **179**: 759–769.

Diamond, J., editor. 1975. *Assembly of Species Communities*. Harvard University Press, Cambridge, MA.

Diamond, J. 1986. Overview: Laboratory experiment, field experiments, and natural experiments. Pages 3–22 in J. Diamond and T. J. Case, editors. *Community Ecology*. Harper and Row, New York.

Diaz, S., A. J. Symstad, F. S. Chapin, D. A. Wardle and L. F. Huenneke. 2003. Functional diversity revealed by removal experiments. Trends in Ecology and Evolution **18**: 140–146.

Diaz, S., J. Fargione, F. S. Chapin and D. Tilman. 2006. Biodiversity loss threatens human well-being. PLoS Biology **4**: 1300–1305.

Dickie, I. A., R. T. Koide and K. C. Steiner. 2002. Influences of established trees on mycorrhizas, nutrition, and growth of *Quercus rubra* seedlings. Ecological Monographs **72**: 505–521.

Dickie, I. A., S. A. Schnitzer, P. B. Reich and S. E. Hobbie. 2005. Spatially disjunct effects of co-occurring competition and facilitation. Ecology Letters **8**: 1191–1200.

Diehl, S. 2007. Paradoxes of enrichment: Effects of increased light versus nutrient supply on pelagic producer-grazer system. American Naturalist **169**: E173–E191.

Diehl, S. and P. Eklöv. 1995. Effects of piscivore-mediated habitat use on resources, diet, and growth of perch. Ecology **76**: 1712–1726.

Dijkstra, F. A., J. B. West, S. E. Hobbie, P. B. Reich and J. Trost. 2007. Plant diversity, CO_2, and N influence inorganic and organic N leaching in grasslands. Ecology **88**: 490–500.

Doak, D. F., D. Bigger, E. K. Harding, M. A. Marvier, R. E. O'Malley and D. Thomson. 1998. The statistical inevitability of stability–diversity relationships in community ecology. American Naturalist **151**: 264–276.

Dobzhansky, Th. 1950. Evolution in the tropics. American Scientist **38**: 209–221.

Dobzhansky, Th. 1964. Biology, molecular and organismic. American Zoologist **4**: 443–452.

Doebeli, M. and N. Knowlton. 1998. The evolution of interspecific mutualisms. Proceedings of the National Academy of Sciences **95**: 8676–8680.

Downing, A. L. 2005. Relative effects of species composition and richness on ecosystem properties in ponds. Ecology **86**: 701–715.

Downing, A. L. and M. A. Leibold. 2010. Species richness facilitates ecosystem resilience in aquatic food webs. Freshwater Biology **55**: 2123–2137.

Drakare, S., J. J. Lennon and H. Hillebrand. 2006. The imprint of the geographical, evolutionary and ecological context on species–area relationships. Ecology Letters **9**: 215–227.

Driscoll, D. A. 2007. How to find a metapopulation. Canadian Journal of Zoology **85**: 1031–1048.

Duffy, J. E., J. P. Richardson and K. E. France. 2005. Ecosystem consequences of diversity depend on food chain length in estuarine vegetation. Ecology Letters **8**: 301–309.

Dugatkin, L. A. 2009. *Principles of Animal Behavior*. Second edition. W. W. Norton & Company, New York.

Dunn, R. R., R. K. Colwell and C. Nilsson. 2006. The river domain: Why are there more species halfway up the river? Ecography **29**: 251–259.

Dunne, J. A. 2006. The network structure of food webs. Pages 27–86 in M. Pascual and J. A. Dunne, editors. *Ecological Networks: Linking Structure to Dynamics in Food Webs*. Oxford University Press, Oxford.

Dunne, J. A., R. J. Williams and N. D. Martinez. 2002. Food web structure and network theory: The role of connectance and size. Proceedings of the National Academy of Sciences **99**: 12917–12922.

Dybzinski, R. and D. Tilman. 2007. Resource use patterns predict long-term outcomes of plant competition for nutrients and light. American Naturalist **170**: 305–318.

Ehrlich, P. R. and L. C. Birch. 1967. The "Balance of Nature" and "Population Control." American Naturalist **101**: 97–107.

Eiserhardt, W. L., S. Bjorholm, J.-C. Svenning, T. F. Rangel and H. Balslev. 2011. Testing the water-energy theory on American palms (Arecaceae) using geographically weighted regression. PLoS ONE **6**(11): e27027. doi: 10.1371/journal.pone.0027027.

Ekström, C. U. 1838. *Kongliga Vetenskaps Academiens Handingar*, p. 213. Cited in C. Brönmark and J. G. Miner. 1992. Predator-induced phenotypical change in body morphology in crucian carp. Science **258**: 1348–1350.

Ellner, S. P., M. A. Geber and N. G. Hairston, Jr. 2011. Does rapid evolution matter? Measuring the rate of contemporary evolution and its impacts on ecological dynamics. Ecology Letters **14**: 603–614.

Ellwood, M. D. F., A. Manica and W. A. Foster. 2009. Stochastic and deterministic processes jointly structure tropical arthropod communities. Ecology Letters **12**: 277–284.

Elmqvist, T., C. Folke, M. Nyström, G. Peterson, J. Bengtsson, B. Walker and J. Norberg. 2003. Response diversity, ecosystem change, and resilience. Frontiers in Ecology and the Environment **1**: 488–494.

Elner, R. W. and R. N. Hughes. 1978. Energy maximization in the diet of the shore crab, *Carcinus maenas*. Journal of Animal Ecology **47**: 103–116.

Elser, J. J. and 11 others. 2000. Nutritional constraints in terrestrial and freshwater food webs. Nature **408**: 578–580.

Elser, J. J. and 9 others. 2007. Global analysis of nitrogen and phosphorus limitation of primary producers in freshwater, marine, and terrestrial ecosystems. Ecology Letters **10**: 1135–1142.

Elton, C. S. 1927. *Animal Ecology*. Reprint 2001, University of Chicago Press, Chicago, IL.

Elton, C. S. 1950. *The Ecology of Animals*. Third edition. Methuen, London.

Elton, C. S. 1958. The Ecology of Invasions by Animals and Plants. Methuen, London.

Emerson, B. C. and R. G. Gillespie. 2008. Phylogenetic analysis of community assembly and structure over space and time. Trends in Ecology and Evolution **23**: 619–630.

Emlen, J. M. 1966. The role of time and energy in food preference. American Naturalist **100**: 611–617.

Emlen, J. M. 1973. *Ecology: An Evolutionary Approach*. Addison-Wesley, New York, NY.

Emmerson, M. C. and D. Raffaelli. 2004. Predator–prey body size, interaction strength and the stability of a real food web. Journal of Animal Ecology **73**: 399–409.

Emmerson, M. and J. M. Yearsley. 2004. Weak interactions, omnivory and emergent food-web properties. Proceedings of the Royal Society of London B **271**: 397–405.

Enquist, B. J. and K. J. Niklas. 2001. Invariant scaling relations across tree-dominated communities. Nature **410**: 655–660.

Eriksson, O. 1997. Clonal life histories and the evolution of seed recruitment. Pages 211–226 in H. de Kroon and V. Groenendael, editors. *The Ecology and Evolution of Clonal Plants*. Backhuys, Leiden.

Ernest, S. K. M., J. H. Brown, K. M. Thibault, E. P. White and J. R. Goheen. 2008. Zero sum, the niche, and metacommunities: Long-term dynamics of community assembly. American Naturalist **172**: E257–E269.

Erwin, D. H. 2001. Lessons from the past: Biotic recoveries from mass extinctions. Proceedings of the National Academy of Sciences **98**: 5399–5403.

Erwin, D. H. 2007. Disparity: Morphological pattern and developmental context. Palaeontology **50**: 57–73.

Estes, J. A. and D. Duggins. 1995. Sea otters and kelp forest in Alaska: Generality and variation in a community ecological paradigm. Ecological Monographs **65**: 75–100.

Estes, J. A. and J. F. Palmisano. 1974. Sea otters: Their role in structuring nearshore communities. Science **185**: 1058–1060.

Estes, J. A., D. O. Duggins and G. B. Rathbun. 1989. The ecology of extinctions in kelp forest communities. Conservation Biology **3**: 252–264.

Estes, J. A., M. T. Tinker, T. M. Williams and D. F. Doak. 1998. Killer whale predation on sea otters linking oceanic and nearshore ecosystems. Science **282**: 473–476.

Estes, J. A. and 23 others. 2011. Trophic downgrading of planet Earth. Science **333**: 301–306.

Etienne, R. S. and D. Alonso. 2005. A dispersal-limited sampling theory for species and alleles. Ecology Letters **8**: 1147–1156.

Evans, K. L. and K. J. Gaston. 2005. Can the evolutionary-rates hypothesis explain species–energy relationships? Functional Ecology **19**: 899–915.

Evans, K. L., J. J. D. Greenwood and K. J. Gaston. 2005. Dissecting the species-energy relationship. Proceedings of the Royal Society of London B **272**: 2155–2163.

Evans, K. L., S. E. Newson, D. Storch, J. J. D. Greenwood and K. J. Gaston. 2008. Spatial scale, abundance and the species-energy relationship in British birds. Journal of Animal Ecology **77**: 395–405.

Ezard, T. H. G., S. D. Côté and F. Pelletier. 2009. Eco-evolutionary dynamics: Disentangling phenotypic, environmental and population fluctuations. Philosophical Transactions of the Royal Society of London B **364**: 1491–1498.

Fagan, W. F. and L. E. Hurd. 1994. Hatch density variation of a generalist arthropod predator: Population consequences and community impact. Ecology **75**: 2022–2032.

Fargione, J. and D. Tilman. 2005. Diversity decreases invasion via both sampling and complementary effects. Ecology Letters **8**: 604–611.

Fargione, J. and 8 others. 2007. From selection to complimentarity: Shifts in the causes of biodiversity–productivity relationships in a long-term biodiversity experiment. Proceedings of the Royal Society of London B **274**: 871–876.

Farnworth, E. G. and F. B. Golley. 1974. Fragile Ecosystems: Evaluation of Research and Applications in the Neotropics. Springer-Verlag, New York.

Farrell, B. D. and C. Mitter. 1993. Phylogenetic determinants of insect/plant community diversity. Pages 253–266 in R. E. Ricklefs and D. Schluter, editors. *Species Diversity in Ecological Communities: Historical and Geographical Perspectives*. University of Chicago Press, Chicago, IL.

Farrell, B. D., C. Mitter and D. J. Futuyma. 1992. Diversification at the insect-plant interface. BioScience **42**: 34–42.

Fauth, J. E., J. Bernardo, M. Camara, W. J. J. Resitarits, J. Van Buskirk and S. A. McCollum. 1996. Simplifying the jargon of community ecology: A conceptual approach. American Naturalist **147**: 282–286.

Fedorov, A. A. 1966. The structure of the tropical rain forest and speciation in the humid tropics. Journal of Ecology **54**: 1–11.

Feldman, T. S., W. F. Morris and W. G. Wilson. 2004. When can two plant species facilitate each other's pollination? Oikos **105**: 197–207.

Felsenstein, J. 1985. Phylogenies and the comparative method. American Naturalist **125**: 1–15.

Fenchel, T. 1975. Character displacement and co-existence in mud snails (Hydrobiidae). Oecologia **20**: 19–32.

Field, R. and 11 others. 2009. Spatial species-richness gradients across scales: A meta-analysis. Journal of Biogeography **36**: 32–147.

Fine, P. E. M. 1975. Vectors and vertical transmission—epidemiologic perspective. Annals of the New York Academy of Sciences **266**: 173–194.

Fine, P. V. A. and R. H. Ree. 2006. Evidence for a time-integrated species–area effect on the latitudinal gradient in tree diversity. American Naturalist **168**: 796–804.

Firn, J., A. P. N. House and Y. M. Buckley. 2010. Alternative states models provide an effective framework for invasive species control and restoration of native communities. Journal of Applied Ecology **47**: 96–105.

Fisher, A. G. 1960. Latitudinal variations in organic diversity. Evolution **14**: 64–81.

Fisher, R. A. 1930. The genetical theory of natural selection. Clarendon Press, Oxford.

Fisher, R. A., A. S. Corbet and C. B. Williams. 1943. The relation between the number of species and the number of individuals in a random sample of an animal population. Journal of Animal Ecology **12**: 42–58.

Fontaine, C. and 7 others. 2011. The ecological and evolutionary implications of merging different types of networks. Ecology Letters **14**: 1170–1181.

Foote, M. 1997. The evolution of morphological diversity. Annual Review of Ecology and Systematics **28**: 129–152.

Forbes, S. A. 1887. The lake as a microcosm. Bulletin of the Scientific Association (Peoria, IL) **1887**: 77–87.

Forister, M. A. and C. R. Feldman. 2011. Phylogenetic cascades and the origins of tropical diversity. Biotropica **43**: 270–278.

Fox, J. W. 2007. Testing the mechanisms by which source-sink dynamics alter competitive outcomes in a model system. American Naturalist **170**: 396–408.

France, K. E. and J. E. Duffy. 2006. Consumer diversity mediates invasion dynamics at multiple trophic levels. Oikos **113**: 515–529.

Francis, A. P. and D. J. Currie. 2003. A globally consistent richness-climate relationship for angiosperms. American Naturalist **161**: 523–536.

Franco, A. C. and P. S. Nobel. 1989. Effect of nurse plants on the microhabitat and growth of cacti. Journal of Ecology **77**: 870–886.

Frank, D. A. and S. J. McNaughton. 1991. Stability increases with diversity in plant communities: Empirical evidence from the 1988 Yellowstone drought. Oikos **62**: 360–362.

Frank, K. T., B. Petrie, J. S. Choi and W. C. Leggett. 2006. Reconciling differences in trophic controls in mid-latitude marine ecosystems. Ecology Letters **9**: 1096–1105.

Frederickson, A. G. and G. Stephanopoulos. 1981. Microbial competition. Science **213**: 972–979.

Freestone, A. L. and R. W. Osman. 2011. Latitudinal variation in local interactions and regional enrichment shape patterns of marine community diversity. Ecology **92**: 208–217.

Freestone, A. L., R. W. Osman, G. M. Ruiz and M. E. Torchin. 2011. Stronger predation in the tropics shapes species richness patterns in marine communities. Ecology **92**: 983–993.

Fridley, J. D. and 8 others. 2007. The invasion paradox: Reconciling pattern and process in species invasions. Ecology **88**: 3–17.

Fryer, G. and T. D. Iles. 1972. The Cichlid Fishes of the Great Lakes of Africa: Their Biology and Evolution. Oliver and Boyd, Edinburgh.

Fryxell, J. M. and P. Lundberg. 1994. Diet choice and predator–prey dynamics. Evolutionary Ecology **8**: 407–421.

Fryxell, J. M. and P. Lundberg. 1998. *Individual Behaviour and Community Dynamics*. Chapman and Hall, New York.

Fung, T., R. M. Seymour and C. R. Johnson. 2011. Alternative stable states and phase shifts in coral reefs under anthropogenic stress. Ecology **92**: 967–982.

Fussmann, G. F., S. P. Ellner, N. G. Hairston, Jr., L. E. Jones, K. W. Shertzer and T. Yoshida. 2005. Ecological and evolutionary dynamics of experimental plankton communities. Advances in Ecological Research **37**: 221–243.

Fussmann, G. F., M. Loreau and P. A. Abrams. 2007. Eco-evolutionary dynamics of communities and ecosystems. Functional Ecology **21**: 465–477.

Futuyma, D. J. 2005. *Evolution*. Sinauer Associates, Sunderland, MA.

Futuyma, D. J. 2010. Evolutionary constraint and ecological consequences. Evolution **64**: 1865–1884.

Gascoigne, J., L. Berec, S. Gregory and F. Courchamp. 2009. Dangerously few liaisons: A review of mate-finding Allee effects. Population Ecology **51**: 355–372.

Gaston, K. J. 2000. Global patterns in biodiversity. Nature **405**: 220–227.

Gaston, K. J. 2003. *The Structure and Dynamics of Geographic Ranges*. Oxford University Press, Oxford.

Gause, G. F. 1934. *The Struggle for Existence*. Williams and Wilkins, Baltimore, MD.

Gause, G. F. and A. A. Witt. 1935. Behavior of mixed populations and the problem of natural selection. American Naturalist **69**: 596–609.

Gavrilets, S. and J. B. Losos. 2009. Adaptive radiation: Contrasting theory with data. Science **323**: 732–737.

Geber, M. A. 2011. Ecological and evolutionary limits to species geographic ranges. American Naturalist **178**: S1–S5.

Gerhold, P. and 8 others. 2011. Phylogenetically poor plant communities receive more alien species, which more easily coexist with natives. American Naturalist **177**: 668–680.

Getz, W. M. and R. G. Haight. 1989. *Population Harvesting: Demographic Models of Fish, Forest and Animal Resources*. Princeton University Press, Princeton, NJ.

Ghazoul, J. 2006. Floral diversity and the facilitation of pollination. Journal of Ecology **94**: 295–304.

Gibson, D. J., J. Connolly, D. C. Hartnett and J. D. Weidenhamer. 1999. Designs for greenhouse studies of interactions between plants. Journal of Ecology **87**: 1–16.

Gilbert, B. and M. J. Lechowicz. 2004. Neutrality, niches, and dispersal in a temperate forest understory. Proceedings of the National Academy of Sciences **101**: 7651–7656.

Gilbert-Norton, L., R. Wilson, J. R. Stevens and K. H. Beard. 2010. A meta-analytic review of corridor effectiveness. Conservation Biology **24**: 660–668.

Gilliam, J. F. 1982. Foraging under mortality risk in size-structured populations. Ph. D. dissertation, Michigan State University, East Lansing, MI.

Gilliam, J. F. and D. F. Fraser. 1987. Habitat selection when foraging under predation hazard: A model and a test with stream-dwelling minnows. Ecology **68**: 1856–1862.

Gillies, C. S. and C. C. St. Clair. 2008. Riparian corridors enhance movement of a forest specialist bird in fragmented tropical forest. Proceedings of the National Academy of Sciences **105**: 19774–19779.

Gillman, L. N. and S. D. Wright. 2006. The influence of productivity on the species richness of plants: A critical assessment. Ecology **87**: 1234–1243.

Gillman, L. N., H. A. Ross, J. D. Keeling and S. D. Wright. 2009. Latitude, elevation and the tempo of molecular evolution in mammals. Proceedings of the Royal Society of London B **276**: 3353–3359.

Gillman, L. N., P. McBride, D. J. Keeling, H. A. Ross and S. D. Wright. 2010. Are rates of molecular evolution in mammals substantially accelerated in warmer environments? Reply. Proceedings of the Royal Society of London B **278**: 1294–1297.

Gilpin, M. E. and F. J. Ayala. 1973. Global models of growth and competition. Proceedings of the National Academy of Sciences **70**: 3590–3593.

Ginzburg, L. R. and H. R. Akçakaya. 1992. Consequences of ratio-dependent predation for steady-state properties of ecosystems. Ecology **73**: 1536–1543.

Gleason, H. A. 1922. On the relation between species and area. Ecology **3**: 158–162.

Gleason, H. A. 1926. The individualistic concept of the plant association. Torrey Botanical Club Bulletin **53**: 7–26.

Gleeson, S. K. and D. S. Wilson. 1986. Equilibrium diet: Optimal foraging and prey coexistence. Oikos **46**: 139–144.

Godfray, H. C. J. and M. P. Hassell. 1992. Long time series reveal density dependence. Nature **359**: 673–674.

Goldberg, D. E. 1987. Neighborhood competition in an old-field plant community. Ecology **68**: 1211–1223.

Goldberg, D. E. and A. M. Barton. 1992. Patterns and consequences of interspecific competition in natural communities: A review of field experiments with plants. American Naturalist **139**: 771–801.

Goldberg, D. E. and S. M. Scheiner. 1993. ANOVA and ANCOVA: Field competition experiments. Pages 69–93 in S. M. Scheiner and J. Gurevitch, editors. *Design and Analysis of Ecological Experiments*. Chapman and Hall, New York.

Goldberg, D. E. and P. A. Werner. 1983. Equivalence of competitors in plant communities—a null hypothesis and a field experimental approach. American Journal of Botany **70**: 1098–1104.

Goldberg, E. E., K. Roy, R. Lande and D. Jablonski. 2005. Diversity, endemism, and age distributions in macroevolutionary sources and sinks. American Naturalist **165**: 623–633.

Gonzalez, A. and M. Loreau. 2009. The causes and consequences of compensatory dynamics in ecological communities. Annual Review of Ecology, Evolution, and Systematics **40**: 393–414.

Gonzalez-Voyer, A., J. M. Padial, S. Castroviejo-Fisher, I. De La Riva and C. Vila. 2011. Correlates of species richness in the largest Neotropical amphibian radiation. Journal of Evolutionary Biology **24**: 931–942.

Gotelli, N. J. 2008. *A Primer of Ecology*. Fourth edition. Sinauer Associates, Sunderland, MA.

Gotelli, N. J. and R. K. Colwell. 2001. Quantifying biodiversity: Procedures and pitfalls in the measurement and comparison of species richness. Ecology Letters **4**: 379–391.

Gotelli, N. J. and G. R. Graves. 1996. *Null Models in Ecology*. Smithsonian Institution Press, Washington, DC.

Gotelli, N. J. and B. J. McGill. 2006. Null versus neutral models: What's the difference? Ecography **29**: 793–800.

Gotelli, N. J. and 20 others. 2009. Patterns and causes of species richness: A general simulation model for macroecology. Ecology Letters **12**: 873–886.

Gouhier, T. C., B. A. Menge and S. D. Hacker. 2011. Recruitment facilitation can promote coexistence and buffer population growth in metacommunities. Ecology Letters **14**: 1201–1210.

Grace, J. B. and R. G. Wetzel. 1981. Habitat partitioning and competitive displacement in cattails (*Typha*)—experimental field studies. American Naturalist **118**: 463–474.

Grace, J. B. and R. G. Wetzel. 1992. Long-term dynamics of *Typha* populations. Aquatic Botany **61**: 137–146.

Grace, J. B. and 12 others. 2007. Does species diversity limit productivity in natural grassland communities? Ecology Letters **10**: 680–689.

Graham, A. 2011. The age and diversification of terrestrial New World ecosystems through Cretaceous and Cenozoic time. American Journal of Botany **98**: 336–351.

Graham, C. H., J. L. Parra, C. Rahbek and J. A. McGuire. 2009. Phylogenetic structure in tropical hummingbird communities. Proceedings of the National Academy of Sciences **106**: 19673–19678.

Grant, P. R. 1972. Convergent and divergent character displacement. Biological Journal of the Linnean Society **4**: 39–68.

Grant, P. R. 1986. *Ecology and Evolution of Darwin's Finches*. Princeton University Press, Princeton, NJ.

Grant, P. R. and B. R. Grant. 2008. *How and Why Species Multiply: The Radiation of Darwin's Finches*. Princeton University Press, Princeton, NJ.

Gravel, D., C. D. Canham, M. Beaudet and C. Messier. 2006. Reconciling niche and neutrality: The continuum hypothesis. Ecology Letters **9**: 399–409.

Gray, R. 1986. Faith and foraging. Pages 69–140 in A. C. Kamil, J. R. Krebs and H. R. Pulliam, editors. *Foraging Behavior*. Plenum Press, New York.

Green, J. L. and 7 others. 2004. Spatial scaling of microbial eukaryote diversity. Nature **432**: 747–750.

Greene, D. F. and E. A. Johnson. 1993. Seed mass and dispersal capacity in wind-dispersed diaspores. Oikos **67**: 69–74.

Greene, D. F. and E. A. Johnson. 1994. Estimating the mean annual seed production of trees. Ecology **75**: 642–647.

Gregory, S. D., C. J. A. Bradshaw, B. W. Brook and F. Courchamp. 2010. Limited evidence for the demographic Allee effect from numerous species across taxa. Ecology **91**: 2151–2161.

Grime, J. P. 1973a. Competitive exclusion in herbaceous vegetation. Nature **242**: 344–347.

Grime, J. P. 1973b. Control of species diversity in herbaceous vegetation. Journal of Environmental Management **1**: 151–167.

Grime, J. P. 1979. *Plant Strategies and Vegetation Processes*. John Wiley and Sons, New York, NY.

Grime, J. P. 1997. Biodiversity and ecosystem function: The debate deepens. Science **277**: 1260–1261.

Grime, J. P., R. Hunt and W. J. Krzanowski. 1987. Evolutionary physiological ecology of plants. Pages 105–125 in P. Calow, editor. *Evolutionary Physiological Ecology*. Cambridge University Press, Cambridge.

Grinnell, J. 1917. The niche-relationship of the California thrasher. Auk **34**: 427–433.

Grman, E., T. M. P. Robinson and C. A. Klausmeier. 2012. Ecological specialization and trade affect the outcome of negotiations in mutualism. American Naturalist.

Gross, K. 2008. Positive interactions among competitors can produce species-rich communities. Ecology Letters **11**: 929–936.

Gross, K. and B. J. Cardinale. 2005. The functional consequences of random vs. ordered species extinctions. Ecology Letters **8**: 409–418.

Gross, K. and B. J. Cardinale. 2007. Does species richness drive community production or vice versa? Reconciling historical and contemporary paradigms in competitive communities. American Naturalist **170**: 207–220.

Gross, K. L. and P. A. Werner. 1982. Colonizing abilities of "biennial" plant species in relation to ground cover:

Implications for their distributions in a successional sere. Ecology **63**: 921–931.

Gross, S. J. and T. D. Price. 2000. Determinants of the northern and southern range limits of a warbler. Journal of Biogeography **27**: 869–878.

Grover, J. P. 1990. Resource competition in a variable environment: Phytoplankton growing according to Mondo's model. American Naturalist **136**: 771–789.

Grover, J. P. 1997. *Resource Competition*. Chapman and Hall, London.

Gruner, D. S. and 11 others. 2008. A cross-system synthesis of consumer and nutrient resource control on producer biomass. Ecology Letters **11**: 740–755.

Guimarães, P. R., Jr., V. Rico-Gray, S. F. dos Reis and J. N. Thompson. 2006. Asymmetries in specialization in ant–plant mutualistic networks. Proceedings of the Royal Society of London B **273**: 2041–2047.

Guimarães, P. R., Jr., C. Sazima, S. F. dos Reis and I. Sazima. 2007. The nested structure of marine cleaning symbiosis: Is it like flowers and bees? Biology Letters **3**: 51–54.

Guimerà, R., D. B. Stouffer, M. Sales-Pardo, E. A. Leicht, M. E. J. Newman and L. A. N. Amaral. 2010. Origin of compartmentalization in food webs. Ecology **91**: 2941–2951.

Gunderson, L. H. 2000. Ecological resilience—in theory and application. Annual Review of Ecology and Systematics **31**: 425–439.

Gurevitch, J., L. L. Morrow, A. Wallace and J. S. Walsh. 1992. A meta-analysis of competition in field experiments. American Naturalist **140**: 539–572.

Gurevitch, J., J. A. Morrison and L. V. Hedges. 2000. The interaction between competition and predation: A meta-analysis of field experiments. American Naturalist **155**: 435–453.

Hacker, S. D. and M. D. Bertness. 1999. Experimental evidence for factors maintaining plant species diversity in a New England salt marsh. Ecology **80**: 2064–2073.

Hairston, N. G., Jr., F. E. Smith and L. B. Slobodkin. 1960. Community

structure, population control, and competition. American Naturalist **94**: 421–425.

Hairston, N. G., Jr., S. P. Ellner, M. A. Geber, T. Yoshida and J. A. Fox. 2005. Rapid evolution and the convergence of ecological and evolutionary time. Ecology Letters **8**: 1114–1127.

Hall, S. R. 2009. Stoichiometrically explicit food webs: Feedbacks between resource supply, elemental constraints, and species diversity. Annual Review of Ecology, Evolution, and Systematics **40**: 503–528.

Hall, S. R., J. B. Shurin, S. Diehl and R. M. Nisbet. 2007. Food quality, nutrient limitation of secondary production, and the strength of trophic cascades. Oikos **116**: 1128–1143.

Hammill, E., O. L. Petchey and B. R. Anholt. 2010. Predator functional response changed by inducible defenses in prey. American Naturalist **176**: 723–731.

Handa, I. T., R. Harmsen and R. L. Jefferies. 2002. Patterns of vegetation change and the recovery potential of degraded areas in a coastal marsh system of the Hudson Bay lowlands. Journal of Ecology **90**: 86–99.

Hanski, I. 1990a. Density dependence, regulation and variability in animal populations. Philosophical Transactions of the Royal Society of London B **330**: 141–150.

Hanski, I. 1990b. Dung and carrion insects. Pages 127–145 in B. Shorrocks and I. R. Swingland, editors. *Living in a Patchy Environment*. Oxford University Press, Oxford.

Hanski, I. 1994. A practical model of metapopulation dynamics. Journal of Animal Ecology **63**: 151–162.

Hanski, I. 1997. Metapopulation dynamics: From concepts and observations to predictive models. Pages 69–91 in I. Hanski and M. Gilpin, editors. *Metapopulation Biology: Ecology, Genetics, and Evolution*. Academic Press, San Diego, CA.

Hanski, I. and M. Gilpin. 1991. Metapopulation dynamics: Brief history and conceptual domain. Biological Journal of the Linnean Society **42**: 3–16.

Hanski, I. and M. Kuussaari. 1995. Butterfly metapopulation dynamics. Pages

149–172 in N. Cappuccino and P. W. Price, editors. *Population Dynamics: New Approaches and Synthesis*. Academic Press, San Diego, CA.

Hanski, I. and E. Ranta. 1983. Coexistence in a patchy environment: Three species of daphnia in rock pools. Journal of Animal Ecology **52**: 263–279.

Hanski, I. and D. Simberloff. 1997. The metapopulation approach, its history, conceptual domain and application to conservation. Pages 5–26 in I. Hanski and M. E. Gilpin, editors. *Metapopulation Biology: Ecology, Genetics, and Evolution*. Academic Press, San Diego, CA.

Hanski, I., M. Kuussaari and M. Niemenen. 1994. Metapopulation structure and migration in the butterfly *Melitaea cinxia*. Ecology **75**: 747–762.

Hardin, G. 1968. The tragedy of the commons. Science **162**: 1243–1248.

Harley, C. D. G. 2011. Climate change, keystone predation, and biodiversity loss. 2011. Science **334**: 1224–1127.

Harmon, L. J., B. Matthews, S. Des Roches, J. M. Chase, J. B. Shurin and D. Schluter. 2009. Evolutionary diversification in stickleback affects ecosystem functioning. Nature **458**: 1167–1170.

Harper, J. L. 1977. *Population Biology of Plants*. Academic Press, London.

Harper, J. L., P. H. Lovell and K. G. Moore. 1970. The shapes and sizes of seeds. Annual Review of Ecology and Systematics **1**: 327–356.

Harpole, W. S. and 10 others. 2011. Nutrient co-limitation of primary producer communities. Ecology Letters **14**: 852–862.

Harrison, S. 1991. Local extinction in a metapopulation context: An empirical evaluation. Biological Journal of the Linnean Society **42**: 73–88.

Harrison, S. 1995. Using density-manipulation experiments to study population regulation. Pages 131–147 in N. Cappuccino and P. W. Price, editors. *Population Dynamics: New Approaches and Synthesis*. Academic Press, San Diego, CA.

Harrison, S. 2008. Commentary on Stohlgren et al. (2008): The myth of plant species saturation. Ecology Letters **11**: 322–324.

Harrison, S. and N. Cappuccino. 1995. Using density-manipulation experiments to study population regulation. Pages 131–147 in N. Cappuccino and P. W. Price, editors. *Population Dynamics: New Approaches and Synthesis*. Academic Press, San Diego, CA.

Harrison, S. and H. V. Cornell. 2008. Towards a better understanding of the regional causes of local community richness. Ecology Letters **11**: 969–979.

Harrison, S. and J. B. Grace. 2007. Biogeographical affinity helps explain productivity–richness relationships at regional and local scales. American Naturalist **170**: S5–S15.

Harrison, S., H. D. Safford, J. B. Grace, J. H. Viers and K. F. Davies. 2006. Regional and local species richness in an insular environment: Serpentine plants in California. Ecological Monographs **76**: 41–56.

Hart, P. J. B. and J. D. Reynolds. 2002. *Handbook of Fish Biology and Fisheries*. Blackwell Publishing, Oxford.

Hartman, A. L., D. M. Lough, D. K. Barupal, O. Fiehn, T. Fishbein, M. Zasloff and J. A. Eisen. 2009. Human gut microbiome adopts an alternative state following small bowel transplantation. Proceedings of the National Academy of Sciences **106**: 17187–17192.

Harvell, C. D. 1990. The ecology and evolution of inducible defenses. Quarterly Review of Biology **65**: 323–340.

Hassell, M. P. 1978. *The Dynamics of Arthropod Predator–Prey Systems*. Princeton University Press, Princeton, NJ.

Hastings, A. 1980. Disturbance, coexistence, history, and competition for space. Theoretical Population Biology **18**: 363–373.

Hastings, A. 2003. Metapopulation persistence with age-dependent disturbance or succession. Science **301**: 1525–1526.

Hawkins, B. A. 1992. Parasitoid–host food webs and donor control. Oikos **65**: 159–162.

Hawkins, B. A. and 11 others. 2003. Energy, water, and broad-scale geographic patterns of species richness. Ecology **84**: 3105–3117.

Hawkins, B. A., J. A. F. Diniz-Filho and A. E. Weis. 2005. The mid-domain effect and diversity gradients: Is there anything to learn? American Naturalist **166**: E140–E143.

Hawkins, B. A., J. A. F. Diniz-Filho, C. A. Jaramillo and S. A. Soeller. 2007. Climate, niche conservatism, and the global bird diversity gradient. American Naturalist **170**: S16–S27.

Hawkins, C. P., R. H. Norris, J. N. Hogue and J. W. Feminella. 2000. Development and evaluation of predictive models for measuring the biological integrity of streams. Ecological Applications **10**: 1456–1477.

Hay, M. E. 1986. Associational plant defenses and the maintenance of species-diversity—turning competitors into accomplices. American Naturalist **128**: 617–641.

He, F., K. J. Gaston, E. F. Conner and D. S. Srivastava. 2005. The local-regional relationship: Immigration, extinction, and scale. Ecology **86**: 360–365.

He, X. and J. F. Kitchell. 1990. Direct and indirect effects of predation on a fish community: A whole-lake experiment. Transactions of the American Fisheries Society **119**: 825–835.

Hector, A. and R. Bagchi. 2007. Biodiversity and ecosystem multifunctionality. Nature **448**: 188–190.

Hector, A. and 33 others. 1999. Plant diversity and productivity experiments in European grasslands. Science **286**: 1123–1127.

Hector, A., E. Bazeley-White, M. Loreau, S. Otway and B. Schmid. 2002. Overyielding in grassland communities: Testing the sampling effect hypothesis with replicated biodiversity experiments. Ecology Letters **5**: 502–511.

Hector, A. and 7 others. 2011. BUGS in the analysis of biodiversity experiments: Species richness and composition are of similar importance for grassland productivity. PLoS One **6(3)**: e17434. doi: 17410.11371/journal.pone.0017434.

Heemsbergen, D. A., M. P. Berg, M. Loreau, J. R. van Hal, J. H. Faber and H. A. Verhoef. 2004. Biodiversity effects on soil processes explained by interspecific functional dissimilarity. Science **306**: 1019–1020.

Hegland, S. J., J.-A. Grytnes and Ø. Totland. 2009. The relative importance of positive and negative interactions for pollinator attraction in a plant community. Ecological Research **24**: 929–936.

Heil, M. and D. McKey. 2003. Protective ant-plant interactions as model systems in ecological and evolutionary research. Annual Review of Ecology, Evolution, and Systematics **34**: 425–453.

Heller, H. C. and D. M. Gates. 1971. Altitudinal zonation of chipmunks (*Eutamias*): Energy budgets. Ecology **52**: 424–433.

Helmus, M. R., K. Savage, M. W. Diebel, J. T. Maxted and A. R. Ives. 2007. Separating the determinants of phylogenetic community structure. Ecology Letters **10**: 917–925.

Hendry, A. P. and M. T. Kinnison. 1999. Perspective: The pace of modern life: Measuring rates of contemporary microevolution. Evolution **53**: 1637–1653.

Herben, T., B. Mandák, K. Bímova and Z. Münzbergova. 2004. Invasibility and species richness of a community: A neutral model and a survey of published data. Ecology **85**: 3223–3233.

Herre, E. A., N. Knowlton, U. G. Mueller and S. A. Rehner. 1999. The evolution of mutualisms: Exploring the paths between conflict and cooperation. Trends in Ecology and Evolution **14**: 49–53.

Hilborn, R. and C. J. Walters. 1992. Quantitative Fisheries Stock Assessment: Choice, Dynamics and Uncertainty. Chapman and Hall, New York.

Hillebrand, H. 2004. On the generality of the latitudinal diversity gradient. American Naturalist **163**: 192–211.

Hillebrand, H. 2005. Regressions of local on regional diversity do not reflect the importance of local interactions or saturation of local diversity. Oikos **110**: 195–198.

Hillebrand, H. and T. Blenckner. 2002. Regional and local impact of species diversity—from pattern to process. Oecologia **132**: 479–491.

Hillebrand, H. and J. B. Shurin. 2005. Biodiversity and aquatic food webs. Pages 184–197 in A. Belgrano, U.

M. Scharler, J. A. Dunne and R. E. Ulanowicz, editors. *Aquatic Food Webs: An Ecosystem Approach*. Oxford University Press, Oxford.

Hillebrand, H. and 10 others. 2007. Consumer versus resource control of producer diversity depends on ecosystem type and producer community structure. Proceedings of the National Academy of Sciences **104**: 10904–10909.

Hoegh-Guldberg, O., L. Hughes, S. McIntyre, D. B. Lindenmayer, C. Parmesan, H. P. Possingham and C. D. Thomas. 2008. Assisted colonization and rapid climate change. Science **321**: 345–346.

Hoeksema, J. D. and M. W. Schwartz. 2003. Expanding comparative-advantage biological market models: Contingency of mutualism on partner's resource requirements and acquisition trade-offs. Proceedings of the Royal Society of London B **270**: 913–919.

Holland, J. N., T. Okuyama and D. L. DeAngelis. 2006. Comment on "Asymmetric coevolutionary networks facilitate biodiversity maintenance." Science **313**: 5795.

Holling, C. S. 1959. The components of predation as revealed by a study of small mammal predation on the European pine sawfly. Canadian Entomologist **91**: 293–320.

Holling, C. S. 1973. Resilience and stability of ecological systems. Annual Review of Ecology and Systematics **4**: 1–23.

Holt, R. D. 1977. Predation, apparent competition, and structure of prey communities. Theoretical Population Biology **12**: 197–229.

Holt, R. D. 1984. Spatial heterogeneity, indirect interactions, and the coexistence of prey species. American Naturalist **124**: 377–406.

Holt, R. D. 1985. Density-independent mortality, non-linear competitive interactions, and species coexistence. Journal of Theoretical Biology **116**: 479–493.

Holt, R. D. 1996. Community modules. Pages 333–348 in M. Begon, A. Gange and V. Brown, editors. *Multitrophic Interactions*. Chapman and Hall, London.

Holt, R. D. 2006. Emergent neutrality. Trends in Ecology and Evolution **21**: 531–533.

Holt, R. D. and T. H. Keitt. 2000. Alternative causes for range limits: A metapopulation perspective. Ecology Letters **3**: 41–47.

Holt, R. D. and J. H. Lawton. 1994. The ecological consequences of shared natural enemies. Annual Review of Ecology and Systematics **25**: 495–520.

Holt, R. D., J. Grover and D. Tilman. 1994. Simple rules of interspecific dominance in systems with exploitative and apparent competition. American Naturalist **144**: 741–771.

Holt, R. D., T. H. Keitt, M. A. Lewis, B. A. Maurer and M. L. Taper. 2005. Theoretical models of species' borders: Single species approaches. Oikos **108**: 18–27.

Holyoak, M. 1993. The frequency of detection of density dependence in insect orders. Ecological Entomology **18**: 339–347.

Holyoak, M. and M. Loreau. 2006. Reconciling empirical ecology with neutral community models. Ecology **87**: 1370–1377.

Holyoak, M., M. A. Leibold, N. Mouquet, R. D. Holt and M. F. Hoopes. 2005. Metacommunities: A framework for large-scale community ecology. Pages 30–31 in M. Holyoak, M. A. Leibold and R. D. Holt, editors. *Metacommunities: Spatial Dynamics and Ecological Communities*. University of Chicago Press, Chicago, IL.

Hone, J. 1999. On rate of increase (*r*): Patterns of variation in Australian mammals and implications for wildlife management. Journal of Applied Ecology **36**: 709–718.

Hooper, D. U. and P. M. Vitousek. 1997. The effects of plant composition and diversity on ecosystem processes. Science **277**: 1302–1305.

Hooper, D. U. and P. M. Vitousek. 1998. Effects of plant composition and diversity on nutrient cycling. Ecological Monographs **68**: 121–149.

Hooper, D. U. and 14 others. 2005. Effects of biodiversity on ecosystem functioning: A consensus of current knowledge. Ecological Monographs **75**: 3–35.

Hoopes, M. F., R. D. Holt and M. Holyoak. 2005. The effects of spatial processes on two species interactions. Pages 35–67 in M. Holyoak, M. A. Leibold and R. D. Holt, editors. *Metacommunities: Spatial Dynamics and Ecological Communities*. University of Chicago Press, Chicago, IL.

Horn, H. S. and R. H. MacArthur. 1972. Competition among fugitive species in a harlequin environment. Ecology **53**: 749–752.

Hortal, J., K. A. Triantis, S. Meiri, E. Thebault and S. Sfenthourakis. 2009. Island species richness increases with habitat diversity. American Naturalist **174**: E205–E217.

Horton, T. R., T. D. Bruns and T. Parker. 1999. Ectomycorrhizal fungi associated with *Arctostaphylos* contribute to *Pseudotsuga menziesii* establishment. Canadian Journal of Botany **77**: 93–102.

Houlahan, J. E. and 17 others. 2007. Compensatory dynamics are rare in natural ecological communities. Proceedings of the National Academy of Sciences **104**: 3273–3277.

Houston, A. I., J. M. MacNamara and J. M. C. Hutchinson. 1993. General results concerning the trade-off between gaining energy and avoiding predators. Philosophical Transactions of the Royal Society of London B **341**: 375–397.

Howe, H. F. and L. C. Westley. 1986. Ecology of pollination and seed dispersal. Pages 185–215 in M. J. Crawley, editor. *Plant Ecology*. Blackwell, London.

Howeth, J. G. and M. A. Leibold. 2010. Species dispersal rates alter diversity and ecosystem stability in pond metacommunities. Ecology **91**: 2727–2741.

Hsu, S.-B., S. P. Hubbell and P. Waltman. 1977. A mathematical theory for single-nutrient competition in continuous cultures of microorganisms. SIAM Journal of Applied Mathematics **32**: 366–383.

Hsu, S.-B., S. P. Hubbell and P. Waltman. 1978. A contribution to the theory of competing predators. Ecological Monographs **48**: 337–349.

Hsu, S.-B., K.-S. Cheng and S. P. Hubbell. 1981. Exploitative competition of microorganisms for two complementary nutrients in continuous cultures. SIAM Journal of Applied Mathematics **41**: 422–444.

Hubbell, S. P. 2001. *The Unified Neutral Theory of Biodiversity and Biogeography*. Princeton University Press, Princeton, NJ.

Hubbell, S. P. 2006. Neutral theory and the evolution of ecological equivalence. Ecology **87**: 1387–1398.

Hubbell, S. P. and R. B. Foster. 1983. Diversity of canopy trees in a Neotropical forest and implications for conservation. Page 25–41 in S. Sutton, T. C. Whitmore and A. Chadwick, editors. *Tropical Rain Forest: Ecology and Management*. Blackwell, Oxford, UK.

Hubbell, S. P. and R. B. Foster. 1986. Biology, chance and history and the structure of tropical rain forest tree communities. Pages 314–329 in J. Diamond and J. M. Case, editors. *Community Ecology*. Harper and Row, New York.

Hughes, A. R., B. D. Inouye, M. T. J. Johnson, N. Underwood and M. Vellend. 2008. Ecological consequences of genetic diversity. Ecology Letters **11**: 609–623.

Hughes, T. P., N. A. J. Graham, J. B. C. Jackson, P. J. Mumby and R. S. Steneck. 2010. Rising to the challenge of sustaining coral reef resilience. Trends in Ecology and Evolution **25**: 633–642.

Huisman, J. and F. J. Weissing. 1999. Biodiversity of plankton by species oscillations and chaos. Nature **402**: 407–410.

Humboldt, A. von. 1808. Ansichten der Natur mit wissenschftlichen Erlauterungen. J. G. Cotta, Tubingen, Germany.

Hunter, M. L. Jr. 2007. Climate change and moving species: Furthering the debate on assisted colonization. Conservation Biology **21**: 1356–1358.

Hurtt, G. C. and S. W. Pacala. 1995. The consequences of recruitment limitation: Reconciling chance, history, and competitive differences between plants. Journal of Theoretical Biology **176**: 1–12.

Huston, M. A. 1979. A general hypothesis of species diversity. American Naturalist **113**: 81–101.

Huston, M. A. 1997. Hidden treatments in ecological experiments: Re-evaluating the ecosystem function of biodiversity. Oecologia **108**: 449–460.

Huston, M. A. and D. L. DeAngelis. 1994. Competition and coexistence: The effects of resource transport and supply rates. American Naturalist **144**: 954–977.

Huston, M. A. and 11 others. 2000. No consistent effect of plant diversity on productivity. Science **289**: 1255.

Hutchings, J. A. and J. D. Reynolds. 2004. Marine fish population collapses: Consequences for recovery and extinction risk. BioScience **54**: 297–309.

Hutchinson, G. E. 1951. Copepodology for the ornithologist. Ecology **32**: 571–577.

Hutchinson, G. E. 1957. Concluding remarks. Cold Spring Harbor Symposium on Quantitative Biology **22**: 415–427.

Hutchinson, G. E. 1959. Homage to Santa Rosalia, or why are there so many kinds of animals? American Naturalist **93**: 145–159.

Hutchinson, G. E. 1961. The paradox of the plankton. American Naturalist **95**: 137–145.

Hutchinson, G. E. 1965. *The Ecological Theater and the Evolutionary Play*. Yale University Press, New Haven, CT.

Hutchinson, G. E. 1978. *An Introduction to Population Ecology*. Yale University Press, New Haven.

Hutto, R. L., J. R. McAuliffe and L. Hogan. 1986. Distributional associates of the saguaro (*Carnegiea gigantea*). Southwest Naturalist **31**: 469–476.

Huxham, M., D. Raffaelli and A. Pike. 1995. Parasites and food web patterns. Journal of Animal Ecology **64**: 168–176.

Ings, T. C. and 16 others. 2009. Ecological networks—beyond food webs. Journal of Animal Ecology **78**: 253–269.

Inouye, B. D. 2001. Response surface experimental designs for investigating interspecific competition. Ecology **82**: 2696–2706.

Isbell, F. and 13 others. 2011. High plant diversity is needed to maintain ecosystem services. Nature **477**: 199–202.

Ives, A. R. and S. R. Carpenter. 2007. Stability and diversity of ecosystems. Science **317**: 58–62.

Ives, A. R. and A. P. Dobson. 1987. Antipredator behavior and the population dynamics of simple predator–prey systems. American Naturalist **130**: 1431–1442.

Ives, A. R., B. J. Cardinale and W. E. Snyder. 2005. A synthesis of subdisciplines: Predator-prey interactions, and biodiversity and ecosystem functioning. Ecology Letters **8**: 102–116.

Jablonski, D. 1993. The tropics as a source of evolutionary novelty through geological time. Nature **364**: 142–144.

Jablonski, D. 2008. Biotic interactions and macroevolution: Extensions and mismatches across scales and levels. Evolution **62**: 715–739.

Jablonski, D., K. Roy and J. W. Valentine. 2006. Out of the tropics: Evolutionary dynamics of the latitudinal diversity gradient. Science **314**: 102–106.

Jabot, F. and J. Chave. 2009. Inferring the parameters of the neutral theory of biodiversity using phylogenetic information and implications for tropical forests. Ecology Letters **12**: 239–248.

Jackson, S. T. and D. F. Sax. 2010. Balancing biodiversity in a changing environment: Extinction debt, immigration credit and species turnover. Trends in Ecology and Evolution **25**: 153–160.

Jaeger, R. G. 1970. Potential extinction through competition between two species of terrestrial salamanders. Evolution **24**: 632–642.

Jakobsson, A. and O. Eriksson. 2003. Trade-offs between dispersal and competitive ability: A comparative study of wind-dispersed Asteraceae forbs. Evolutionary Ecology **17**: 233–246.

Jansson, R. and T. J. Davies. 2008. Global variation in diversification rates of flowering plants: Energy vs. climate change. Ecology Letters **11**: 173gy v.

Janzen, D. H. 1966. Coevolution of mutualism between ants and acacias in Central America. Evolution **20**: 249–275.

Janzen, D. H. 1967. Why mountain passes are higher in the tropics. American Naturalist **101**: 233–249.

Janzen, D. H. 1970. Herbivores and the number of tree species in tropical forests. American Naturalist **104**: 501–528.

Jaramillo, C., J. J. Rueda and G. Mora. 2006. Cenozoic plant diversity in the Neotropics. Science **311**: 1893–1896.

Jenkins, D. G. and A. L. J. Buikema. 1998. Do similar communities develop in similar sites? A test with zooplankton structure and function. Ecological Monographs **68**: 421–443.

Jeschke, J. M., M. Kopp and R. Tollrian. 2004. Consumer-food systems: Why type I functional responses are exclusive to filter feeders. Biological Reviews **79**: 337–349.

Jetz, W. and C. Rahbek. 2001. Geometric constraints explain much of the species richness pattern in African birds. Proceedings of the National Academy of Sciences **98**: 5661–5666.

Jiang, L. and Z. Pu. 2009. Different effects of species diversity on temporal stability in single-trophic and multitrophic communities. American Naturalist **174**: 651–659.

Jiang, L., H. Joshi and S. N. Patel. 2009. Predation alters relationships between biodiversity and temporal stability. American Naturalist **173**: 389–399.

Johnson, D. M. and P. H. Crowley. 1980. Habitat and seasonal segregation among coexisting odonate larvae. Odonatologica **9**: 297–308.

Johnson, M. T. J., M. Vellend and J. R. Stinchcombe. 2009. Evolution in plant populations as a driver of ecological changes in arthropod communities. Philosophical Transactions of the Royal Society of London B **364**: 1593–1605.

Johnson, N. C. 2010. Resource stoichiometry elucidates the structure and function of arbuscular mycorrhizas across scales. New Phytologist **185**: 631–647.

Johnson, N. C., J. H. Graham and F. A. Smith. 1997. Functioning of mycorrhizal associations along the mutualism-parasitism continuum. New Phytologist **135**: 575–585.

Jones, C. G., J. H. Lawton and M. Shachak. 1994. Organisms as ecosystem engineers. Oikos **69**: 373–386.

Jones, C. G., J. H. Lawton and M. Shachak. 1997. Ecosystem engineering by organisms: Why semantics matters. Trends in Ecology and Evolution **12**: 275.

Jones, L. E. and S. P. Ellner. 2007. Effects of rapid prey evolution on predator–prey cycles. Journal of Mathematical Biology **55**: 541–573.

Jones, L. E., L. Becks, S. P. Ellner, N. G. Hairston, Jr., T. Yoshida and G. F. Fussmann. 2009. Rapid contemporary evolution and clonal food web dynamics. Philosophical Transactions of the Royal Society of London B **364**: 1579–1591.

Jordano, P. 2000. Fruits and frugivory. Pages 125–166 in M. Fenner, editor. *Seeds: The Ecology of Regeneration in Natural Plant Communities*. Commonwealth Agricultural Bureau International, Wallingford, UK.

Jump, A. S. and J. Peñuelas. 2005. Running to stand still: Adaptation and the response of plants to rapid climate change. Ecology Letters **81**: 1010–1020.

Kalmar, A. and D. J. Currie. 2006. A global model of island biogeography. Global Ecology and Biogeography **15**: 72–81.

Kalmar, A. and D. J. Currie. 2007. A unified model of avian species richness on islands and continents. Ecology **88**: 1309–1321.

Kaufman, D. M. 1998. The structure of mammalian faunas in the New World: From continents to communities. Master's thesis, University of New Mexico, Albuquerque, NM.

Kaunzinger, C. M. K. and P. J. Morin. 1998. Productivity controls food-chain properties in microbial communities. Nature **395**: 495–497.

Keddy, P. A. 1989. *Competition*. Chapman and Hall, London.

Kéfi, S. 2008. Reading the signs: Spatial vegetation patterns, arid ecosystems, and desertification. Ph.D. dissertation, Utrecht University, The Netherlands.

Kéfi, S., M. Rietkerk, C. L. Alados, Y. Pueyo, V. P. Papanastasis, A. Elaich and P. C. de Ruiter. 2007a. Spatial vegetation patterns and imminent desertification in Mediterranean arid ecosystems. Nature **449**: 213–217.

Kéfi, S., M. Rietkerk, M. van Baalen and M. Loreau. 2007b. Local facilitation, bistability and transitions in arid ecosystems. Theoretical Population Biology **71**: 367–379.

Kermack, W. O. and A. G. McKendrick. 1927. A contribution to the mathematical theory of epidemics. Proceedings of the Royal Society of London A **115**: 700–721.

Kerr, J. T., M. Perring and D. J. Currie. 2006. The missing Madagascan mid-domain effect. Ecology Letters **9**: 149–159.

Keymer, J. E., P. A. Marquet, J. X. Velasco-Hernandez and S. A. Levin. 2000. Extinction thresholds and meta-population persistence in dynamic landscapes. American Naturalist **156**: 478–494.

Kiessling, W., C. Simpson and M. Foote. 2010. Reefs as cradles of evolution and sources of biodiversity in the Phanerozoic. Science **327**: 196–198.

Kilham, S. S. 1986. Dynamics of Lake Michigan natural phytoplankton communities in continuous cultures along a Si:P loading gradient. Canadian Journal of Fisheries and Aquatic Sciences **43**: 351–360.

Kimura, M. 1968. Evolutionary rate at the molecular level. Nature **217**: 624–626.

Kimura, M. and T. Ohta. 1971. *Theoretical Aspects of Population Genetics*. Princeton University Press, Princeton, NJ.

Kingsland, S. 1985. *Modeling Nature*. University of Chicago Press, Chicago, IL.

Kingsolver, J. G. and S. E. Diamond. 2011. Phenotypic selection in natural populations: What limits directional selection? American Naturalist **177**: 346–357.

Kingsolver, J. G. and 8 others. 2001. The strength of phenotypic selection in natural populations. American Naturalist **157**: 245–261.

Kinzig, A. P., R. Ryan, M. Etienne, H. Allison, T. Elmqvist and B. H. Walker. 2006. Resilience and regime shifts:

Assessing cascading effects. Ecology and Society **11**: 20 (online).

Kislaliogu, M. and R. N. Gibson. 1976. Prey handling time and its importance in food selection by the 15-spined stickleback *Spinachia spinachia*. Journal of Experimental Marine Biology and Ecology **25**: 151–158.

Kleidon, A. and H. A. Mooney. 2000. A global distribution of biodiversity inferred from climatic constraints: Results for a process-based modeling study. Global Change Biology **6**: 507–523.

Kneitel, J. M. and J. M. Chase. 2004. Trade-offs in community ecology: Linking spatial scales and species coexistence. Ecology Letters **7**: 69–80.

Kneitel, J. M. and T. E. Miller. 2003. Dispersal rates affect species composition in metacommunities of *Sarracenia purpurea* inquilines. American Naturalist **162**: 165–171.

Koch, A. L. 1974. Competitive coexistence of two predators utilizing the same prey under constant environmental conditions. Journal of Theoretical Biology **44**: 387–395.

Kondoh, M. and K. Ninomiya. 2009. Food-chain length and adaptive foraging. Proceedings of the Royal Society of London B **276**: 3113–3121.

Kondoh, M., S. Kato and Y. Sakato. 2010. Food webs are built up with nested subwebs. Ecology **91**: 3123–3130.

Kraft, N. J. B. and D. D. Ackerly. 2010. Functional trait and phylogenetic tests of community assembly across spatial scales in an Amazonian forest. Ecological Monographs **80**: 401–422.

Kraft, N. J. B., W. K. Cornwell, C. O. Webb and D. D. Ackerly. 2007. Trait evolution, community assembly, and the phylogenetic structure of ecological communities. American Naturalist **170**: 271–283.

Kramer, A. M., O. Sarnelle and R. A. Knapp. 2008. Allee effect limits colonization success of sexually reproducing zooplankton. Ecology **89**: 2760–2769.

Kramer, A. M., B. Dennis, A. M. Liebhold and J. M. Drake. 2009. The evidence for Allee effects. Population Ecology **51**: 341–354.

Krause, A. E., K. A. Frank, D. M. Mason, R. E. Ulanowicz and W. W. Taylor. 2003. Compartments revealed in food-web structure. Nature **426**: 282–285.

Krebs, C. J. 1995. Two paradigms of population regulation. Wildlife Research **22**: 1–10.

Krebs, J. R. 1978. Optimal foraging: Decision rules for predators. Pages 22–63 in J. R. Krebs and N. B. Davies, editors. *Behavioural Ecology*. Blackwell Scientific, Oxford.

Krebs, J. R., J. T. Erichsen, M. I. Webber and E. L. Charnov. 1977. Optimal prey-selection by the great tit (*Parus major*). Animal Behaviour **25**: 30–38.

Kreft, H. and W. Jetz. 2010. A framework for delineating biogeographical regions based on species distributions. Journal of Biogeography **37**: 2029–2053.

Křivan, V. and J. Eisner. 2006. The effect of the Holling type II functional response on apparent competition. Theoretical Population Biology **70**: 421–430.

Krug, A. Z., D. Jablonski and J. W. Valentine. 2009. Signature of the End-Cretaceous mass extinction in the modern biota. Science **323**: 767–771.

Kuang, J. J. and P. L. Chesson. 2008. Predation-competition interactions for seasonally recruiting species. American Naturalist **171**: E119–E133.

Kuang, J. J. and P. L. Chesson. 2010. Interacting coexistence mechanisms in annual plant communities: Frequency-dependent predation and the storage effect. Theoretical Population Biology **77**: 56–70.

Kuris, A. M. and 17 others. 2008. Ecosystem energetic implications of parasite and free-living biomass in three estuaries. Nature **454**: 515–518.

Lack, D. 1947. *Darwin's Finches*. Cambridge University Press, Cambridge.

Lafferty, K. D., A. P. Dobson and A. M. Kuris. 2006. Parasites dominate food web links. Proceedings of the National Academy of Sciences **103**: 11211–11216.

Lafferty, K. D. and 17 others. 2008. Parasites in food webs: The ultimate missing links. Ecology Letters **11**: 533–546.

Lampert, W. 1987. Vertical migration of freshwater zooplankton: Indirect effects of vertebrate predators on algal communities. Pages 291–299 in W. C. Kerfoot and A. Sih, editors. *Predation: Direct and Indirect Impacts on Aquatic Communities*. University Press of New England, Hanover, NH.

Lande, R. 1987. Extinction thresholds in demographic models of territorial populations. American Naturalist **130**: 624–635.

Lande, R. 1988a. Demographic models of the northern spotted owl. Oecologia **75**: 601–607.

Lande, R. 1988b. Genetics and demography in biological conservation. Science **241**: 1455–1460.

Lande, R., S. Engen and B. E. Saether. 2003. *Stochastic Population Dynamics in Ecology and Conservation*. Oxford University Press, Oxford.

Larkin, P. 1977. An epitaph for the concept of maximum sustained yield. Transactions of the American Fisheries Society **106**: 1–11.

Laska, M. S. and J. T. Wootton. 1998. Theoretical concepts and empirical approaches to measuring interaction strength. Ecology **79**: 461–476.

Latham, R. E. and R. E. Ricklefs. 1993. Continental comparisons of temperate-zone tree species diversity. Pages 294–314 in R. E. Ricklefs and D. Schluter, editors. *Species Diversity in Ecological Communities: Historical and Geographical Perspectives*. University of Chicago Press, Chicago, IL.

Lavergne, S., N. Mouquet, W. Thuiller and O. Ronce. 2010. Biodiversity and climate change: Integrating evolutionary and ecological responses of species and communities. Annual Review of Ecology, Evolution, and Systematics **41**: 321–350.

Laverty, T. M. 1992. Plant interactions for pollinator visits: A test of the magnet species effect. Oecologia **89**: 502–508.

Law, R. and R. D. Morton. 1996. Permanence and the assembly of ecological communities. Ecology **77**: 762–775.

Lawrence, P. A. 2011. "The heart of research is sick." Interview by J. Garwood. Lab Times **2-2011**: 24–31.

Lawton, J. H. 1999. Are there general laws in ecology? Oikos **84**: 177–192.

Lawton, J. H., S. Nee, A. J. Letcher and P. H. Harvey. 1994. Animal distributions; patterns and processes. Pages 41–58 in P. J. Edwards, R. M. May and N. R. Webb, editors. *Large-Scale Ecology and Conservation Biology*. Blackwell Scientific, Oxford.

Lehman, C. and D. Tilman. 2000. Biodiversity, stability, and productivity in competitive communities. American Naturalist **156**: 534–552.

Leibold, M. A. 1989. Resource edibility and the effects of predators and productivity on the outcome of trophic interactions. American Naturalist **134**: 922–949.

Leibold, M. A. 1996. A graphical model of keystone predators in food webs: Trophic regulation of abundance, incidence, and diversity patterns in communities. American Naturalist **147**: 784–812.

Leibold, M. A. 1997. Do nutrient-competition models predict nutrient availabilities in limnetic ecosystems? Oecologia **110**: 132–142.

Leibold, M. A. and M. A. McPeek. 2006. Coexistence of the niche and neutral perspectives in community ecology. Ecology **87**: 1399–1410.

Leibold, M. A. and J. Norberg. 2004. Biodiversity in metacommunities: Plankton as complex adaptive systems? Limnology and Oceanography **49**: 1278–1289.

Leibold, M. A. and J. T. Wootton. 2001. Introduction to *Animal Ecology*, by Charles Elton. University of Chicago Press, Chicago, IL.

Leibold, M. A., J. M. Chase, J. B. Shurin and A. L. Downing. 1997. Species turnover and the regulation of trophic structure. Annual Review of Ecology and Systematics **28**: 467–494.

Leibold, M. A. and 11 others. 2004. The metacommunity concept: A framework for multi-scale community ecology. Ecology Letters **7**: 601–613.

Leibold, M. A., R. D. Holt and M. Holyoak. 2005. Adaptive and coadaptive dynamics in metacommunities. Pages 439–464 in M. Holyoak, M. A. Leibold and R. D. Holt, editors. *Metacommunities: Spatial Dynamics and*

Ecological Communities. University of Chicago Press, Chicago, IL.

Leibold, M. A., E. P. Economo and P. Peres-Neto. 2010. Metacommunity phylogenetics: Separating the roles of environmental filters and historical biogeography. Ecology Letters **13**: 1290–1299.

Leigh, E. G. 1981. The average lifetime of a population in a varying environment. Journal of Theoretical Biology **90**: 213–239.

Leigh, E. G. 1999. *Tropical Forest Ecology*. Oxford University Press, Oxford.

Leigh, E. G. 2007. Neutral theory: A historical perspective. Journal of Evolutionary Biology **20**: 2075–2091.

Leigh, E. G. and T. E. Rowell. 1995. The evolution of mutualism and other forms of harmony at various levels of biological organization. Ecologie **26**: 131–158.

Leigh, E. G., P. Davidar, C. W. Dick, J. P. Puyravaud, J. Terborgh, H. ter Steege and S. J. Wright. 2004. Why do some tropical forests have so many species of trees? Biotropica **36**: 447–473.

León, J. A. and D. B. Tumpson. 1975. Competition between two species for two complementary or substitutable resources. Journal of Theoretical Biology **50**: 185–201.

Leslie, H. M. 2005. Positive intraspecific effects trump negative effects in high-density barnacle aggregations. Ecology **86**: 2716–2725.

Levin, S. A. 1974. Dispersion and population interaction. American Naturalist **108**: 207–228.

Levin, S. A. 1998. Ecosystems and the biosphere as complex adaptive systems. Ecosystems **1**: 431–436.

Levine, J. M. 2000. Species diversity and biological invasions: Relating local process to community pattern. Science **288**: 852–854.

Levine, J. M. and J. HilleRisLambers. 2009. The importance of niches for the maintenance of species diversity. Nature **461**: 254–257.

Levine, J. M. and M. Rees. 2002. Coexistence and relative abundance in annual plant assemblages: The roles of competition and colonization. American Naturalist **160**: 452–467.

Levine, J. M., P. B. Adler and S. G. Yelenik. 2004. A meta-analysis of biotic resistance to exotic plant invasions. Ecology Letters **7**: 975–989.

Levins, R. 1968. *Evolution in Changing Environments*. Princeton University Press, Princeton, NJ.

Levins, R. 1969. Some demographic and genetic consequences of environmental heterogeneity for biological control. Bulletin of the Entomological Society of America **15**: 237–240.

Levins, R. 1970. Extinction. Pages 75–107 in M. Gerstenhaber, editor. *Some Mathematical Problems in Biology*. American Mathematical Society, Providence, RI.

Levins, R. and D. Culver. 1971. Regional coexistence of species and competition between rare species. Proceedings of the National Academy of Sciences **68**: 1246–1248.

Levitan, C. 1987. Formal stability analysis of a planktonic freshwater community. Pages 71–100 in W. C. Kerfoot and A. Sih, editors. *Predation: Direct and Indirect Impacts on Aquatic Communities*. University Press of New England, Hanover, NH.

Lewin, R. 1989. Biologists disagree over bold signature of nature. Science **244**: 527–528.

Lewinsohn, T. M., P. I. Prado, P. Jordano, J. Bascompte and J. M. Olesen. 2006. Structure in plant-animal interaction assemblages. Oikos **113**: 174–184.

Lewontin, R. C. 1969. The meaning of stability. Brookhaven Symposium on Biology **22**: 13–23.

Lewontin, R. C. 2000. *The Triple Helix, Gene, Organisms, and Environment*. Harvard University Press, Cambridge, MA.

Lima, S. 1998a. Nonlethal effects in the ecology of predator–prey interactions. BioScience **48**: 25–34.

Lima, S. 1998b. Stress and decision making under the risk of predation: Recent developments from behavioral, reproductive, and ecological perspectives. Advances in the Study of Behavior **27**: 215–290.

Lindeman, R. L. 1942. The trophic-dynamic aspect of ecology. Ecology **23**: 399–417.

Loehle, C. 1998. Height growth rate tradeoffs determine northern and southern range limits for trees. Journal of Biogeography **25**: 735–742.

Loeuille, N. and M. Loreau. 2005. Evolutionary emergence of size-structured food webs. Proceedings of the National Academy of Sciences **102**: 5761–5766.

Long, Z. T., J. F. Bruno and J. E. Duffy. 2007. Biodiversity mediates productivity through different mechanisms at adjacent trophic levels. Ecology **88**: 2821–2829.

Lonsdale, W. M. 1999. Global patterns of plant invasions and the concept of invasibility. Ecology **80**: 1522–1536.

Loreau, M. 2000. Biodiversity and ecosystem functioning: Recent theoretical advances. Oikos **91**: 3–17.

Loreau, M. 2010a. *From Populations to Ecosystems*. Princeton University Press, Princeton, NJ.

Loreau, M. 2010b. Linking biodiversity and ecosystems: Towards a unifying ecological theory. Philosophical Transactions of the Royal Society of London B **365**: 49–60.

Loreau, M. and C. de Mazancourt. 2008. Species synchrony and its drivers: Neutral and nonneutral community dynamics in fluctuating environments. American Naturalist **172**: E48–E66.

Loreau, M. and A. Hector. 2001. Partitioning selection and complementarity in biodiversity experiments. Nature **412**: 72–76.

Loreau, M. and 11 others. 2001. Biodiversity and ecosystem functioning: Current knowledge and future challenges. Science **294**: 804–808.

Loreau, M., N. Mouquet and A. Gonzalez. 2003. Biodiversity as spatial insurance in heterogeneous landscapes. Proceedings of the National Academy of Sciences **100**: 12765–12770.

Lortie, C. J. and R. M. Callaway. 2006. Reanalysis of meta-analysis: Support for the stress-gradient hypothesis. Journal of Ecology **94**: 7–16.

Losos, J. B. 2008a. Phylogenetic niche conservatism, phylogenetic signal and the relationship between phylogenetic relatedness and ecological

similarity between species. Ecology Letters **11**: 995–1007.

Losos, J. B. 2008b. Rejoinder to Wiens (2008): Phylogenetic niche conservatism, its occurrence and importance. Ecology Letters **11**: 1005–1007.

Losos, J. B. 2009. Lizards in an Evolutionary Tree: Ecology and Adaptive Radiation of Anoles. University of California Press, Berkeley, CA.

Losos, J. B. 2010. Adaptive radiation, ecological opportunity, and evolutionary determinism. American Naturalist **175**: 623–639.

Losos, J. B. 2011. Seeing the forest for the trees: The limitations of phylogenetics in comparative biology. American Naturalist **177**: 709–727.

Losos, J. B. and D. B. Miles. 2002. Testing the hypothesis that a clade has adaptively radiated: Iguanid lizard clades as a case study. American Naturalist **160**: 147–157.

Losos, J. B. and R. E. Ricklefs. 2009. Adaptation and diversification on islands. Nature **457**: 830–836.

Losos, J. B., T. R. Jackman, A. Larson, K. de Queiroz and L. Rodriguez-Schettino. 1998. Contingency and determinism in replicated adaptive radiations of island lizards. Science **279**: 2115–2118.

Loss, S. R., L. A. Terwilliger and A. C. Peterson. 2010. Assisted colonization: Integrating conservation strategies in the face of climate change. Biological Conservation **144**: 92–100.

Lotka, A. J. 1925. *Elements of Physical Biology*. Williams and Wilkins, Baltimore, MD.

Louette, G. and L. De Meester. 2005. High dispersal capacity of cladoceran zooplankton in newly founded communities. Ecology **86**: 353–359.

Louette, G. and L. De Meester. 2007. Predation and priority effects in experimental zooplankton communities. Oikos **116**: 419–426.

Louette, G., L. De Meester and S. Declerck. 2008. Assembly of zooplankton communities in newly created ponds. Freshwater Biology **53**: 2309–2320.

Lovejoy, T. E. 1986. Species leave the ark one by one. Pages 13–27 in B. G. Norton, editor. *The Preservations of Species: The Value of Biological*

Diversity. Princeton University Press, Princeton, NJ.

Lubchenco, J. 1978. Plant species diversity in a marine intertidal community: Importance of herbivore food preferences and algal competitive abilities. American Naturalist **112**: 23–39.

Ludwig, D. and L. Rowe. 1990. Life history strategies for energy gain and predator avoidance under time constraints. American Naturalist **135**: 686–707.

Ludwig, D., R. Hilborn and C. Walters. 1993. Uncertainty, resource exploitation, and conservation: Lessons from history. Science **260**: 17 and 36.

Ma, B. O., P. A. Abrams and C. E. Brassil. 2003. Dynamic versus instantaneous models of diet choice. American Naturalist **162**: 668–684.

MacArthur, R. H. 1955. Fluctuations of animal populations and a measure of community stability. Ecology **36**: 533–536.

MacArthur, R. H. 1968. The theory of the niche. Pages 159–176 in R. C. Lewontin, editor. *Population Biology and Evolution*. Syracuse University Press, Syracuse, NY.

MacArthur, R. H. 1969. Species packing, and what interspecies competition minimizes. Proceedings of the National Academy of Sciences **64**: 1369–1371.

MacArthur, R. H. 1970. Species packing and competitive equilibrium for many species. Theoretical Population Biology **1**: 1–11.

MacArthur, R. H. 1972. *Geographical Ecology*. Harper and Row, New York.

MacArthur, R. H. and R. Levins. 1967. The limiting similarity, convergence, and divergence of coexisting species. American Naturalist **101**: 377–385.

MacArthur, R. H. and E. R. Pianka. 1966. On optimal use of a patchy environment. American Naturalist **100**: 603–609.

MacArthur, R. H. and E. O. Wilson. 1967. *The Theory of Island Biogeography*. Princeton University Press, Princeton, NJ.

MacArthur, R. H., J. Diamond and J. M. Karr. 1972. Density compensation in island faunas. Ecology **53**: 330–342.

Mace, G. M., H. Masundire and J. E. M. Baillie. 2005. Biodiversity. Pages 77–122 in B. Scholes and R. Hassan, editors. *Ecosystems and Human Well-Being: Current State and Trends*. Island Press, Washington, DC.

Mack, M. C. and C. M. D'Antonio. 1998. Impacts of biological invasions on disturbance regimes. Trends in Ecology and Evolution **13**: 195–198.

Mack, M. C., C. M. D'Antonio and R. E. Ley. 2001. Alteration of ecosystem nitrogen dynamics by exotic plants: A case study of C_4 grasses in Hawaii. Ecological Applications **11**: 1323–1335.

Mackay, R. L. and D. J. Currie. 2001. The diversity–disturbance relationship: Is it generally strong and peaked? Ecology **82**: 3479–3492.

MacLean, W. P. and R. D. Holt. 1979. Distributional patterns in St. Croix, *Sphaerodactylus* lizards: Taxon cycle in action. Biotropica **11**: 189–195.

Madenjian, C. P., D. W. Schloesser and K. A. Krieger. 1998. Population models of burrowing mayfly recolonization in Western Lake Erie. Ecological Applications **8**: 1206–1212.

Maestre, F. T. and J. Cortina. 2004. Do positive interactions increase with abiotic stress? A test from a semi-arid steppe. Proceedings of the Royal Society of London B Supplement **271**: S331–S333.

Maestre, F. T., F. Valladares and J. F. Reynolds. 2005. Is the change of plant-plant interactions with abiotic stress predictable? A meta-analysis of field results in arid environments. Journal of Ecology **93**: 748–757.

Maestre, F. T., F. Valladares and J. F. Reynolds. 2006. The stress-gradient hypothesis does not fit all relationships between plant-plant interactions and abiotic stress: Further insights from arid environments. Journal of Ecology **94**: 17–22.

Maestre, F. T., R. M. Callaway, F. Valladares and C. J. Lortie. 2009. Refining the stress-gradient hypothesis for competition and facilitation in plant communities. Journal of Ecology **97**: 199 205.

Maguire, B. 1973. Niche response to structure and the analytical potentials of its relationship to habitat. American Naturalist **107**: 213–246.

Magurran, A. E. 2004. *Measuring Biological Diversity*. Blackwell Publishing, Oxford.

Magurran, A. E. 2005. Species abundance distributions: Pattern or process? Functional Ecology **19**: 177–181.

Magurran, A. E. and B. J. McGill. 2011. *Biological Diversity: Frontiers in Measurement and Assessment*. Oxford University Press, Oxford.

Mahler, D. L., L. J. Revell, R. E. Glor and J. B. Losos. 2010. Ecological opportunity and the rate of morphological evolution in the diversification of Greater Antillean anoles. Evolution **64**: 2731–2745.

Malthus, T. R. 1798. *An Essay on the Principle of Population*. London: J. Johnson.

Mangan, S. A., S. A. Schnitzer, E. A. Heere, K. M. L. Mack. M. C. Valencia, E. I. Sanchez and J. D. Bever. 2010. Negative plant–soil feedback predicts tree species relative abundance in a tropical forest. Nature **466**: 752–755.

Manly, B. 1974. A model for certain types of selection experiments. Biometrics **30**: 281–294.

Manly, B. 1985. The Statistics of Natural Selection on Animal Populations. Chapman and Hall, London.

Mäntylä, E., T. Klemola and T. Laaksonen. 2011. Birds help plants: A meta-analysis of top-down trophic cascades caused by avian predators. Oecologia **165**: 143–151.

Marcogliese, D. J. and D. K. Cone. 1997. Food webs: A plea for parasites. Trends in Ecology and Evolution **12**: 320–325.

Maron, J. and M. Marler. 2007. Native plant diversity resists invasion at both low and high resource levels. Ecology **88**: 2651–2661.

Maron, J., M. Marler, J. N. Klironomos and C. C. Cleveland. 2011. Soil fungal pathogens and the relationship between plant diversity and productivity. Ecology Letters **14**: 36–41.

Marquard, E. and 8 others. 2009. Plant species richness and functional composition drive overyielding in a six-year grassland experiment. Ecology **90**: 3290–3302.

Marshall, S. D., S. E. Walker and A. L. Rypstra. 2000. A test for a differential colonization and competitive ability in two generalist predators. Ecology **81**: 3341–3349.

Martin, K. L. and L. K. Kirkman. 2009. Management of ecological thresholds to re-establish disturbance-maintained herbaceous wetlands of the south-eastern USA. Journal of Applied Ecology **46**: 906–914.

Martinez, N. D. 1992. Constant connectance in community food webs. American Naturalist **139**: 1208–1218.

Matsuda, H. and P. A. Abrams. 2006. Maximal yields from multispecies fisheries systems: Rules for systems with multiple trophic levels. Ecological Applications **16**: 225–237.

May, R. M. 1971. Stability in multispecies community models. Mathematical Biosciences **12**: 59–79.

May, R. M. 1972. Will a large complex system be stable? Nature **238**: 413–414.

May, R. M. 1973a. *Stability and Complexity in Model Ecosystems*. Princeton University Press, Princeton, NJ.

May, R. M. 1973b. Stability in randomly fluctuating versus deterministic environments. American Naturalist **107**: 621–650.

May, R. M. 1975. Patterns of species abundance and diversity. Pages 81–120 in M. L. Cody and J. Diamond, editors. *Ecology and Evolution of Communities*. Belknap Press of Harvard University, Cambridge, MA.

May, R. M. 1977a. Mathematical models and ecology: Past and future. Pages 189–202 in C. E. Goulden, editor. *Changing Scenes in the Natural Sciences*. Special Publication no. 12. Academy of Natural Sciences, Philadelphia, PA.

May, R. M. 1977b. Thresholds and breakpoints in ecosystems with a multiplicity of stable states. Nature **269**: 471–477.

May, R. M. 1981. Models for two interacting populations. Pages 78–104 in R. M. May, editor. *Theoretical Ecology: Principles and Applications*. Sinauer Associates, Sunderland, MA.

May, R. M. 2010. Foreword. In J. B. Losos and R. E. Ricklefs, editors. *The Theory of Island Biogeography Revisited*. Princeton University Press, Princeton, NJ.

May, R. M. and R. H. MacArthur. 1972. Niche overlap as a function of environmental variability. Proceedings of the National Academy of Sciences **69**: 1109–1113.

May, R. M. and J. Seger. 1986. Ideas in ecology. American Scientist **74**: 256–267.

May, R. M., J. R. Beddington, C. W. Clark, S. J. Holt and R. M. Laws. 1979. Management of multispecies fisheries. Science **205**: 267–277.

Mayfield, M. M. and J. M. Levine. 2010. Opposing effects of competitive exclusion on the phylogenetic structure of communities. Ecology Letters **13**: 1085–1093.

Maynard Smith, J. 1974. *Models in Ecology*. Cambridge University Press, Cambridge.

McCain, C. M. 2007. Area and mammalian elevational diversity. Ecology **88**: 76–86.

McCann, K. S. 2000. The diversity–stability debate. Nature **405**: 228–233.

McCann, K. S., A. Hastings and G. R. Huxel. 1998. Weak trophic interactions and the balance of nature. Nature **395**: 794–798.

McCauley, E. and F. Briand. 1979. Zooplankton grazing and phytoplankton species richness: Field tests of the predation hypothesis. Limnology and Oceanography **24**: 243–252.

McCauley, E., W. W. Murdoch and S. Watson. 1988. Simple models and variation in plankton densities among lakes. American Naturalist **132**: 383–403.

McCook, L. J. 1999. Macroalgae, nutrients and phase shifts on coral reefs: Scientific issues and management consequences for the Great Barrier Reef. Coral Reefs **18**: 357–367.

McGill, B. J. 2003. A test of the unified neutral theory of biodiversity. Nature **422**: 881–885.

McGill, B. J. 2005. A mechanistic model of a mutualism and its ecological and evolutionary dynamics. Ecological Modelling **187**: 413–425.

McGill, B. J. 2006. A renaissance in the study of abundance. Science **314**: 770–772.

McGill, B. J. 2011. Linking biodiversity patterns to autocorrelated random sampling. American Journal of Botany **98**: 481–502.

McGill, B. J. and G. G. Mittelbach. 2006. An allometric vision and motion model to predict prey encounter rates. Evolutionary Ecology Research **8**: 1–11.

McGill, B. J. and J. C. Nekola. 2010. Mechanisms in macroecology: AWOL or purloined letter? Towards a pragmatic view of mechanism. Oikos **119**: 591–603.

McGill, B. J., B. A. Maurer and M. D. Weiser. 2006. Empirical evaluation of neutral theory. Ecology **87**: 1411–1423.

McGrady-Steed, J., P. M. Harris and P. J. Morin. 1997. Biodiversity regulates ecosystem predictability. Nature **390**: 162–165.

McHugh, P., A. R. McIntosh and P. Jellyman. 2010. Dual influences of ecosystem size and disturbance on food chain length in streams. Ecology Letters **13**: 881–890.

McIntosh, R. P. 1980. The background and some current problems of theoretical ecology. Synthese **43**: 195–255.

McIntosh, R. P. 1985. *The Background of Ecology: Concept and Theory*. Cambridge University Press, Cambridge.

McIntosh, R. P. 1987. Pluralism in ecology. Annual Review of Ecology and Systematics **18**: 321–341.

McKenna, D. D. and B. D. Farrell. 2006. Tropical forests are both evolutionary cradles and museums of leaf beetle diversity. Proceedings of the National Academy of Sciences **103**: 10947–10951.

McKinney, M. L. 1997. Extinction vulnerability and selectivity: Combining ecological and paleontological views. Annual Review of Ecology and Systematics **28**: 495–516.

McLachlan, J. S., J. J. Hellman and M. W. Schwartz. 2007. A framework for debate of assisted migration in an era of climate change. Conservation Biology **21**: 297–302.

McLain, C. R., E. P. White and A. H. Hurlbert. 2007. Challenges in the application of geometric constraint models.

Global Ecology and Biogeography **16**: 257–264.

McNamara, J. M. and A. I. Houston. 1994. The effect of a change in foraging options on intake rate and predation rate. American Naturalist **144**: 978–1000.

McNaughton, S. J. 1977. Diversity and stability of ecological communities: A comment on the role of empiricism in ecology. American Naturalist **111**: 515–525.

McNaughton, S. J. 1993. Biodiversity and stability of grazing ecosystems. Pages 361–383 in E.-D. Shulze and H. A. Mooney, editors. *Biodiversity and Ecosystem Function.* Springer-Verlag, Berlin.

McPeek, M. A. 1998. The consequences of changing the top predator in a food web. Ecological Monographs **68**: 1–23.

McPeek, M. A. 2004. The growth/predation risk trade-off: So what is the mechanism? American Naturalist **163**: E88–E111.

McPeek, M. A. 2007. The macroevolutionary consequences of ecological differences among species. Palaeontology **50**: 111–129.

McPeek, M. A. 2008. The ecological dynamics of clade diversification and community assembly. American Naturalist **172**: E270–E284.

McPeek, M. A. and B. L. Peckarsky. 1998. Life histories and the strengths of species interactions: Combining mortality, growth, and fecundity effects. Ecology **79**: 867–879.

McPeek, M. A., L. Shen and H. Farid. 2008. The correlated evolution of 3-dimensional reproductive structures between male and female damselflies. Evolution **63**: 73–83.

Meiners, S. J. 2007. Apparent competition: An impact of exotic shrub invasion on tree regeneration. Biological Invasions **9**: 849–855.

Meire, P. M. and A. Ervynck. 1986. Are oystercatchers (*Haematopus ostralegus*) selecting the most profitable mussels (*Mytilus edulis*)? Animal Behaviour **34**: 1427–1435.

Melián, C. J. and J. Bascompte. 2002. Complex networks: Two ways to be robust. Ecology Letters **5**: 705–708.

Melián, C. J. and J. Bascompte. 2004. Food web cohesion. Ecology **85**: 352–358.

Melián, C. J., J. Bascompte, P. Jordano and V. Křivan. 2009. Diversity in a complex ecological network with two interaction types. Oikos **118**: 122–130.

Memmott, J., N. M. Waser and M. V. Price. 2004. Tolerance of pollination networks to species extinctions. Proceedings of the Royal Society of London B **271**: 2605–2611.

Menge, B. A. 1995. Indirect effects in marine rocky intertidal interaction webs: Patterns and importance. Ecological Monographs **65**: 21–74.

Menge, B. A. and W. J. Sutherland. 1987. Community regulation: Variation in disturbance, competition, and predation in relation to environmental stress and recruitment. American Naturalist **130**: 730–757.

Menge, B. A., E. L. Berlow, C. A. Blanchette, S. A. Navarrete and S. B. Yamada. 1994. The keystone species concept—variation in interaction strength in a rocky intertidal habitat. Ecological Monographs **64**: 249–286.

Menge, B. A. and 7 others. 2011. Potential impact of climate-related changes is buffered by differential responses to recruitment and interactions. Ecological Monographs **81**: 493–509.

Menges, E. S. 1990. Population viability analysis for an endangered plant. Conservation Biology **4**: 52–62.

Merriam, G. 1991. Corridors in restoration of fragmented landscapes. Pages 71–87 in D. A. Saunders, R. J. Hobbs and P. R. Ehrlich, editors. *Nature Conservation 3: Reconstruction of Fragmented Ecosystems.* Surrey Beatty and Sons, NSW, Australia.

Michels, E., K. Cottenie, L. Neys and L. De Meester. 2001. Zooplankton on the move: First results on the quantification of dispersal in a set of interconnected ponds. Hydrobiologia **442**: 117–126.

Miki, T., T. Yokokawa, T. Nagata and N. Yamamura. 2008. Immigration of prokaryotes to local environments enhances remineralization efficiency of sinking particles: A metacommunity model. Marine Ecology Progress Series **366**: 1–14.

Milchunas, D. G. and W. K. Lauenroth. 1993. Quantitative effects of grazing on vegetation and soils over a global range of environments. Ecological Monographs **63**: 327–366.

Miller, T. E. and 7 others. 2005. A critical review of twenty years' use of the resource-ratio theory. American Naturalist **165**: 339–448.

Miller, T. E. and 7 others. 2007. Evaluating support for the resource-ratio hypothesis: A reply to Wilson et al. American Naturalist **169**: 707–708.

Miriti, M. N. 2006. Ontogenetic shift from facilitation to competition in a desert shrub. Journal of Ecology **94**: 973–979.

Mitchell, W. A. and T. J. Valone. 1990. The optimization research program: Studying adaptations by their function. Quarterly Review of Biology **65**: 43–52.

Mittelbach, G. G. 1981. Foraging efficiency and body size: A study of optimal diet and habitat use by bluegills. Ecology **62**: 1370–1386.

Mittelbach, G. G. 1988. Competition among refuging sunfishes and effects of fish density on littoral zone invertebrates. Ecology **69**: 614–623.

Mittelbach, G. G. 2002. Fish foraging and habitat choice: A theoretical perspective. Pages 251–266 in P. J. B. Hart and J. D. Reynolds, editors. *Handbook of Fish Biology and Fisheries*, Vol. 1, *Fish Biology.* Blackwell, Oxford.

Mittelbach, G. G. 2010. Understanding species richness-productivity relationships: The importance of meta-analysis. Ecology **91**: 2540–2544.

Mittelbach, G. G. and P. L. Chesson. 1987. Predation risk: Indirect effects on fish populations. Pages 315–332 in W. C. Kerfoot and A. Sih, editors. *Predation: Direct and Indirect Impacts on Aquatic Communities.* University Press of New England, Hanover, NH.

Mittelbach, G. G. and C. W. Osenberg. 1993. Stage-structured interactions in bluegill: Consequences of adult resource variation. Ecology **74**: 2381–2394.

Mittelbach, G. G., C. W. Osenberg and M. A. Leibold. 1988. Trophic relations and ontogenetic niche shifts in aquatic organisms. Pages 217–235 in B. Ebenman and L. Persson, editors. *Size-Structured Populations.* Springer-Verlag, Berlin.

Mittelbach, G. G., A. M. Turner, D. J. Hall, J. E. Rettig and C. W. Osenberg. 1995. Perturbation and resilience in an aquatic community: A long-term study of the extinction and reintroduction of a top predator. Ecology **76**: 2347–2360.

Mittelbach, G. G. and 8 others. 2001. What is the observed relationship between species richness and productivity? Ecology **82**: 2381–2396.

Mittelbach, G. G., S. M. Scheiner and C. F. Steiner. 2003. What is the observed relationship between species richness and productivity? Reply. Ecology **84**: 3390–3395.

Mittelbach, G. G., E. A. Garcia and Y. Taniguchi. 2006. Fish reintroductions reveal smooth transitions between lake community states. Ecology **87**: 312–318.

Mittelbach, G. G. and 21 others. 2007. Evolution and the latitudinal diversity gradient: Speciation, extinction and biogeography. Ecology Letters **10**: 315–331.

Moles, A. T., D. S. Falster, M. R. Leishman and M. Westoby. 2004. Small-seeded species produce more seeds per square metre of canopy per year, but not per individual per lifetime. Journal of Ecology **92**: 384–396.

Monod, J. 1950. La technique de culture continue, théorie et applications. Annales d'Institut Pasteur **79**: 390–410.

Montoya, J. M., S. L. Pimm and R. V. Solé. 2006. Ecological networks and their fragility. Nature **442**: 259–264.

Montoya, J. M., G. Woodward, M. Emmerson and R. V. Solé. 2009. Press perturbations and indirect effects in real food webs. Ecology **90**: 2426–2433.

Moore, J. 2002. *Parasites and the Behavior of Animals*. Oxford University Press, Oxford.

Morales, C. L. and A. Traveset. 2009. A meta-analysis of impacts of alien vs. native plants on pollinator visitation and reproductive success of co-flowering native plants. Ecology Letters **12**: 716–728.

Motomura, I. 1932. On the statistical treatment of communities (in Japanese). Zoological Magazine Tokyo **44**: 379–383.

Mouquet, N. and M. Loreau. 2003. Community patterns in source-sink metacommunities. American Naturalist **162**: 544–557.

Mouquet, N., M. F. Hoopes and P. Amarasekare. 2005. The world is patchy and heterogeneous! Trade-off and source-sink dynamics in competitive metacommunities. Pages 237–262 in M. Holyoak, M. A. Leibold and R. D. Holt, editors. *Metacommunities: Spatial Dynamics and Ecological Communities*. University of Chicago Press, Chicago, IL.

Mullen, S. P., W. K. Savage, N. Wahlberg and K. R. Willmott. 2011. Rapid diversification and not clade age explains high diversity in Neotropical *Adelpha* butterflies. Proceedings of the Royal Society of London B **278**: 1777–1785.

Müller, C. B. and H. C. J. Godfray. 1997. Apparent competition between two aphid species. Journal of Animal Ecology **66**: 57–64.

Muller-Landau, H. C. 2008. Colonization-related trade-offs in tropical forests and their role in the maintenance of plant species diversity. Pages 182–195 in W. P. Carson and S. A. Schnitzer, editors. *Tropical Forest Community Ecology*. Wiley-Blackwell, Chichester.

Munger, J. C. 1984. Long-term yield from harvester ant colonies: Implications for horned lizard foraging strategy. Ecology **65**: 1077–1086.

Murdoch, W. W. 1966. "Community structure, population control, and competition"—a critique. American Naturalist **100**: 219–226.

Murdoch, W. W. 1969. Switching in general predators: Experiments on predator specificity and stability of prey populations. Ecological Monographs **39**: 335–354.

Murdoch, W. W. 1970. Population regulation and population inertia. Ecology **51**: 497–502.

Murdoch, W. W. 1991. The shift from an equilibrium to a non-equilibrium paradigm in ecology. Bulletin of the Ecological Society of America **72**: 49–51.

Murdoch, W. W. 1994. Population regulation in theory and practice. Ecology **75**: 271–287.

Murdoch, W. W. and S. J. Walde. 1989. Analysis of insect population dynamics. Pages 113–140 in P. J. Grubb and J. B. Whittaker, editors. *Towards a More Exact Ecology*. Blackwell, Oxford.

Murdoch, W. W., C. J. Briggs and R. M. Nisbet. 2003. *Consumer-Resource Dynamics*. Princeton University Press, Princeton, NJ.

Mutshinda, C. M., R. B. O'Hara and I. P. Woiwod. 2010. What drives community dynamics? Proceedings of the Royal Society of London B **276**: 2923–2929.

Naeem, S. 2008. Green with complexity. Science **319**: 913–914.

Naeem, S. and S. Li. 1997. Biodiversity and ecosystem reliability. Nature **390**: 507–509.

Naeem, S., L. J. Thompson, S. P. Lawler, J. H. Lawton and R. M. Woodfin. 1994. Declining biodiversity can alter the performance of ecosystems. Nature **368**: 734–737.

Naeslund, B. and J. Norberg. 2006. Ecosystem consequences of the regional species pool. Oikos **115**: 504–512.

Nagel, L. and D. Schluter. 1998. Body size, natural selection, and speciation in sticklebacks. Evolution **52**: 209–218.

Navarrete, S. A. and B. A. Menge. 1996. Keystone predation and interaction strength: Interactive effects of predators on their main prey. Ecological Monographs **66**: 409–429.

Nee, S. 1994. How populations persist. Nature **367**: 123–124.

Nee, S. 2005. The neutral theory of biodiversity: Do the numbers add up? Functional Ecology **19**: 173–176.

Nee, S. and R. M. May. 1992. Dynamics of metapopulations: Habitat destruction and competitive coexistence. Journal of Animal Ecology **61**: 37–40.

Nee, S., R. M. May and M. P. Hassell. 1997. Two-species metapopulation models. Pages 123–147 in I. Hanski and M. Gilpin, editors. *Metapopulation Biology*. Academic Press, San Diego, CA.

Neff, J. C., S. E. Hobbie and P. M. Vitousek. 2003. Nutrient and mineralogical control on dissolved organic C, N and P fluxes and stoichiometry in Hawaiian soils. Biogeochemistry **51**: 283–302.

Neill, W. E. 1974. The community matrix and interdependence of the competition coefficients. American Naturalist **108**: 399–408.

Nekola, J. C. and P. S. White. 1999. The distance decay of similarity in biogeography and ecology. Journal of Biogeography **26**: 867–878.

Nicholson, A. J. 1957. The self-adjustment of populations to change. Cold Spring Harbor Symposia on Quantitative Biology **22**: 153–172.

Niklaus, P. A., E. Kandeler, P. W. Leadley, B. Schmid, D. Tscherko and C. Körner. 2001. A link between plant diversity, elevated CO_2 and soil nitrate. Oecologia **127**: 540–548.

Nisbet, R. M. and W. S. C. Gurney. 1982. *Modeling Fluctuating Populations*. John Wiley & Sons, New York.

Noë, R. and P. Hammerstein. 1994. Biological markets: Supply and demand determine the effect of partner choice in cooperation, mutualism and mating. Behavioral Ecology and Sociobiology **35**: 1–11.

Noonburg, E. G. and J. E. Byers. 2005. More harm than good: When invader vulnerability to predators enhances impact on native species. Ecology **86**: 2555–2560.

Norberg, J. 2000. Resource-niche complementarity and autotrophic compensation determines ecosystem-level responses to increased cladoceran species richness. Oecologia **112**: 264–272.

Norberg, J., D. P. Swaney, J. Dushoff, J. Lin, R. Casagrandi and S. A. Levin. 2001. Phenotypic diversity and ecosystem functioning in changing environments: A theoretical framework. Proceedings of the National Academy of Sciences **98**: 11376–11381.

Norbury, G. 2001. Conserving dryland lizards by reducing predator-mediated apparent competition and direct competition with introduced rabbits. Journal of Applied Ecology **38**: 1350–1361.

Northfield, T. D., G. B. Snyder, A. R. Ives and W. E. Snyder. 2010. Niche saturation reveals resource partitioning

among consumers. Ecology Letters **13**: 338–348.

Nuñez, C. I., E. Raffaele, M. A. Nuñez and F. Cuassolo. 2009. When do nurse plants stop nursing? Temporal changes in water stress levels in *Austrocedrus chilensis* growing within and outside shrubs. Journal of Vegetation Science **20**: 1064–1071.

Nyström, M., N. A. J. Graham, J. Lokrantz and A. Norström. 2008. Capturing the cornerstones of coral reef resilience: Linking theory to practice. Coral Reefs **27**: 795–809.

O'Brien, W. J. 1974. The dynamics of nutrient limitation of phytoplankton algae: A model reconsidered. Ecology **50**: 930–938.

Odum, E. P. 1953. *Fundamentals of Ecology*. Saunders, Philadelphia, PA.

Odum, E. P. 1959. *Fundamentals of Ecology*. Saunders, Philadelphia, PA.

Oelmann, Y. and 7 others. 2007. Soil and plant nitrogen pools as related to plant diversity in an experimental grassland. Soil Science Society of America Journal **71**: 720–729.

O'Gorman, E. J. and M. C. Emmerson. 2009. Perturbations to trophic interactions and the stability of complex food webs. Proceedings of the National Academy of Sciences **106**: 13393–13398.

Oksanen, L., S. D. Fretwell, J. Arruda and P. Niemelä. 1981. Exploitation ecosystems in gradients of primary productivity. American Naturalist **118**: 240–261.

Okuyama, T. and B. M. Bolker. 2007. On quantitative measures of indirect interactions. Ecology Letters **10**: 264–271.

Oliver, M., J. J. Luque-Larena and X. Lambin. 2009. Do rabbits eat voles? Apparent competition, habitat heterogeneity and large-scale coexistence under mink predation. Ecology Letters **12**: 1201–1209.

Orrock, J. L., M. S. Witter and O. J. Reichman. 2008. Apparent competition with an exotic plant reduces native plant establishment. Ecology **89**: 1168–1174.

Osenberg, C. W. and G. G. Mittelbach. 1996. The relative importance of

resource limitation and predator limitation in food chains. Pages 134–148 in G. A. Polis and K. O. Winemiller, editors. *Food Webs: Integration of Patterns and Dynamics*. Chapman and Hall, New York.

Osenberg, C. W., O. Sarnelle and S. D. Cooper. 1997. Effect size in ecological experiments: The application of biological models in meta-analysis. American Naturalist **150**: 798–812.

Osenberg, C. W., O. Sarnelle, S. D. Cooper and R. D. Holt. 1999. Resolving ecological questions through meta-analysis: Goals, metrics, and models. Ecology: 1105–1117.

Östman, Ö. and A. R. Ives. 2003. Scale-dependent indirect interactions between two prey species through a shared predator. Oikos **102**: 505–514.

Pacala, S. W. and D. Tilman. 1994. Limiting similarity in mechanistic and spatial models of plant competition in heterogeneous environments. American Naturalist **143**: 222–257.

Paine, R. T. 1966. Food web complexity and species diversity. American Naturalist **100**: 65–75.

Paine, R. T. 1969. A note on trophic complexity and community stability. American Naturalist **103**: 91–93.

Paine, R. T. 1980. Food webs, linkage interaction strength, and community infrastructure. Journal of Animal Ecology **49**: 667–685.

Paine, R. T. 1992. Food-web analysis through field measurement of per capita interaction strength. Nature **355**: 73–75.

Pajunen, V. I. and I. Pajunen. 2003. Long-term dynamics in rock pool *Daphnia* metapopulations. Ecography **26**: 731–738.

Pake, C. E. and D. L. Venable. 1995. Is coexistence of Sonoran Desert annuals mediated by temporal variability in reproductive success? Ecology **76**: 246–261.

Palkovacs, E. P. and D. M. Post. 2009. Experimental evidence that phenotypic divergence in predators drives community divergence in prey. Ecology **90**: 300–305.

Palmer, T. M., M. L. Stanton, T. P. Young, J. R. Goheen, R. M. Pringle and R. Karban. 2008. Breakdown of an ant-plant mutualism follows the loss of large herbivores from an African savanna. Science **319**: 192–195.

Pangle, K. L., S. D. Peacor and O. E. Johannsson. 2007. Large nonlethal effects of an invasive invertebrate predator on zooplankton population growth rate. Ecology **88**: 402–412.

Parker, S. and A. Huryn. 2006. Food web structure and function in two arctic streams with contrasting disturbance. Freshwater Biology **51**: 1249–1263.

Parmesan, C. 2006. Ecological and evolutionary responses to recent climate change. Annual Review of Ecology, Evolution, and Systematics **37**: 637–669.

Parmesan, C. and G. Yohe. 2003. A globally coherent fingerprint of climate change impacts across natural systems. Nature **421**: 37–42.

Parmesan, C., S. Gaines, L. Gonzalez, D. M. Kaufman, J. Kingsolver, A. Townsend Peterson and R. Sagarin. 2005. Empirical perspectives on species borders: From traditional biogeography to global change. Oikos **108**: 58–75.

Partel, M. and M. Zobel. 2007. Dispersal limitation may result in the unimodal productivity–diversity relationship: A new explanation for a general pattern. Journal of Ecology **95**: 90–94.

Partel, M., L. Laanisto and M. Zobel. 2007. Contrasting plant productivity–diversity relationships across latitude: The role of evolutionary history. Ecology **88**: 1091–1097.

Pascual, M. and J. A. Dunne, editors. 2006. *Ecological Networks: Linking Structure to Dynamics in Food Webs*. Oxford University Press, Oxford.

Pauly, D. and 7 others. 2002. Towards sustainability in world fisheries. Nature **418**: 689–695.

Pausas, J. G. and M. Verdú. 2010. The jungle of methods for evaluating phenotypic and phylogenetic structure of communities. BioScience **60**: 614–625.

Peacor, S. D. 2002. Positive effect of predators on prey growth rate through induced modifications of prey behaviour. Ecology Letters **5**: 77–85.

Peacor, S. D. and E. E. Werner. 1997. Trait mediated indirect interactions in a simple aquatic food web. Ecology **78**: 1146–1156.

Pearman, P. B., A. Guisan, O. Broennimann and C. F. Randin. 2008. Niche dynamics in space and time. Trends in Ecology and Evolution **23**: 149–158.

Peckarsky, B. L., C. A. Cowan, M. A. Penton and C. Anderson. 1993. Sublethal consequences of predator-avoidance by stream-dwelling mayfly larvae to adult fitness. Ecology **74**: 1836–1846.

Peckarsky, B. L., B. W. Taylor, A. R. McIntosh, M. A. McPeek and D. A. Lytle. 2001. Variation in mayfly size at metamorphosis as a developmental response to risk of predation. Ecology **82**: 740–757.

Peckarsky, B. L., A. R. McIntosh, B. W. Taylor and J. Dahl. 2002. Predator chemicals induce changes in mayfly life history traits: A whole-stream manipulation. Ecology **83**: 612–618.

Pedruski, M. T. and S. E. Arnott. 2011. The effects of habitat connectivity and regional heterogeneity on artificial pond metacommunities. Oecologia **166**: 221–228.

Pels, B., A. M. de Roos and M. W. Sabelis. 2002. Evolutionary dynamics of prey exploitation in a metapopulation of predators. American Naturalist **159**: 172–189.

Pennings, S. C., E. R. Seling, L. T. Houser and M. D. Bertness. 2003. Geographic variation in positive and negative interactions among salt marsh plants. Ecology **84**: 1527–1538.

Pereira, H. M. and 22 others. 2010. Scenarios for global biodiversity in the 21st century. Science **330**: 1496–1501.

Perlman, D. L. and G. Adelson. 1997. *Biodiversity: Exploring Values and Priorities in Conservation*. Blackwell Scientific, Cambridge, MA.

Persson, L. 1993. Predator-mediated competition in prey refuges: The importance of habitat dependent prey resources. Oikos **68**: 12–22.

Persson, L., P.-A. Amundsen, A. M. De Roos, A. Klemetsen, R. Knudsen and R. Primicerio. 2007. Culling prey promotes predator recovery—alternative states in a whole-lake experiment. Science **316**: 1743–1746.

Pervez, A. and Omkar. 2005. Functional responses of coccinellid predators: An illustration of a logistic approach. Journal of Insect Science **5**: 5; available online at insectscience.org/5.5.

Petchey, O. L., A. L. Downing, G. G. Mittelbach, L. Persson, C. F. Steiner, P. H. Warren and G. Woodward. 2004. Species loss and the structure and functioning of multitrophic aquatic systems. Oikos **104**: 467–478.

Petchey, O. L., A. P. Beckerman, J. O. Riede and P. H. Warren. 2008. Size, foraging, and food web structure. Proceedings of the National Academy of Sciences **105**: 4191–4196.

Petchey, O. L., A. P. Beckerman, J. O. Riede and P. H. Warren. 2011. Fit, efficiency, and biology: Some thoughts on judging food web models. Journal of Theoretical Biology **279**: 169–171.

Peters, R. H. 1983. *The Ecological Implications of Body Size*. Cambridge University Press, Cambridge.

Pettersson, L. and C. Brönmark. 1997. Trading off safety against food: State dependent habitat choice and foraging in crucian carp. Oecologia **95**: 353–357.

Pfisterer, A. B. and B. Schmid. 2002. Diversity-dependent production can decrease the stability of ecosystem functioning. Nature **416**: 84–86.

Phillimore, A. B. and T. D. Price. 2008. Density-dependent cladogenesis in birds. PLoS Biology **6(3)**: e71. doi: 10.1371/journal.pbio.0060071.

Phillips, O. M. 1973. The equilibrium and stability of simple marine biological systems. I. Primary nutrient consumers. American Naturalist **107**: 73–93.

Phillips, O. M. 1974. The equilibrium and stability of simple marine systems. II. Herbivores. Archiv fur Hydrobiologie **73**: 310–333.

Pierce, G. J. and J. G. Ollason. 1987. Eight reasons why optimal foraging theory is a complete waste of time. Oikos **49**: 111–118.

Pimm, S. L. 1984. The complexity and stability of ecosystems. Nature **307**: 321–326.

Pimm, S. L. 1992. *Food Webs*. Chapman and Hall, London.

Pimm, S. L. and J. H. Lawton. 1977. On the number of trophic levels. Nature **268**: 329–331.

Pimm, S. L., H. L. Jones and J. Diamond. 1988. On the risk of extinction. American Naturalist **132**: 757–785.

Pimm, S. L., J. H. Lawton and J. E. Cohen. 1991. Food web patterns and their consequences. Nature **350**: 669–674.

Pitman, N. C. A. and 7 others. 2002. A comparison of tree species diversity in two upper Amazonian forests. Ecology **83**: 3210–3224.

Platt, W. J. and I. M. Weiss. 1977. Resource partitioning and competition within a guild of fugitive prairie plants. American Naturalist **111**: 479–513.

Polis, G. 1991. Complex trophic interactions in deserts: An empirical critique of food web theory. American Naturalist **138**: 123–155.

Polis, G. A. and D. R. Strong. 1996. Food web complexity and community dynamics. American Naturalist **147**: 813–846.

Pollard, E., K. H. Lakhani and P. Rothery. 1987. The detection of density-dependence from a series of annual censuses. Ecology **68**: 2046–2055.

Post, D. M. 2002. The long and short of food-chain length. Trends in Ecology and Evolution **17**: 269–277.

Post, D. M. and E. P. Palkovacs. 2009. Eco-evolutionary feedbacks in community and ecosystem ecology: Interactions between the ecological theatre and the evolutionary play. Philosophical Transactions of the Royal Society of London B **364**: 1629–1640.

Post, D. M., M. Pace and N. G. Hairston, Jr. 2000. Ecosystem size determines food-chain length in lakes. Nature **405**: 1047–1049.

Post, D. M., E. P. Palkovacs, E. G. Schielke and S. I. Dodson. 2008. Intraspecific phenotypic variation in a predator affects community structure and cascading trophic interactions. Ecology **89**: 2019–2032.

Powell, M. G. 2007. Latitudinal diversity gradients for brachiopod genera during late Palaeozoic time: Links between climate, biogeography and evolutionary rates. Global Ecology and Biogeography **16**: 519–528.

Powell, M. G. 2009. The latitudinal diversity gradient of brachiopods over the past 530 million years. Journal of Geology **117**: 585–594.

Power, M. E. 1990. Effects of fish in river food webs. Science **250**: 811–814.

Power, M. E. 1997. Ecosystem engineering by organisms: Why semantics matters—reply. Trends in Ecology and Evolution **12**: 275–276.

Power, M. E., W. J. Mathews and A. J. Stewart. 1985. Grazing minnows, piscivorous bass, and stream algae: Dynamics of a strong interaction. Ecology **66**: 1448–1456.

Power, M. E. and 9 others. 1996. Challenges in the quest for keystones. BioScience **46**: 609–620.

Pratt, H. D. 2005. *The Hawaiian Honeycreepers; Drepanididae*. Oxford University Press, Oxford.

Preisser, E. I., D. I. Bolnick and M. F. Benard. 2005. Scared to death? The effects of intimidation and consumption in predator–prey interactions. Ecology **86**: 501–509.

Preston, F. W. 1948. The commonness, and rarity, of species. Ecology **29**: 254–283.

Preston, F. W. 1960. Time and space and the variation of species. Ecology **41**: 611–627.

Preston, F. W. 1962. The canonical distribution of commonness and rarity: Part I. Ecology **43**: 185–215.

Price, P. W. 1980. *Evolutionary Biology of Parasites*. Princeton University Press, Princeton, NJ.

Price, T. D. and M. Kirkpatrick. 2009. Evolutionary stable range limits set by interspecific competition. Proceedings of the Royal Society of London B **276**: 1429–1434.

Price, T. D., D. Mohan, D. T. Tietze, D. M. Hooper, C. D. L. Orme and P. C. Rasmussen. 2011. Determinants of northerly range limits along the Himalayan bird diversity gradient. American Naturalist **178**, Suppl. 1: S97–S108.

Proulx, R. and 33 others. 2010. Diversity promotes temporal stability across levels of ecosystem organization in experimental grasslands. PLoS One **5(10)**: e13382. doi: 10.1371/journal.pone.0013382.

Ptacnik, R., T. Andersen, P. Brettum, L. Lepistö and E. Willén. 2010. Regional species pools control community saturation in lake phytoplankton. Proceedings of the Royal Society of London B **277**: 3755–3764.

Pulliam, H. R. 1974. On the theory of optimal diets. American Naturalist **108**: 59–75.

Pulliam, H. R. 1988. Sources, sinks and population regulation. American Naturalist **132**: 652–661.

Purvis, A., J. L. Gittleman, G. Cowlishaw and G. M. Mace. 2000. Predicting extinction risk in declining species. Proceedings of the Royal Society of London B **267**: 1947–1952.

Pyke, G. H. 1984. Optimal foraging theory: A critical review. Annual Review of Ecology and Systematics **15**: 523–575.

Rabosky, D. L. 2009. Ecological limits on clade diversification in higher taxa. American Naturalist **173**: 662–674.

Rabosky, D. L. 2010. Extinction rates should not be estimated from molecular phylogenies. Evolution **64**: 1816–1824.

Rabosky, D. L. and M. E. Alfaro. 2010. Evolutionary bangs and whimpers: Methodological advances and conceptual frameworks for studying exceptional diversification. Systematic Biology **59**: 615–618.

Rabosky, D. L. and I. J. Lovette. 2008. Density-dependent diversification in North American wood warblers. Proceedings of the National Academy of Sciences **275**: 2363–2371.

Raffaelli, D. and S. R. Hall. 1996. Assessing the relative importance of trophic links in food webs. Pages 185–191 in G. A. Polis and K. O. Winemiller, editors. *Food Webs: Integration of Patterns and Dynamics*. Chapman and Hall, New York.

Rahbek, C., N. J. Gotelli, R. K. Colwell, G. L. Entsminger, T. Rangel and G. R. Graves. 2007. Predicting continental-scale patterns of bird species richness with spatially explicit models. Proceedings of the Royal Society of London B **274**: 165–174.

Rand, T. A. and S. M. Louda. 2004. Exotic weed invasion increases the susceptibility of native plants attack by a biocontrol herbivore. Ecology **85**: 1548–1554.

Ranta, E., V. Kaitala, M. S. Fowler, J. Laakso, L. Ruokolainen and R. O'Hara. 2008. Detecting compensatory dynamics in competitive communities under environmental forcing. Oikos **117**: 1907–1911.

Rasmussen, J. B., D. J. Rowan, R. S. Lean and J. H. Carey. 1990. Food chain structure in Ontario lakes determines PCB levels in lake trout (*Salvelinus namaycush*) and other pelagic fish. Canadian Journal of Fisheries and Aquatic Sciences **47**: 2030–2038.

Rathcke, B. 1988. Competition and facilitation among plants for pollination. Pages 305–329 in L. Real, editor. *Pollination Biology*. Academic Press, New York.

Read, D. J., H. K. Koucheki and J. R. Hodgson. 1976. Vesicular-arbuscular mycorrhiza in natural vegetation systems. I. The occurrence of infection. New Phytologist **77**: 641–654.

Reed, D. H., J. J. O'Grady, B. W. Brook, J. D. Ballou and R. Frankham. 2003. Estimates of minimum viable population sizes for vertebrates and factors influencing those estimates. Biological Conservation **113**: 23–34.

Rees, M. 1995. Community structure in sand dune annuals: Is seed weight a key quantity? Journal of Ecology **83**: 857–864.

Rees, M., P. J. Grubb and D. Kelly. 1996. Quantifying the impact of competition and spatial heterogeneity on the structure and dynamics of a four-species guild of winter annuals. American Naturalist **147**: 1–32.

Reichman, O. J. and E. W. Seabloom. 2002. Ecosystem engineering: A trivialized concept? Trends in Ecology and Evolution **17**: 308.

Rejmanek, M. 2003. The rich get richer—responses. Frontiers in Ecology and the Environment **1**: 122–123.

Resetarits, W. J. J. and J. Bernardo. 1998. *Experimental Ecology: Ideas and Perspectives*. Oxford University Press, New York.

Rex, M. A., J. A. Crame, C. T. Stuart and A. Clark. 2005. Large-scale biogeographic patterns in marine mollusks: A confluence of history and productivity? Ecology **86**: 2288–2297.

Reynolds, J. D. and C. A. Peres. 2006. Overexploitation. Pages 253–277 in M. J. Groom, G. K. Meffe and C. R. Carroll, editors. *Principles of Conservation Biology*. Third ed. Sinauer Associates, Sunderland, MA.

Rezende, E. L., J. E. Lavabre, P. R. Guimarães, Jr., P. Jordano and J. Bascompte. 2007. Non-random coextinctions in phylogenetically structured mutualistic networks. Nature **448**: 925–928.

Rezende, E. L., E. M. Albert, M. A. Fontuna and J. Bascompte. 2009. Compartments in a marine food web associated with phylogeny, body mass, and habitat structure. Ecology Letters **12**: 779–788.

Reznick, D. N. and C. K. Ghalambor. 2005. Selection in nature: Experimental manipulations of natural populations. Integrative Comparative Biology **45**: 456–462.

Reznick, D. N. and R. E. Ricklefs. 2009. Darwin's bridge between microevolution and macroevolution. Nature **457**: 837–842.

Ricciardi, A. and D. Simberloff. 2009. Assisted colonization is not a viable conservation strategy. Trends in Ecology and Evolution **24**: 248–253.

Richardson, H. and N. A. M. Verbeek. 1986. Diet selection and optimization by Northwestern crows feeding on Japanese little-neck clams. Ecology **67**: 1219–1226.

Ricklefs, R. E. 1987. Community diversity: Relative roles of regional and local processes. Science **235**: 167–171.

Ricklefs, R. E. 2003. Global diversification rates of passerine birds. Proceedings of the Royal Society of London B **270**: 2285–2291.

Ricklefs, R. E. 2004. A comprehensive framework for global patterns in biodiversity. Ecology Letters **7**: 1–15.

Ricklefs, R. E. 2005. Phylogenetic perspectives on patterns of regional and local species richness. Pages 16–40 in E. Bermingham, C. Dick and C. Moritz, editors. *Rainforest: Past, Present and Future*. University of Chicago Press, Chicago, IL.

Ricklefs, R. E. 2006a. Evolutionary diversification and the origin of the diversity–environment relationship. Ecology **87**: S3–S13.

Ricklefs, R. E. 2006b. Global variation in the diversification rate of passerine birds. Ecology **87**: 2468–2478.

Ricklefs, R. E. 2006c. The unified neutral theory of biodiversity: Do the numbers add up? Ecology **87**: 1424–1431.

Ricklefs, R. E. 2007. Estimating diversification rates from phylogenetic information. Trends in Ecology and Evolution **22**: 601–610.

Ricklefs, R. E. 2011. A biogeographic perspective on ecological systems: Some personal reflections. Journal of Biogeography **38**: 2045–2056.

Ricklefs, R. E. and D. G. Jenkins. 2011. Biogeography and ecology: Towards the integration of two disciplines. Philosophical Transactions of the Royal Society of London B **366**: 2438–2448.

Ricklefs, R. E. and S. S. Renner. 1994. Species richness within families of flowering plants. Evolution **48**: 1619–1636.

Ricklefs, R. E. and D. Schluter, editors. 1993. *Species Diversity in Ecological Communities: Historical and Geographical Perspectives*. University of Chicago Press, Chicago, IL.

Rietkerk, M., S. C. Dekker, P. C. deRuiter and J. van de Koppel. 2004. Self-organized patchiness and catastrophic shifts in ecosystems. Science **305**: 1926–1929.

Ripple, W. J. and R. Beschta. 2007. Hardwood tree decline following large carnivore loss on the Great Plains. Frontiers in Ecology and the Environment **5**: 241–246.

Robinson, G. R., J. F. Quinn and M. L. Stanton. 1995. Invasibility of experimental habitat islands in a California winter annual grassland. Ecology **76**: 786–794.

Roelants, K., A. Hass and F. Bossuyt. 2011. Anuran radiations and the evolution of tadpole morphospace. Proceedings of the National Academy of Sciences **108**: 8731–8736.

Rohde, K. 1992. Latitudinal gradients in species diversity: The search for the primary cause. Oikos **65**: 514–527.

Rooney, N. and K. S. McCann. 2012. Integrating food web diversity, structure and stability. Trends in Ecology and Evolution **27**: 40–46.

Rooney, N., K. S. McCann, G. Gellner and J. C. Moore. 2006. Structural asymmetry and stability of diverse food webs. Nature **442**: 265–269.

Root, R. B. 1967. The niche exploitation pattern of the blue-gray gnatcatcher. Ecological Monographs **37**: 317–350.

Rosemond, A. D., P. J. Mulholland and J. W. Elwood. 1993. Top-down and bottom-up control of stream periphyton—effects of nutrients and herbivores. Ecology **74**: 1264–1280.

Rosenzweig, M. L. 1969. Why the prey curve has a hump. American Naturalist **103**: 81–87.

Rosenzweig, M. L. 1971. Paradox of enrichment: Destabilization of exploitation ecosystems in ecological time. Science **171**: 385–387.

Rosenzweig, M. L. 1973. Exploitation in three trophic levels. American Naturalist **107**: 275–294.

Rosenzweig, M. L. 1977. Aspects of biological exploitation. Quarterly Review of Biology **52**: 371–380.

Rosenzweig, M. L. 1995. *Species Diversity in Space and Time*. Cambridge University Press, Cambridge.

Rosenzweig, M. L. and R. H. MacArthur. 1963. Graphical representation and stability conditions of predator–prey interactions. American Naturalist **97**: 209–223.

Rosindell, J. and S. J. Cornell. 2007. Species–area relationships from a spatially explicit neutral model in an infinite landscape. Ecology Letters **10**: 586–595.

Rosindell, J., S. J. Cornell, S. P. Hubbell and R. S. Etienne. 2010. Protracted speciation revitalizes the neutral theory of biodiversity. Ecology Letters **13**: 716–727.

Rossberg, A. G., H. Matsuda, T. Amemiya and K. Itoh. 2006. Some properties of the speciation model for food-web structure—mechanisms for degree distribution and intervality. Journal of Theoretical Biology **238**: 401–415.

Rothhaupt, K. O. 1988. Mechanistic resource competition theory applied to laboratory experiments with zooplankton. Nature **333**: 660–662.

Roughgarden, J. 1975. Evolution of a marine symbiosis—a simple cost–benefit model. Ecology **56**: 1201–1208.

Roughgarden, J. 1979. Theory of Population Genetics and Evolutionary Ecology: An Introduction. MacMillan, New York.

Roughgarden, J. 1997. Production functions from ecological populations: A survey with emphasis on spatially implicit models. Pages 296–317 in D. Tilman and P. Kareiva, editors. *Spatial Ecology*. Princeton University Press, Princeton, NJ.

Roughgarden, J. and M. Feldman. 1975. Species packing and predation pressure. Ecology **56**: 489–492.

Rowe, L. and D. Ludwig. 1991. Size and timing of metamorphosis in complex life cycles: Time constraints and variation. Ecology **72**: 413–427.

Roxburgh, S. H., K. Shea and J. B. Wilson. 2004. The intermediate disturbance hypothesis: Patch dynamics and mechanisms of species coexistence. Ecology **85**: 359–371.

Roy, K., D. Jablonski and J. W. Valentine. 1996. Higher taxa in biodiversity studies: Patterns from Eastern Pacific marine molluscs. Philosophical Transactions of the Royal Society of London B **351**: 1605–1613.

Rudolf, V. H. W. and K. D. Lafferty. 2011. Stage structure alters how complexity affects stability of ecological networks. Ecology Letters **14**: 75–79.

Runkle, J. R. 1989. Synchrony of regeneration, gaps, and latitudinal differences in tree species diversity. Ecology **79**: 546–547.

Sabo, J., J. Finlay, T. Kennedy and D. M. Post. 2010. The role of discharge variation in scaling of drainage area and food chain length in rivers. Science **330**: 965–967.

Sachs, J. L., U. G. Mueller, T. P. Wilcox and J. J. Bull. 2004. The evolution of cooperation. Quarterly Review of Biology **79**: 136–160.

Sala, E. and M. H. Graham. 2002. Community-wide distribution of predator–prey interaction strength in kelp forests. Proceedings of the National Academy of Sciences **99**: 3678–3683.

Sale, P. F. 1977. Maintenance of high diversity in coral reef fish communities. American Naturalist **111**: 337–359.

Sandel, B., L. Arge, B. Dalsgaard, R. G. Davies, K. J. Gaston, W. J. Sutherland and J.-C. Svenning. 2011. The influence of late quaternary climate-change velocity on species endemism. Science **334**: 660–664.

Sanders, H. L. 1969. Benthic marine diversity and the stability-time hypothesis. Pages 71–81 in G. M. Woodwell and H. H. Smith, editors. *Diversity and Stability in Ecological Systems*. Brookhaven Symposium no. 22. U.S. Department of Commerce, Springfield, VA.

Sarnelle, O. and R. A. Knapp. 2004. Zooplankton recovery after fish removal: Limitations of the egg bank. Limnology and Oceanography **49**: 1382–1392.

Sarnelle, O. and A. E. Wilson. 2008. Type III functional response in *Daphnia*. Ecology **89**: 1723–1732.

Sax, D. F. and S. D. Gaines. 2003. Species diversity: From global decreases to local increases. Trends in Ecology and Evolution **18**: 561–566.

Sax, D. F., S. D. Gaines and J. H. Brown. 2002. Species invasions exceed extinctions on islands world-wide: A comparative study of plants and birds. American Naturalist **160**: 766–783.

Schaffer, W. M. and M. Kot. 1985. Do strange attractors govern ecological systems? BioScience **35**: 342–350.

Scheffer, M. 1998. *Ecology of Shallow Lakes*. Chapman and Hall, London.

Scheffer, M. 2009. *Critical Transitions in Nature and Society*. Princeton University Press, Princeton, NJ.

Scheffer, M. 2010. Alternative states in ecosystems. Pages 287–298 in J. Terborgh and J. A. Estes, editors. *Trophic Cascades: Predators, Prey, and the Changing Dynamics of Nature*. Island Press, Washington, DC.

Scheffer, M. and S. R. Carpenter. 2003. Catastrophic regime shifts in ecosystems: Linking theory to observation. Trends in Ecology and Evolution **18**: 648–656.

Scheffer, M. and E. H. Van Nes. 2007. Shallow lakes theory revisited: Various alternative regimes driven by climate, nutrients, depth and lake size. Hydrobiologia **584**: 455–466.

Scheffer, M., S. R. Carpenter, J. A. Foley, C. Folke and B. Walker. 2001. Catastrophic shifts in ecosystems. Nature **413**: 591–596.

Scheffer, M. and 9 others. 2009. Early-warning signals for critical transitions. Nature **461**: 53–59.

Scheiner, S. M. and J. Gurevitch. 1993. *Design and Analysis of Ecological Experiments*. Chapman and Hall, New York.

Scheiner, S. M., S. B. Cox, M. Willig, G. G. Mittelbach, C. W. Osenberg and M. Kaspari. 2000. Species richness, species–area curves and Simpson's paradox. Evolutionary Ecology Research **2**: 791–802.

Schemske, D. W. 1981. Floral convergence and pollinator sharing in two bee-pollinated tropical herbs. Ecology **62**: 946–954.

Schemske, D. W. 2002. Tropical diversity: Patterns and processes. Pages 163–173 in R. Chazdon and T. Whitmore, editors. *Ecological and Evolutionary Perspectives on the Origins of Tropical Diversity: Key Papers and Commentaries.* University of Chicago Press, Chicago, IL.

Schemske, D. W., G. G. Mittelbach, H. V. Cornell, J. M. Sobel and K. Roy. 2009. Is there a latitudinal gradient in the importance of biotic interactions? Annual Review of Ecology, Evolution, and Systematics **40**: 245–269.

Scherer-Lorenzen, M., C. Palmborg, A. Prinz and E. D. Schulze. 2003. The role of plant diversity and composition for nitrate leaching in grasslands. Ecology **84**: 1539–1552.

Schindler, D. E., R. Hilborn, B. Chasco, C. P. Boatright, T. P. Quinn, L. A. Rogers and M. S. Webster. 2010. Population diversity and the portfolio effect in an exploited species. Nature **465**: 609–612.

Schlesinger, W. H., J. F. Reynolds, G. L. Cunningham, L. F. Huenneke, W. M. Jarrell, R. A. Virginia and W. G. Whitford. 1990. Biological feedbacks in global desertification. Science **247**: 1043–1048.

Schluter, D. 1990. Species-for-species matching. American Naturalist **136**: 560–568.

Schluter, D. 1996. Ecological causes of adaptive radiation. American Naturalist **148**: S40–S64.

Schluter, D. 2000. *The Ecology of Adaptive Radiation*. Oxford University Press, New York.

Schluter, D. and J. D. McPhail. 1992. Ecological character displacement and speciation in sticklebacks. American Naturalist **140**: 85–108.

Schluter, D., T. D. Price and P. R. Grant. 1985. Ecological character displacement in Darwin finches. Science **227**: 1056–1059.

Schmid, B. 2002. The species richness–productivity controversy. Trends in Ecology and Evolution **17**: 113–114.

Schmitt, R. J. 1987. Indirect interactions between prey: Apparent competition, predator aggregation, and habitat segregation. Ecology **68**: 1887–1897.

Schmitt, R. J. and S. J. Holbrook. 2003. Mutualism can mediate competition and promote coexistence. Ecology Letters **6**: 898–902.

Schmitz, O. 1998. Direct and indirect effects of predation and predation risk in old-field interactions webs. American Naturalist **151**: 327–342.

Schmitz, O. 2003. Top predator control of plant biodiversity and productivity in an old-field ecosystem. Ecology Letters **6**: 156–163.

Schmitz, O. 2004. Perturbation and abrupt shift in trophic control of biodiversity and productivity. Ecology Letters **7**: 403–409.

Schmitz, O. 2006. Predators have large effects on ecosystem properties by changing plant diversity not plant biomass. Ecology **87**: 1432–1437.

Schmitz, O. 2008. Effects of predator hunting mode on grassland ecosystem function. Science **319**: 952–954.

Schmitz, O. 2010. *Resolving Ecosystem Complexity*. Princeton University Press, Princeton, NJ.

Schmitz, O., P. A. Hambäck and A. P. Beckerman. 2000. Trophic cascades in terrestrial systems: A review of the effects of carnivore removals on plants. American Naturalist **155**: 141–153.

Schmitz, O., V. Kivan and O. Ovadia. 2004. Trophic cascades: The primacy of trait-mediated indirect interactions. Ecology Letters **7**: 153–163.

Schmitz, O., E. L. Kalies and M. G. Booth. 2006. Alternative dynamic regimes and trophic control of plant succession. Ecosystems **9**: 659–672.

Schnitzer, S. A. and 11 others. 2011. Soil microbes drive the classic plant diversity–productivity pattern. Ecology **92**: 296–303.

Schoener, T. W. 1971. Theory of feeding strategies. Annual Review of Ecology and Systematics **2**: 369–404.

Schoener, T. W. 1974. Competition and the form of habitat shift. Theoretical Population Biology **6**: 265–307.

Schoener, T. W. 1976. Alternative to Lotka-Volterra competition: Models of intermediate complexity. Theoretical Population Biology **10**: 309–333.

Schoener, T. W. 1983. Field experiments on interspecific competition. American Naturalist **122**: 240–285.

Schoener, T. W. 1986. Kinds of ecological communities: Ecology becomes pluralistic. Pages 467–479 in J. Diamond and T. J. Case, editors. *Community Ecology*. New York: Harper & Row.

Schoener, T. W. 1989. Food webs from the small to the large. Ecology **70**: 1559–1589.

Schoener, T. W. 1993. On the relative importance of direct versus indirect effects in ecological communities. Pages 365–411 in H. Kawanabe, J. E. Cohen and K. Iwasaki, editors. *Mutualism and Community Organization: Behavioral, Theoretical and Food Web Approaches*. Oxford University Press, Oxford.

Schoener, T. W. 2011. The newest synthesis: Understanding the interplay of evolutionary and ecological dynamics. Science **331**: 426–429.

Schoener, T. W. and D. Spiller. 1999. Indirect effects in an experimentally staged invasion by a major predator. American Naturalist **153**: 347–358.

Schooler, S. S., B. Salau, M. H. Julien and A. R. Ives. 2011. Alternative stable states explain unpredictable biological control of *Salvinia molesta* in Kakadu. Nature **470**: 86–89.

Schröder, A., L. Persson and A. M. De Roos. 2005. Direct experimental evidence for alternative stable states: A review. Oikos **110**: 3–19.

Schröder, A., L. Persson and A. M. De Roos. 2012. Complex shifts between food web states in response to whole-ecosystem manipulations. Oikos **121**: 417–427.

Schtickzelle, N. and T. P. Quinn. 2007. A metapopulation perspective for salmon and other anadromous fish. Fish and Fisheries **8**: 297–314.

Schupp, E. W. 1995. Seed–seedling conflicts, habitat choice, and patterns of plant recruitment. American Journal of Botany **82**: 399–409.

Schwartz, M. W. and J. D. Hoeksema. 1998. Specialization and resource trade: Biological markets as a model of mutualisms. Ecology **79**: 1029–1038.

Schwartz, M. W., C. A. Brigham, J. D. Hoeksema, K. G. Lyons, M. H. Mills and P. J. van Mantgem. 2000. Linking biodiversity to ecosystem function: Implications for conservation ecology. Oecologia **122**: 297–305.

Scurlock, J. M. O. and R. J. Olson. 2002. Terrestrial net primary productivity—A brief history and a new worldwide database. Environmental Reviews **10**: 91–109.

Seabloom, E. W., W. S. Harpole, O. J. Reichman and D. Tilman. 2003. Invasion, competitive dominance, and resource use by exotic and native California grassland species. Proceedings of the National Academy of Sciences **100**: 13384–13389.

Sears, A. L. W. and P. L. Chesson. 2007. New methods for quantifying the spatial storage effect: An illustration with desert annuals. Ecology **88**: 2240–2247.

Seehausen, O. 2006. African cichlid fish: A model system in adaptive radiation research. Proceedings of the Royal Society of London B **273**: 1987–1998.

Seifert, R. P. and F. H. Seifert. 1976. A community matrix analysis of *Heliconia* insect communities. American Naturalist **110**: 461–483.

Sepkoski, J. J., Jr. 1978. A kinetic model of Phanerozoic taxonomic diversity. I. Analysis of marine orders. Paleobiology **4**: 223–251.

Sexton, J. P., P. J. McIntyre, A. L. Angert and K. J. Rice. 2009. Evolution and ecology of species range limits. Annual Review of Ecology, Evolution, and Systematics **40**: 415–436.

Shea, K., S. H. Roxburgh and E. S. J. Rauschert. 2004. Moving from pattern to process: Coexistence mechanisms under intermediate disturbance regimes. Ecology Letters **7**: 491–508.

Shelford, V. E. 1913. *Animal Communities in Temperate America*. University of Chicago Press, Chicago, IL.

Shipley, B. and J. Dion. 1992. The allometry of seed production in herbaceous angiosperms. American Naturalist **139**: 467–483.

Shipley, B., D. Vile and E. Garnier. 2006. From plant traits to plant communities: A statistical mechanistic approach to biodiversity. Science **314**: 812–814.

Shixiao, Y. L., Z.-P. Xie and C. Staehelin. In press. Analysis of a negative plant-soil feedback in a subtropical monsoon forest. Journal of Ecology. doi: 10.1111/j.1365-2745.2012.01953.x.

Shmida, A. and S. P. Ellner. 1984. Coexistence of plant species with similar niches. Vegetatio **58**: 29–55.

Shmida, A. and M. V. Wilson. 1985. Biological determinants of species diversity. Journal of Biogeography **12**: 1–20.

Shreve, F. 1951. *Vegetation of the Sonoran Desert*. Publication no. 591. Carnegie Institution, Washington, DC.

Shurin, J. B. 2000. Dispersal limitation, invasion resistance, and the structure of pond zooplankton communities. Ecology **81**: 3074–3086.

Shurin, J. B. and D. S. Srivastava. 2005. New perspectives on local and regional diversity: Beyond saturation. Pages 399–417 in M. Holyoak, M. A. Leibold and R. D. Holt, editors. *Metacommunities: Spatial Dynamics and Ecological Communities*. University of Chicago Press, Chicago, IL.

Shurin, J. B., J. E. Havel, M. A. Leibold and B. Pinel-Alloul. 2000. Local and regional zooplankton species richness: A scale-independent test for saturation. Ecology **81**: 3062–3073.

Shurin, J. B. and 7 others. 2002. A cross-ecosystem comparison of the strength of trophic cascades. Ecology Letters **5**: 785–791.

Shurin, J. B., K. Cottenie and H. Hillebrand. 2009. Spatial autocorrelation and dispersal limitation in freshwater organisms. Oecologia **159**: 151–159.

Shurin, J. B., R. W. Markel and B. Matthews. 2010. Comparing trophic cascades across ecosystems. Pages 319–335 in J. Terborgh and J. A. Estes, editors. *Trophic Cascades: Predators, Prey, and the Changing Dynamics of Nature*. Island Press, Washington, DC.

Sibly, R. M. and J. Hone. 2002. Population growth rate and its determinants: An overview. Philosophical Transactions of the Royal Society of London B **357**: 1153–1170.

Sidorovich, V., H. Kruuk and D. W. MacDonald. 1999. Body size and interactions between European and American mink (*Mustela lutreola* and *M. vison*) in Eastern Europe. Journal of Zoology **248**: 521–527.

Siepielski, A. M. and M. A. McPeek. 2010. On the evidence for species coexistence: A critique of the coexistence program. Ecology **91**: 3153–3164.

Siepielski, A. M., K.-L. Hung, E. E. B. Bein and M. A. McPeek. 2010. Experimental evidence for neutral community dynamics governing an insect assemblage. Ecology **91**: 847–857.

Sih, A. 1980. Optimal behavior: Can foragers balance two conflicting demands? Science **210**: 1041–1043.

Sih, A. and B. Christensen. 2001. Optimal diet theory: When does it work and why does it fail? Animal Behaviour **61**: 379–390.

Silvertown, J. 1987. *Introduction to Plant Population Ecology*. Second edition. Longman, New York.

Simberloff, D. 1976. Experimental zoogeography of islands: Effects of island size. Ecology **57**: 629–648.

Simberloff, D. and W. Boecklen. 1981. Santa Rosalia reconsidered: Size ratios and competition. Evolution **35**: 1206–1228.

Simberloff, D. and E. F. Conner. 1981. Missing species combinations. American Naturalist **118**: 215–239.

Simberloff, D. and J. Cox. 1987. Consequences and costs of conservation corridors. Conservation Biology **1**: 63–71.

Simberloff, D., J. A. Farr, J. Cox and D. W. Mehlman. 1992. Movement corridors: Conservation bargains or poor investments? Conservation Biology **6**: 493–504.

Simms, E. L. and D. L. Taylor. 2002. Partner choice in nitrogen-fixation mutualisms of legumes and rhizobia. Integrative and Comparative Biology **42**: 369–380.

Simms, E. L., D. L. Taylor, J. Povich, R. P. Shefferson, J. L. Sachs, M. Urbina and Y. Tausczik. 2006. An empirical test of partner choice mechanisms in a wild legume–rhizobium interaction. Proceedings of the Royal Society of London B **273**: 77–81.

Šímová, I., D. Storch, P. Keil, B. Boyle, O. L. Phillips and B. J. Enquist. 2011. Global species-energy relationship in forest plots: Role of abundance, temperature and species climatic tolerances. Global Ecology and Biogeography **20**: 842–856.

Simpson, G. G. 1953. *The Major Features of Evolution*. Columbia University Press, New York.

Sinclair, T. R. E., S. Mduma and J. S. Brashares. 2003. Patterns of predation in a diverse predator–prey system. Nature **425**: 288–290.

Sjögren, P. 1988. Metapopulation biology of *Rana lessonae* Camerano on the northern periphery of its range. Acta Universitatis Upsaliensis **157**: 1–35.

Sjögren Gulve, P. 1991. Extinctions and isolation gradients in metapopulations: The case of the pool frog (*Rana lessonae*). Biological Journal of the Linnean Society **42**: 135–147.

Skellam, J. G. 1951. Random dispersal in theoretical populations. Biometrika **38**: 196–218.

Skelly, D. K. and E. E. Werner. 1990. Behavioral and life historical responses of larval American toads to an odonate predator. Ecology **71**: 2313–2322.

Slatkin, M. 1993. Isolation by distance in equilibrium and non-equilibrium populations. Evolution **47**: 264–279.

Slobodkin, L. B. 1961. *Growth and Regulation of Animal Populations*. Holt, Rinehart and Winston, New York.

Slobodkin, L. B. 1968. How to be a predator. American Zoologist **8**: 43–51.

Slobodkin, L. B. 1974. Prudent predation does not require group selection. American Naturalist **108**: 665–678.

Sluijs, A. and 14 others. 2006. Subtropical Arctic Ocean temperatures during the Palaeocene/Eocene thermal maximum. Nature **441**: 610–613.

Smith, F. A., E. J. Grace and S. E. Smith. 2009. More than a carbon economy: Nutrient trade and ecological sustainability in facultative arbuscular mycorrhizal symbiosis. New Phytologist **182**: 347–358.

Smith, M. D. and A. K. Knapp. 2003. Dominant species maintain ecosystem function with non-random species loss. Ecology Letters **6**: 509–517.

Sobel, J. M., G. F. Chen, L. R. Watt and D. W. Schemske. 2009. The biology of speciation. Evolution **64**: 295–315.

Soininen, J., R. McDonald and H. Hillebrand. 2007. The distance decay of similarity in ecological communities. Ecography **30**: 3–12.

Solan, M., B. J. Cardinale, A. L. Downing, K. A. M. Engelhardt, J. L. Ruesink and D. S. Srivastava. 2004. Extinction and ecosystem function in the marine benthos. Science **306**: 1177–1180.

Sommer, U. 1986. Phytoplankton competition along a gradient of dilution rates. Oecologia **68**: 503–506.

Sommer, U. 1989. The role of resource competition in phytoplankton succession. Pages 57–170 in U. Sommer, editor. *Plankton Ecology: Succession in Plankton Communities*. Springer, New York.

Sousa, W. P. 1979a. Disturbance in marine intertidal boulder fields: The non-equilibrium maintenance of species diversity. Ecology **60**: 1225–1239.

Sousa, W. P. 1979b. Experimental investigations of disturbance and ecological succession in a rocky intertidal community. Ecological Monographs **49**: 227–254.

Spehn, E. M. and 12 others. 2002. The role of legumes as a component of biodiversity in a cross-European study of grassland biomass nitrogen. Oikos **98**: 205–218.

Spiller, D. and T. W. Schoener. 1990. A terrestrial field experiment showing the impact of eliminating top predators on foliage damage. Nature **347**: 469–472.

Springer, A. and 7 others. 2003. Sequential megafaunal collapse in the North Pacific Ocean: An ongoing legacy of industrial whaling? Proceedings of the National Academy of Sciences **100**: 12233–12228.

Srinivasan, U. T., J. A. Dunne, J. Harte and N. D. Martinez. 2007. Response of complex food webs to realistic extinction sequences. Ecology **88**: 671–682.

Stachowicz, J. J. 2001. Mutualism, facilitation, and the structure of ecological communities. BioScience **51**: 235–246.

Stachowicz, J. J., R. B. Whitlatch and R. W. Osman. 1999. Species diversity and invasion resistance in a marine ecosystem. Science **286**: 1577–1578.

Stachowicz, J. J., H. Fried, R. W. Osman and R. B. Whitlatch. 2002. Biodiversity, invasion resistance, and marine ecosystem function: Reconciling pattern and process. Ecology **83**: 2575–2590.

Stachowicz, J. J., J. F. Bruno and J. E. Duffy. 2007. Understanding the effects of marine biodiversity on communities and ecosystems. Annual Review of Ecology, Evolution, and Systematics **38**: 739–766.

Stebbins, G. L. 1974. *Flowering Plants: Evolution above the Species Level*. The Belknap Press of Harvard University Press, Cambridge, MA.

Steiner, C. F. 2001. The effects of prey heterogeneity and consumer identity on the limitation of trophic-level biomass. Ecology **82**: 2495–2506.

Steiner, C. F., Z. T. Long, J. A. Krumins and P. J. Morin. 2006. Population and community resilience in multitrophic communities. Ecology **87**: 996–1007.

Steneck, R. S., M. H. Grahman, B. J. Bourque, D. Corbett, J. M. Erlandson, J. A. Estes and M. J. Tegner. 2002. Kelp forest ecosystems: Biodiversity, stability, resilience and future. Environmental Conservation **29**: 436–459.

Stephens, D. W. and J. R. Krebs. 1986. *Foraging Theory*. Princeton University Press, Princeton, NJ.

Stephens, D. W., J. S. Brown and R. C. Ydenberg. 2007. *Foraging: Behavior and Ecology*. University of Chicago Press, Chicago, IL.

Stephens, P. A., W. J. Sutherland and R. P. Freckleton. 1999. What is the Allee effect? Oikos **87**: 185–190.

Sterner, R. W. and J. J. Elser. 2002. *Ecological Stoichiometry*. Princeton University Press, Princeton, NJ.

Stich, H. B. and W. Lampert. 1981. Predator evasion as an explanation of diurnal vertical migration by zooplankton. Nature **293**: 396–398.

Stohlgren, T. J. and 9 others. 1999. Exotic plant species invade hot spots of native plant diversity. Ecological Monographs **69**: 25–46.

Stohlgren, T. J., C. Flather, C. S. Jarnevich, D. T. Barnett and J. Kartesz. 2008. Rejoinder to Harrison (2008): The myth of plant species saturation. Ecology Letters **11**: 324–326.

Stokstad, E. 2011. Open-source ecology takes root across the world. Science **334**: 308–309.

Stomp, M. and 8 others. 2004. Adaptive divergence in pigment composition promotes phytoplankton biodiversity. Nature **432**: 104–107.

Stomp, M., J. Huisman, G. G. Mittelbach, E. Litchman and C. A. Klausmeier. 2011. Large-scale biodiversity patterns in freshwater phytoplankton. Ecology **92**: 2096–2107.

Storch, D. and 12 others. 2006. Energy, range dynamics and global species richness patterns: Reconciling mid-domain effects and environmental determinants of avian diversity. Ecology Letters **9**: 1308–1320.

Stouffer, D. B. 2010. Scaling from individuals to networks in food webs. Functional Ecology **24**: 44–51.

Stouffer, D. B. and J. Bascompte. 2011. Compartmentalization increases food-web persistence. Proceedings of the National Academy of Sciences **108**: 3648–3652.

Strauss, S. Y., J. A. Lau and S. P. Carroll. 2006a. Evolutionary responses of natives to introduced species: What do introductions tell us about natural communities? Ecology Letters **9**: 357–374.

Strauss, S. Y., C. O. Webb and N. Salamin. 2006b. Exotic taxa less related to native species are more invasive. Proceedings of the National Academy of Sciences **103**: 5841–5845.

Strauss, S. Y., J. A. Lau, T. W. Schoener and P. Tiffin. 2008. Evolution in ecological field experiments: Implications for effect size. Ecology Letters **11**: 199–207.

Strong, D. R. 1986. Density vagueness: Abiding the variance in the demography of real populations. Pages 257–268. in J. Diamond and T. J. Case, editors. *Community Ecology*. Harper and Row, New York.

Strong, D. R. 1992. Are trophic cascades all wet? Differentiation and donor-control in speciose ecosystems. Ecography **73**: 747–754.

Strong, D. R., L. A. Szyska and D. Simberloff. 1979. Test of community-wide character displacement against null hypotheses. Evolution **33**: 897–913.

Suding, K. N. and R. J. Hobbs. 2009. Threshold models in restoration and conservation: A developing framework. Trends in Ecology and Evolution **24**: 271–279.

Suding, K. N., K. L. Gross and G. R. Houseman. 2004. Alternative states and positive feedbacks in restoration ecology. Trends in Ecology and Evolution **19**: 46–53.

Suding, K. N., A. E. Miller, H. Bechtold and W. D. Bowman. 2006. The consequence of species loss to ecosystem nitrogen cycling depends on community compensation. Oecologia **149**: 141–149.

Sugihara, G. 1980. Minimal community structure: An explanation of species abundance patterns. American Naturalist **116**: 770–787.

Sutherland, J. P. 1974. Multiple stable points in natural communities. American Naturalist **108**: 859–873.

Swenson, N. G. 2011. The role of evolutionary processes in producing biodiversity patterns, and the interrelationships between taxonomic, functional and phylogenetic biodiversity. American Journal of Botany **98**: 472–480.

Swenson, N. G. and B. J. Enquist. 2007. Ecological and evolutionary determinants of a key plant functional trait: Wood density and its community-wide variation across latitude and elevation. American Journal of Botany **94**: 451–495.

Swenson, N. G., B. J. Enquist, J. Pither, J. Thompson and J. K. Zimmerman. 2006. The problem and promise of scale dependency in community phylogenetics. Ecology **87**: 2418–2424.

Swenson, N. G., B. J. Enquist, J. Thompson and J. K. Zimmerman. 2007. The influence of spatial and size scales on phylogenetic relatedness in tropical forest communities. Ecology **88**: 1770–1780.

Symstad, A. J. and D. Tilman. 2001. Diversity loss, recruitment limitation, and ecosystem functioning: Lessons learned from a removal experiment. Oikos **82**: 424–435.

Symstad, A. J. and 7 others. 2003. Long-term and large-scale perspectives on the relationship between biodiversity and ecosystem functioning. BioScience **53**: 89–98.

Takimoto, G., D. Spiller and D. M. Post. 2008. Ecosystem size, but not disturbance, determines food-chain length on islands of the Bahamas. Ecology **89**: 3001–3007.

Tanner, J., T. P. Hughes and J. H. Connell. 2009. Community-level density dependence: An example from a shallow coral assemblage. Ecology **90**: 506–516.

Tansley, A. G. 1923. Practical Plant Ecology: A Guide for Beginners in Field Study of Plant Communities. Allen and Unwin, London.

Tansley, A. G. 1939. *The British Islands and Their Vegetation*. Cambridge University Press, Cambridge.

Taylor, B. W. and R. E. Irwin. 2004. Linking economic activities to the distribution of exotic plants. Proceedings of the National Academy of Sciences **101**: 17725–17730.

Taylor, E. B. and J. D. McPhail. 2000. Historical contingency and ecological determinism interact to prime speciation in sticklebacks, *Gasterosteus*. Proceedings of the Royal Society of London B **267**: 2375–2384.

Tegner, M. J. and P. K. Dayton. 1991. Sea urchins, El Niños, and the long term stability of southern California kelp forest communities. Marine Ecology Progress Series **77**: 49–63.

Temperton, V. M., P. N. Mwangi, M. Scherer-Lorenzen, B. Schmid and N. Buchmann. 2007. Positive interactions between nitrogen-fixing legumes and four different neighboring species in a biodiversity experiment. Oecologia **151**: 190–205.

Terborgh, J. 1973. On the notion of favorableness in plant ecology. American Naturalist **107**: 481–501.

Terborgh, J. and J. A. Estes, editors. 2010. *Trophic Cascades: Predators, Prey, and the Changing Dynamics of Nature.* Island Press, Washington, DC.

Terborgh, J., R. B. Foster and V. Nuñez. 1996. Tropical tree communities: A test of the nonequilibrium hypothesis. Ecology **77**: 561–567.

Terborgh, J. and 10 others. 2001. Ecological meltdown in predator-free forest fragments. Science **294**: 1923–1926.

Terborgh, J., R. D. Holt and J. A. Estes. 2010. Trophic cascades: What they are, how they work, and why they matter. Pages 1–18 in J. Terborgh and J. A. Estes, editors. *Trophic Cascades: Predators, Prey, and the Changing Dynamics of Nature.* Island Press, Washington DC.

terHorst, C. P., T. E. Miller and D. R. Levitan. 2010. Evolution of prey in ecological time reduces the effect size of predators in experimental microcosms. Ecology **91**: 629–636.

Tewksbury, J. J. and J. D. Lloyd. 2001. Positive interactions under nurse-plants: Spatial scale, stress gradients and benefactor size. Oecologia **127**: 425–434.

Thebault, E. and C. Fontaine. 2010. Stability of ecological communities and the architecture of mutualistic and trophic networks. Science **329**: 853–856.

Thebault, E., V. Huber and M. Loreau. 2007. Cascading extinctions and ecosystem functioning: Contrasting effects of diversity depending on food web structure. Oikos **116**: 163–173.

Thierry, A., O. L. Petchey, A. P. Beckerman, P. H. Warren and R. J. Williams. 2011. The consequences of size dependent foraging for food web topology. Oikos **120**: 493–502.

Thomas, C. D., E. J. Bodsworth, R. J. Wilson, A. D. Simmons, Z. G. Davies, M. Musche and L. Conradt. 2001. Ecological and evolutionary processes at expanding range margins. Nature **411**: 577–581.

Thompson, J. N. 1998. Rapid evolution as an ecological process. Trends in Ecology and Evolution **13**: 329–332.

Thompson, K., L. C. Rickard, D. J. Hodkinson and M. Rees. 2002. Seed dispersal, the search for trade-offs. Pages 152–172 in J. M. Bullock, R. E. Kenward and R. S. Hails, editors. *Dispersal Ecology.* Blackwell, Oxford.

Thompson, R. M., K. N. Mouritsen and R. Poulin. 2005. Importance of parasites and their life cycle characteristics in determining the structure of a large marine food web. Journal of Animal Ecology **74**: 77–85.

Thomson, J. D. 1978. Effects of stand composition on insect visitation in two-species mixtures of *Hieracium.* American Midland Naturalist **100**: 431–440.

Thuiller, W., S. Lavorel, M. B. Araújo, M. T. Sykes and I. C. Prentice. 2005. Climate change threats to plant diversity in Europe. Proceedings of the National Academy of Sciences **102**: 8245–8250.

Tielbörger, K. and R. Kadmon. 2000. Temporal environmental variation tips the balance between facilitation and interference in desert plants. Ecology **81**: 1544–1553.

Tilman, D. [published as D. Titman]. 1976. Ecological competition between algae: Experimental confirmation of resource-based competition theory. Science **192**: 463–466.

Tilman, D. 1977. Resource competition between planktonic algae: Experimental and theoretical approach. Ecology **58**: 338–348.

Tilman, D. 1980. A graphical-mechanistic approach to competition and predation. American Naturalist **116**: 362–393.

Tilman, D. 1981. Tests of resource competition theory using four species of Lake Michigan algae. Ecology **62**: 802–825.

Tilman, D. 1982. *Resource Competition and Community Structure.* Princeton University Press, Princeton, NJ.

Tilman, D. 1987. The importance of mechanisms of interspecific competition. American Naturalist **129**: 769–774.

Tilman, D. 1994. Competition and biodiversity in spatially structured habitats. Ecology **75**: 2–16.

Tilman, D. 1999. The ecological consequences of changes in biodiversity: A search for general principles. Ecology **80**: 1455–1474.

Tilman, D. 2001. An evolutionary approach to ecosystem functioning. Proceedings of the National Academy of Sciences **98**: 10979–10980.

Tilman, D. 2004. Niche tradeoffs, neutrality, and community structure: A stochastic theory of resource competition, invasion, and community assembly. Proceedings of the National Academy of Sciences **101**: 10854–10861.

Tilman, D. 2007. Interspecific competition and multispecies coexistence. Pages 84–97 in R. M. May and A. McLean, editors. *Theoretical Ecology: Principles and Applications.* Third edition. Oxford University Press, Oxford.

Tilman, D. and A. L. Downing. 1994. Biodiversity and stability in grasslands. Nature **367**: 363–365.

Tilman, D. and S. Pacala. 1993. The maintenance of species richness in plant communities. Pages 13–25 in R. E. Ricklefs and D. Schluter, editors. *Species Diversity in Ecological Communities: Historical and Geographical Perspectives.* University of Chicago Press, Chicago, IL.

Tilman, D. and D. Wedin. 1991a. Plant traits and resource reduction for five grasses growing on a nitrogen gradient. Ecology **72**: 685–700.

Tilman, D. and D. Wedin. 1991b. Dynamics of nitrogen competition between successional grasses. Ecology **72**: 1038–1049.

Tilman, D. and D. Wedin. 1991c. Components of plant competition along an experimental gradient of nitrogen availability. Ecology **72**: 1050–1065.

Tilman, D., R. M. May, C. L. Lehman and M. A. Nowak. 1994. Habitat destruction and the extinction debt. Nature **371**: 65–66.

Tilman, D., D. Wedin and J. M. H. Knops. 1996. Productivity and sustainability influenced by biodiversity in grassland ecosystems. Nature **379**: 718–720.

Tilman, D., J. Knops, D. Wedin, P. B. Reich, M. Ritchie and E. Siemann. 1997. The influence of functional diversity and composition on ecosystem processes. Science **277**: 1300–1302.

Tilman, D., C. L. Lehman and C. E. Bristow. 1998. Diversity–stability relationships: Statistical inevitability or ecological consequence? American Naturalist **151**: 277–282.

Tilman, D., P. B. Reich, J. Knops, D. Wedin, T. Mielke and C. Lehman. 2001. Diversity and productivity in a long-term grassland experiment. Science **294**: 843–845.

Tilman, D., P. B. Reich and J. M. H. Knops. 2006. Biodiversity and ecosystem stability in a decade-long grassland experiment. Nature **441**: 629–632.

Tokeshi, M. 1999. *Species Coexistence*. Blackwell Scientific, Oxford.

Tollrian, R. and C. D. Harvell, editors. 1999. *The Ecology and Evolution of Inducible Defenses*. Princeton University Press, Princeton, NJ.

Townsend, C. R., R. M. Thompson, A. R. McIntosh, C. Kilroy, E. Edwards and M. Scarsbrook. 1998. Disturbance, resource supply, and food-web architecture in streams. Ecology Letters **1**: 200–209.

Tracy, C. R. and T. L. George. 1992. On the determinants of extinction. American Naturalist **139**: 102–122.

Tuomisto, H. 2010a. A diversity of beta diversities: Straightening up a concept gone awry. Part 1. Defining beta diversity as a function of alpha and gamma diversity. Ecography **33**: 2–22.

Tuomisto, H. 2010b. A diversity of beta diversities: Straightening up a concept gone awry. Part 2. Quantifying beta diversity and related phenomena. Ecography **33**: 23–45.

Tuomisto, H., K. Ruokolainen and M. Yli-Halla. 2003. Dispersal, environment, and floristic variation of western Amazonian forests. Science **299**: 241–244.

Turchin, P. 1990. Rarity of density dependence or population regulation with lags? Nature **344**: 660–663.

Turchin, P. 1999. Population regulation: A synthetic view. Oikos **84**: 153–159.

Turchin, P. 2003. *Complex Population Dynamics: A Theoretical/Empirical Synthesis*. Princeton University Press, Princeton, NJ.

Turchin, P. and A. D. Taylor. 1992. Complex dynamics in ecological time series. Ecology **73**: 289–305.

Turgeon, J., R. Stokes, R. A. Thum, J. M. Brown and M. A. McPeek. 2005. Simultaneous Quaternary radiations of three damselfly clades across the Holarctic. American Naturalist **165**: E78–E107.

Turnbull, L. A., M. Rees and M. J. Crawley. 1999. Seed mass and the competition/colonization trade-off: A sowing experiment. Journal of Ecology **87**: 899–912.

Turner, A. M. and G. G. Mittelbach. 1990. Predator avoidance and community structure: Interactions among piscivores, planktivores, and plankton. Ecology **71**: 2241–2254.

Ulrich, W., M. Ollik and K. I. Ugland. 2010. A meta-analysis of species-abundance distributions. Oikos **119**: 1149–1155.

Underwood, A. J. 1986. The analysis of competition by field experiments. Pages 240–268 in J. Kikkawa and D. J. Anderson, editors. *Community Ecology: Pattern and Process*. Blackwell, Melbourne.

Urban, M. C. and 13 others. 2008. The evolutionary ecology of metacommunities. Trends in Ecology and Evolution **23**: 311–317.

Valentine, J. W. and D. Jablonski. 2010. Origins of marine patterns of biodiversity: Some correlates and applications. Palaeontology **53**: 1203–1210.

Valone, T. J. and M. R. Schutzenhofer. 2007. Reduced rodent biodiversity destabilizes plant populations. Ecology **88**: 26–31.

Vamosi, S. M., S. B. Heard, J. C. Vamosi and C. O. Webb. 2009. Emerging patterns in the comparative analysis of phylogenetic community structure. Molecular Ecology **18**: 572–592.

Van Bael, S. and S. Pruett-Jones. 1996. Exponential population growth of Monk Parakeets in the United States. Wilson Bulletin **108**: 584–588.

Vance, R. R. 1978. Predation and resource partitioning in one predator–two prey model communities. American Naturalist **112**: 797–813.

van der Heijden, M. G. A., R. D. Bardgett and N. M. van Straalen. 2008. The unseen majority: Soil microbes as drivers of plant diversity and productivity in terrestrial ecosystems. Ecology Letters **11**: 296–310.

van der Heijden, M. G. A. and T. R. Horton. 2009. Socialism in soil? The importance of mycorrhizal fungal networks for facilitation in natural ecosystems. Journal of Ecology **97**: 1139–1150.

Vandermeer, J. H. 1970. The community matrix and the number of species in a community. American Naturalist **104**: 73–83.

Vandermeer, J. H. 1972. On the covariance of the community matrix. Ecology **53**: 187–189.

Vandermeer, J. H. 1979. The community matrix and the number of species in a community. American Naturalist **104**: 73–83.

Vandermeer, J. H. 1989. *The Ecology of Intercropping*. Cambridge University Press, Cambridge.

Vandermeer, J. H. and D. E. Goldberg. 2003. *Population Ecology: First Principles*. Princeton University Press, Princeton, NJ.

Van Der Windt, H. J. and J. A. A. Swart. 2008. Ecological corridors, connecting science and politics: The case of the Green River in the Netherlands. Journal of Applied Ecology **45**: 124–132.

Vander Zanden, M. J. and W. Fetzer. 2007. Global patterns of aquatic food chain length. Oikos **116**: 1378–1388.

Vander Zanden, M. J. and J. B. Rasmussen. 1996. A trophic position model of pelagic food webs: Impact on contaminant bioaccumulation in lake trout. Ecological Monographs **66**: 451–477.

Vander Zanden, M. J., B. Shuter, N. Lester and J. B. Rasmussen. 1999. Patterns of food chain length in lakes: A stable isotope study. American Naturalist **154**: 405–416.

Van Nes, E. H. and M. Scheffer. 2004. Large species shifts triggered by small forces. American Naturalist **164**: 255–266.

Vanormelingen, P., K. Cottenie, E. Michels, K. Muylaert, W. Vyverman and L. De Meester. 2008. The relative importance of dispersal and local processes in structuring phytoplankton communities in a set of highly interconnected ponds. Freshwater Biology **53**: 2170–2183.

Vanschoenwinkel, B., C. DeVries, M. Seaman and L. Brendonck. 2007. The role of metacommunity processes in shaping invertebrate rock pool communities along a dispersal gradient. Oikos **116**: 1255–1266.

van Veen, F. J., R. J. Morris and H. C. J. Godfray. 2006. Apparent competition, quantitative food webs, and the structure of phytophagous insect communities. Annual Review of Entomology **51**: 187–208.

van Veen, F. J. F., C. B. Muller, J. K. Pell and H. C. J. Godfray. 2008. Food web structure of three guilds of natural enemies: Predators, parasitoids and pathogens of aphids. Journal of Animal Ecology **77**: 191–200.

Vázquez, D. P., W. F. Morris and P. Jordano. 2005. Interaction frequency as a surrogate for the total effect of animal mutualists on plants. Ecology Letters **8**: 1088–1094.

Vellend, M. 2010. Conceptual synthesis in community ecology. Quarterly Review of Biology **85**: 183–206.

Verboom, J., A. Schotman, P. Opdam and J. A. J. Metz. 1991. European nuthatch metapopulations in a fragmented agricultural landscape. Oikos **61**: 149–156.

Verhulst, P. F. 1838. Notice sur la loi que la population suit dans son accroissement. Correspondances Mathématiques et Physiques **10**: 113–121.

Violle, C., D. R. Nemergut, Z. Pu and L. Jiang. 2011. Phylogenetic limiting similarity and competitive exclusion. Ecology Letters **14**: 782–787.

Vogt, R. J., T. N. Romanuk and J. Kolasa. 2006. Species richness–variability relationships in multi-trophic aquatic microcosms. Oikos **113**: 55–66.

Volkov, I., J. R. Banavar, S. P. Hubbell and A. Maritan. 2003. Neutral theory and relative species abundance in ecology. Nature **424**: 1035–1037.

Volterra, V. 1926. Variazioni e fluttuazioni del numero d'individui in specie animali conviventi. Memoria della Regia Accademia Nazionale dei Lincei **ser 6,2 (1926)**: 31–113. Reprinted 1931 in R. N. Chapman, *Animal Ecology*, McGraw-Hill, New York.

Volterra, V. 1931. Variations and fluctuations of the number of individuals in animal species living together. Pages 409–448 in R. N. Chapman, editor. *Animal Ecology*. McGraw-Hill Publishers, New York.

Vos, M., A. Vershoor, B. Kooi, F. Wäckers, D. DeAngelis and W. Mooij. 2004. Inducible defenses and trophic structure. Ecology **85**: 2783–2794.

Wade, P. R., J. M. ver Hoef and D. P. DeMaster. 2009. Mammal-eating killer whales and their prey—trend data for pinnipeds and sea otters in the North Pacific Ocean do not support the sequential megafaunal collapse hypothesis. Marine Mammal Science **25**: 737–747.

Wake, D. and V. Vredenburg. 2008. Are we in the midst of the 6th mass extinction? A view from the world of amphibians. Proceedings of the National Academy of Sciences **105**: 11466–11473.

Walker, B. H. 1993. Rangeland ecology: Understanding and managing change. Ambio **22**: 2–3.

Walker, L. R. and R. del Moral. 2003. *Primary Succession and Landscape Restoration*. Cambridge University Press, Cambridge.

Walker, M. K. and R. M. Thompson. 2010. Consequences of realistic patterns of biodiversity loss: An experimental test from the intertidal zone. Marine and Freshwater Research **61**: 1015–1022.

Wallace, A. R. 1876. *The Geographical Distribution of Animals*. Harper and Brothers, New York.

Wallace, A. R. 1878. *Tropical Nature and Other Essays*. Macmillan, New York.

Walters, A. and D. M. Post. 2008. An experimental disturbance alters fish size structure but not food chain length in streams. Ecology **89**: 3261–3267.

Walters, C. J. and S. J. Martell. 2004. *Fisheries Ecology and Management*. Princeton University Press, Princeton, NJ.

Wardle, D. A. 1999. Is "sampling effect" a problem for experiments investigating biodiversity ecosystem function relationships? Oikos **87**: 403–407.

Wardle, D. A. and O. Zackrisson. 2005. Effects of species and functional group loss on island ecosystem properties. Nature **435**: 806–810.

Wardle, D. A., R. D. Bardgett, R. M. Callaway and W. H. Van der Putten. 2011. Terrestrial ecosystem responses to species gains and losses. Science **332**: 1273–1277.

Warren, D. L., R. E. Glor and M. Turelli. 2008. Environmental niche equivalency versus conservatism: Quantitative approaches to niche evolution. Evolution **62**: 2868–2883.

Warren, P. H. 1996. Dispersal and destruction in a multiple habitat system: An experimental approach using protist communities. Oikos **77**: 317–325.

Watkins, J. E., C. Cardelus, R. K. Colwell and R. C. Moran. 2006. Species richness and distribution of ferns along an elevational gradient in Costa Rica. American Journal of Botany **93**: 73–83.

Watson, H. C. 1859. Cybele Britannica, or British Plants and their Geographical Relations. Longman and Company, London.

Watson, S., E. McCauley and J. A. Downing. 1992. Sigmoid relationships between production and biomass among lakes. Canadian Journal of Fisheries and Aquatic Sciences **45**: 915–920.

Watts, D. J. and S. H. Strogatz. 1998. Collective dynamics of "small-world" networks. Nature **393**: 440–442.

Webb, C. O., D. D. Ackerly, M. A. McPeek and M. J. Donoghue. 2002. Phylogenies and community ecology. Annual Review of Ecology and Systematics **33**: 475–505.

Weiher, E. and P. A. Keddy. 1995. Assembly rules, null models, and trait dispersion: New questions from old patterns. Oikos **74**: 159–164.

Weir, J. T. and D. Schluter. 2007. The latitudinal gradient in recent speciation and extinction rates of birds and mammals. Science **315**: 1574–1576.

Went, F. W. 1942. The dependence of certain annual plants on shrubs in Southern California deserts. Bulletin of the Torrey Botanical Club **69**: 100–114.

Werner, E. E. 1986. Amphibian metamorphosis: Growth rate, predation risk, and the optimal size at transformation. American Naturalist **128**: 319–341.

Werner, E. E. and B. R. Anholt. 1993. Ecological consequences of the trade-off between growth and mortality rates mediated by foraging activity. American Naturalist **142**: 242–272.

Werner, E. E. and J. F. Gilliam. 1984. The ontogenetic niche and species interactions in size-structured populations. Annual Review of Ecology and Systematics **15**: 393–425.

Werner, E. E. and D. J. Hall. 1974. Optimal foraging and the size selection of prey by the bluegill sunfish (*Lepomis macrochirus*). Ecology **55**: 1042–1052.

Werner, E. E. and M. A. McPeek. 1994. The roles of direct and indirect effects on the distributions of two frog species along an environmental gradient. Ecology **75**: 1368–1382.

Werner, E. E. and S. D. Peacor. 2003. A review of trait-mediated indirect interactions in ecological communities. Ecology **84**: 1083–1100.

Werner, P. A. and W. J. Platt. 1976. Ecological relationships of co-occurring goldenrods (*Solidago*: Compositae). American Naturalist **110**: 959–971.

Werner, E. E., J. F. Gilliam, D. J. Hall and G. G. Mittelbach. 1983. An experimental test of the effects of predation risk on habitat use in fish. Ecology **64**: 1540–1548.

West, S. A., E. T. Kiers, E. L. Simms and R. F. Denison. 2002. Nitrogen fixation and the stability of the legume–rhizobium mutualism. Proceedings of the Royal Society of London B **269**: 685–694.

Westoby, M., M. R. Leishman and J. Lord. 1996. Comparative ecology of seed size and dispersal. Philosophical Transactions of the Royal Society of London B **351**: 1309–1318.

Whittaker, R. H. 1956. Vegetation of the Great Smoky Mountains. Ecological Monographs **26**: 1–80.

Whittaker, R. H. 1960. Vegetation of the Siskiyou Mountains, Oregon and California. Ecological Monographs **30**: 279–338.

Whittaker, R. H. 1972. Evolution and measurement of species diversity. Taxon **21**: 213–251.

Whittaker, R. J. 2010. Meta-analysis and mega-mistakes: Calling time on meta-analysis of the species richness–productivity relationship. Ecology **91**: 2522–2533.

Wiens, J. A. 1977. On competition and variable environments: Populations may experience "ecological crunches" in variable climates, nullifying the assumptions of competition theory and limiting the usefulness of short-term studies of population patterns. American Scientist **65**: 590–597.

Wiens, J. J. 2007. Global patterns of diversification and species richness in amphibians. American Naturalist **170**: S86–S106.

Wiens, J. J. 2008. Commentary on Losos (2008): Niche conservatism deja vu. Ecology Letters **11**: 1004–1005.

Wiens, J. J. 2011. The causes of species richness patterns across space, time, and clades and the role of "ecological limits." Quarterly Review of Biology **86**: 75–96.

Wiens, J. J. and M. J. Donoghue. 2004. Historical biogeography, ecology and species richness. Trends in Ecology and Evolution **19**: 639–644.

Wiens, J. J. and C. H. Graham. 2005. Niche conservatism: Integrating evolution, ecology, and conservation biology. Annual Review of Ecology, Evolution, and Systematics **36**: 519–539.

Wiens, J. J., C. H. Graham, D. S. Moen, S. A. Smith and T. W. Reeder. 2005. Evolutionary and ecological causes of the latitudinal diversity gradient in hylid frogs: Treefrog trees unearth the roots of high tropical diversity. American Naturalist **168**: 579–596.

Wiens, J. J., J. Sukumaran, R. A. Pyron and R. M. Brown. 2009. Evolutionary and biogeographic origins of high tropical diversity in old world frogs (Ranidae). Evolution **63**: 1217–1231.

Wiens, J. J. and 13 others. 2010. Niche conservatism as an emerging principle in ecology of evolutionary biology. Ecology Letters **13**: 1310–1324.

Williams, C. B. 1964. *Patterns in the Balance of Nature*. Academic Press, London.

Williams, R. J. and N. D. Martinez. 2000. Simple rules yield complex food webs. Nature **404**: 180–183.

Williams, T. M., J. A. Estes, D. F. Doak and A. Springer. 2004. Killer appetites: Assessing the role of predators in ecological communities. Ecology **85**: 3373–3384.

Williamson, M. 1972. *The Analysis of Biological Populations*. Edward Arnold, London.

Williamson, M. H. 1981. *Island Populations*. Oxford University Press, Oxford.

Willig, M., D. M. Kaufman and R. D. Stevens. 2003. Latitudinal gradients of biodiversity: Pattern, process, scale, and synthesis. Annual Review of Ecology and Systematics **34**: 273–310.

Willis, A. J. and J. Memmott. 2005. The potential for indirect effects between a weed, one of its biocontrol agents and native herbivores: A food web approach. Biological Control **35**: 299–306.

Wilson, D. S. 1992. Complex interactions in metacommunities, with implications for biodiversity and higher levels of selection. Ecology **73**: 1984–2000.

Wilson, E. O. 2002. *The Future of Life*. Knopf Publishing Group, New York.

Wilson, J. B. and A. D. Q. Agnew. 1992. Positive feedback switches in plant communities. Advances in Ecological Research **23**: 263–336.

Wilson, J. B., E. Spijkerman and J. Huisman. 2007. Is there really insufficient support for Tilman's R^* concept? A comment on Miller et al. American Naturalist **169**: 700–706.

Wiser, S. K., R. B. Allen, P. W. Clinton and K. H. Platt. 1998. Community structure and forest invasion by an exotic herb over 23 years. Ecology **79**: 2071–2081.

Woiwod, I. P. and I. Hanski. 1992. Patterns of density dependence in moths and aphids. Journal of Animal Ecology **61**: 619–629.

Wojdak, J. M. 2005. Relative strength of top-down, bottom-up, and consumer species richness effects on pond ecosystems. Ecological Monographs **75**: 489–504.

Wojdak, J. M. and G. G. Mittelbach. 2007. Consequences of niche overlap for ecosystem functioning: An experimental test with pond grazers. Ecology **88**: 2072–2083.

Wolda, H. 1989. The equilibrium concept and density dependence tests: What does it all mean? Oecologia **81**: 430–432.

Wolda, H. and B. Dennis. 1993. Density dependent tests, are they? Oecologia **95**: 581–591.

Wollkind, D. J. 1976. Exploitation in three trophic levels: An extension allowing intraspecies carnivore interaction. American Naturalist **110**: 431–447.

Woodward, G., B. Ebenman, M. Emmerson, J. M. Montoya, J. M. Olesen, A. Valido and P. H. Warren. 2005. Body size in ecological networks. Trends in Ecology and Evolution **20**: 402–409.

Woodward, G. and 7 others. 2010. Individual-based food webs: Species identity, body size and sampling effects. Advances in Ecological Research **43**: 209–265.

Wootton, J. T. 1994. Putting the pieces together: Testing the independence of interactions among organisms. Ecology **75**: 1544–1551.

Wootton, J. T. 1997. Estimates and tests of per capita interaction strength: Diet, abundance, and impact of intertidally foraging birds. Ecological Monographs **67**: 45–64.

Wootton, J. T. 2005. Field parameterization and experimental test of the neutral theory of biodiversity. Nature **433**: 309–312.

Wootton, J. T. and A. L. Downing. 2003. Understanding the effects of reduced biodiversity: A comparison of two approaches. Pages 85–104 in P. Kareiva and S. A. Levin, editors. *The Importance of Species*. Princeton University Press, Princeton, NJ.

Wootton, J. T. and M. Emmerson. 2005. Measurement of interaction strength in nature. Annual Review of Ecology and Systematics **36**: 419–444.

Wootton, J. T. and M. E. Power. 1993. Productivity, consumers, and the structure of a river food chain. Proceedings of the National Academy of Sciences **90**: 1384–1387.

Worm, B. and 20 others. 2010. Rebuilding global fisheries. Science **325**: 578–585.

Wright, D. H. 1983. Species-energy theory: An extension of species-area theory. Oikos **41**: 496–506.

Wright, D. H., D. J. Currie and B. A. Maurer. 1993. Energy supply and patterns of species richness on local and regional scales. Pages 66–74 in R. E. Ricklefs and D. Schluter, editors. *Species Diversity in Ecological Communities: Historical and Geographical Perspectives*. University of Chicago Press, Chicago, IL.

Wright, D. H., J. Keeling and L. Gilman. 2006. The road from Santa Rosalia: A faster tempo of evolution in tropical climates. Proceedings of the National Academy of Sciences **103**: 7718–7722.

Wright, J. P. and C. G. Jones. 2006. The concept of organisms as ecosystem engineers ten years on: Progress, limitations, and challenges. BioScience **56**: 203–209.

Wright, S., R. D. Gray and R. C. Gardner. 2003. Energy and the rate of evolution: Inferences from plant rDNA substitution rates in the western Pacific. Evolution **57**: 2893–2898.

Yang, L. H. and V. H. W. Rudolf. 2010. Phenology, ontogeny and the effects of climate change on the timing of species interactions. Ecology Letters **13**: 1–10.

Yodzis, P. 1977. Harvesting and limiting similarity. American Naturalist **111**: 833–843.

Yodzis, P. 1978. Competition for Space and the Structure of Ecological Communities. Springer-Verlag, Berlin.

Yodzis, P. 1988. The indeterminacy of ecological interactions as perceived through perturbation experiments. Ecology **69**: 508–515.

Yodzis, P. 1989. *Introduction to Theoretical Ecology*. Harper and Row, New York.

Yodzis, P. 1998. Local trophodynamics and the interaction of marine mammals and fisheries in the Benguela ecosystem. Journal of Animal Ecology **67**: 635–658.

Yodzis, P. 2001. Must top predators be culled for the sake of fisheries? Trends in Ecology and Evolution **16**: 78–84.

Yoon, I., R. J. Williams, E. Levine, S. Yoon, J. A. Dunne and N. D. Martinez. 2004. Webs on the web (WoW): 3D visualization of ecological networks on the WWW for collaborative research and education. Proceedings of the IS&T/SPIE Symposium on Electronic Imaging, Visualization and Data Analysis **5295**: 124–132.

Yoshida, T., L. E. Jones, S. P. Ellner, G. F. Fussmann and N. G. Hairston, Jr. 2003. Rapid evolution drives ecological dynamics in a predator–prey system. Nature **424**: 303–306.

Yu, D. W., H. B. Wilson and N. E. Pierce. 2001. An empirical model of species coexistence in a spatially structured environment. Ecology **82**: 1761–1771.

Yu, D. W., H. B. Wilson, M. E. Frederickson, W. Palomino, R. de la Colina, D. P. Edwards and A. A. Balareso. 2004. Experimental demonstration of species coexistence enabled by dispersal limitation. Journal of Animal Ecology **73**: 1102–1114.

Zachos, J. C., G. R. Dickens and R. E. Zeebe. 2008. An early Cenozoic perspective on greenhouse warming and carbon-cycle dynamics. Nature **451**: 279–283.

Zillio, T. and R. Condit. 2007. The impact of neutrality, niche differentiation and species input on diversity and abundance distributions. Oikos **116**: 931–940.

Zook, A. E., A. Eklof, U. Jacob and S. Allesina. 2011. Food webs: Ordering species according to body size yields high degree of intervality. Journal of Theoretical Biology **271**: 106–113.

Zuckerkandl, E. and L. Pauling. 1965. Evolutionary divergence and convergence in proteins. Pages 97–166 in V. Bryson and H. J. Vogel, editors. *Evolving Genes and Proteins*. Academic Press, New York.

Author Index

Subject Index